Numerical Solutions of the Euler Equations for Steady Flow Problems

by Albrecht Eberle, Arthur Rizzi, and Ernst Heinrich Hirschel

Notes on Numerical Fluid Mechanics (NNFM) Volume 34

Series Editors: Ernst Heinrich Hirschel, München
Kozo Fujii, Tokyo
Bram van Leer, Ann Arbor
Keith William Morton, Oxford
Maurizio Pandolfi, Torino
Arthur Rizzi, Stockholm
Bernard Roux, Marseille

(Adresses of the Editors: see last page)

Volumes 1 to 3, 5 to 7, 10, and 24 are out of print.

Numerical Solutions of the Euler Equations for Steady Flow Problems

by
Albrecht Eberle,
Arthur Rizzi,
and Ernst Heinrich Hirschel

vieweg

Die Deutsche Bibliothek – CIP-Einheitsaufnahme

Eberle, Albrecht:
Numerical solutions of the Euler equations for steady flow problems / by Albrecht Eberle, Arthur Rizzi, and Ernst Heinrich Hirschel. – Braunschweig; Wiesbaden: Vieweg, 1992
(Notes on numerical fluid mechanics; vol. 34)
ISBN 3-528-07634-8

NE: Rizzi, Arthur:; Hirschel, Ernst Heinrich:; GT

Vieweg is a subsidiary company of the Bertelsmann Publishing Group International.

Produced by W. Langelüddecke, Braunschweig
Printed on acid-free paper
Printed in Germany

ISSN 0179-9614
ISBN 3-528-07634-8

Foreword

The last decade has seen a dramatic increase of our abilities to solve numerically the governing equations of fluid mechanics. In design aerodynamics the classical potential-flow methods have been complemented by higher modelling-level methods. Euler solvers, and for special purposes, already Navier-Stokes solvers are in use.

The authors of this book have been working on the solution of the Euler equations for quite some time. While the first two of us have worked mainly on algorithmic problems, the third has been concerned off and on with modelling and application problems of Euler methods.

When we started to write this book we decided to put our own work at the center of it. This was done because we thought, and we leave this to the reader to decide, that our work has attained over the years enough substance in order to justify a book. The problem which we soon faced, was that the field still is moving at a fast pace, for instance because hypersonic computation problems became more and more important.

The book thus reflects mainly our own work up to the year 1990, and is less a review of the field. It puts emphasis on the finite-volume concept: Chapter IV, on centered differencing: Chapter V, on upwinding: Chapter VI, and on the modelling of vortex flows: Chapter X. In order to demonstrate the versatility of present-day Euler methods, the large Chapter XI on methods in practical applications was added. In this chapter also results from other solution approaches are presented.

Naturally, the responsibility for each of the main chapters was in the hand of only one of us. The initiated reader of course can tell how the relations are. We have aimed for a high degree of commonality in the nomenclature, however we have not enforced it strictly. Hopefully this will not impede the reading of the book.

Our aim has been to give an introduction into the numerical solution of the Euler equations for the beginner, but also to give a coherent picture of the field for the expert. The aerodynamic designer who is not an expert in solution methods, should find all the background information he needs to know about his tool, even if it is a finite-difference method. Especially he will get an idea about the potential and the limitations of the Euler approach in design work, and also about the possibilities for instance to take into account viscous effects. He can be assured that it is possible to compute vortical flows with Euler methods, and that the solutions are right in principle, if certain pitfalls are avoided. Finally the researcher in the field will see that many important problems still exist, and that the field is far from being finished. The biggest challenge, however, will be future interdisciplinary applications, part of which are already in work in the area of hypersonic aerothermodynamics.

Spring 1991

A. Eberle
A. Rizzi
E.H. Hirschel

Acknowledgements

Many persons and institutions have provided us with information and material, or have given us their permission to reproduce graphs and figures.

We wish to thank particularly the following colleagues for material which they made available for us: N.C. Bissinger, L. Fornasier, F. Monnoyer, Ch. Mundt, M.A. Schmatz, W. Schwarz, W. Staudacher, K.M. Wanie (all from MBB-Flugzeuge), M. Hartmann, M. Pfitzner, C. Weiland (all from MBB-Space Division), H. Felici, I.L. Ryhming, A. Saxer (all from IMHEF/EPFL), A. Jameson (Princeton University), H. Viviand (Principia), G. Krukow (BMW), L.-E. Eriksson (Volvo Flygmotor), B. Engqvist (UCLA), B. Gustafsson (Uppsala University).

Thanks are due to Academic Press, American Institute for Aeronautics and Astronautics, Cambridge University Press, Elsevier Science Publications, Pitman Publishing, Royal Aeronautical Society, Springer Verlag, Zeitschrift für Flugwissenschaften und Weltraumforschung, and last not least the Vieweg Verlag with the Notes on Numerical Fluid Mechanics, for the permission to reprint various figures.

We state that copyright rests with the American Institute of Aeronautics and Astronautics of Figs. 10.8 to 10.11, 11.11 11.46 to 11.50, 11.52 and 11.53, 11.69 and 11.70, 11.83 and 11.84. The original versions of Figs. 11.37 to 11.39, and 11.41 to 11.43 were first published by the Advisory Group for Aerospace Research and Development, North Atlantic Treaty Organisation (AGARD/NATO) in Conference Proceedings CP437 Vol. 1 "Validation of Computational Fluid Dynamics" in December 1988. We gratefully acknowledge the permission to reprint these figures.

We acknowledge further the permission from FFA and MBB-Flugzeuge to use graphs and figures to illustrate the application potential of Euler methods.

FFA has very kindly taken the responsibility for the production of the camera-ready manuscript, and a number of people have devotedly engaged themselves in this over a number of years. For the typing of the manuscript we wish to extend our gratitude to Ms. Birgitta Arhed, Ms. Inez Engström, and Ms. Ingrid Jonason, who prepared the text and cheerfully accepted our many changes and revisions. We are equally grateful to Ms. Irmili Viitala-Larsson for her very conscientious drawing of the original figures and diagrams in the book. The book would not have been possible without the help of these four generous ladies.

A.E
A.R
E.H.H

CONTENTS

I HISTORICAL ORIGINS OF THE INVISCID MODEL

This book concerns the development and application of numerical methods to the solution of the Euler equations. A brief sketch of the evolution of the ideas that led to the formulation of these equations, the problems to which they were applied, and the efforts to solve them before computers were available then sets the proper perspective for an appreciation of the material that follows.

The histories of the exact sciences usually show a familiar and easily recognized pattern. Initially there is a period of unguided and unrelated experimentation when the simpler and more striking facts are discovered, often in a dramatic or unexpected fashion. At this stage the experimenters are working in the dark and any theories which may be formulated are necessarily speculative and, as often as not, utterly wrong. This period is generally followed by an era of consolidation, particularly on the experimental side, with results becoming more precise and numerical, if less spectacular, than before and a definite technique of experimentation begins to emerge. This stage is almost always characterized by an abundance of empirical formulae which have been invented, in default of a general theory, as a convenient way of summarizing results and as guides for future work. The science reaches maturity when the experimental probings have gone deep enough to reveal the true theoretical basis of the subject; the empirical formulae give place to exact mathematical theorems, which may ultimately attain the status of "natural laws", the truth of which is accepted without question by all concerned.

Fluid mechanics, of which aerodynamics is a part, has not entirely followed this familiar pattern of development. In mechanics and astronomy, and later in electricity and magnetism, optics, heat and thermodynamics, and other branches of mathematical physics, the calculus of Newton and Leibniz opened the way of swift and sure progress with theory and observation only rarely in conflict. In the eighteenth century Euler and Bernoulli applied the calculus to problems of fluid motion and so founded the classical hydrodynamics, which deals with motion in a hypothetical medium called an "ideal fluid". The classical hydrodynamics became a subject of immense attraction to mathematicians who made it so highly abstract as almost to deserve the name of "pure" mathematics. To engineers it had considerably less appeal, for they found its results either unintelligible or completely at variance with reality when applied to real fluids. In the eighteenth and nineteenth centuries the classical hydrodynamics, following its aim of producing a logical and consistent theory capable of yielding exact solutions of idealized problems, became a largely academic study and as such made little effective contribution to the problem of flight until it was realized that viscosity played an important role in very thin layers close to the wing surface, called boundary layers. If the places where these layers separate from the surface could be determined from some external criteria, the idealized theory of hydrodynamics supplemented, e.g. by the Kutta condition at a sharp trailing edge, could then become very practical indeed.

Presently we are in the midst of a debate about the understanding of "inviscid" separation observed in numerical solutions to the Euler equations and the creation of vorticity even when no Kutta condition is enforced explicitly. We shall present some of the current arguments of this debate in a later chapter. But first we think it is important to have an appreciation of the reasoning and thought processes that established our current understanding of the mechanics of fluids, all before the age of numerical solutions.

1.1 From Antiquity to the Renaissance

The study of fluids can be traced back to antiquity when Archimedes founded hydrostatics. The earliest serious analysis of the motion of a projectile in flight occurs in the Physica of Aristotle (384-322 B.C.) as a part of his proof of the impossibility of a vacuum. The "proof" that led to his "medium theory" is worthy of consideration because it is a vivid illustration of the gulf which sometimes lies between the ancient Greek mind and our own in the domain of physical science.

Briefly, Aristotle argues that a body such as an arrow can continue in motion only as long as force is continually applied to it and that any withdrawal of the force would immediately cause the body to stop. Further, since it was held to be impossible to conceive of anything in the nature of action at a distance, the force required some material medium in contact with the body for its transmission. Thus a projectile could not move in a vacuum, which is absurd, and hence a vacuum cannot exist.

The argument clearly requires that the air sustains the projectile in its flight, but Aristotle did not linger over this, to us the real problem. He suggested that the atmosphere might push the arrow along by rushing to fill the vacuum in its rear, or (rather more obscurely) that motion, once started, would be maintained by the air as a consequence of its "fluidity". In both cases the air sustained and did not retard the flight of the arrow.

In these arguments lie the characteristic features of the strength and the weakness of ancient Greek science. The strength lies in the power of conceiving a philosophical problem as an abstraction from a material phenomenon, a process which is still an essential feature of all mathematical physics. The whole argument, however, collapses and proves nothing because its premises are at variance with the essential physical facts, and this because of the absence of experimental data from Greek science. Such, however, was the authority of the Aristotelian doctrines that these views held sway, or were given serious consideration, right up to the middle ages of European culture.

The centuries which lie between Aristotle and the birth of modern science with Galileo and Newton hold little of interest for aerodynamics until we come to Leonardo da Vinci (1452-1519) who discarded the central concept of the Aristotelian scheme, that the air assisted the motion, and instead considered the atmosphere as a resisting medium[1]. This is the essential step, without which all aerodynamic theory would be in vain. Leonardo's concept is that of making a surface support a weight by the application of power to the resistance of the air, but he went wrong in his subsequent development of the idea. He supposed that the flapping motion of the wing of a bird causes the air in contact with it to "condense" and behave as a rigid body on which the bird is supported, the motion of the wing being sufficiently rapid to ensure that the stroke is completed before the local "condensation" is passed on to other layers of air. Soaring flight, in which the wings are held nearly motionless, Leonardo explained on the same hypothesis by saying (quite correctly) that what mattered was the relative motion of the air and the wing so that, given a favourable wind, the bird can soar without beating its wings.

A "condensation" process resembling that envisaged by Leonardo does occur in nature, but becomes appreciable only at very high speeds. What Leonardo did not know was that to attain any noticeable amount of local compression in a free atmosphere it would be necessary for the wings to be moving through the air at extremely high speed, so that his explanation could not possibly apply to, say, the flight of an eagle or any other bird which uses slow powerful wing beats. On the other hand his conception of the bird being able to fly because, and only because, the air offers resistance is essentially correct and marks a great step forward on the Aristotelian doctrine. He effectively posed that the fundamental problem for a theory is to explain quantitatively the details of resistance to motion. In the later Middle Ages machines came into use in small manufacturing, public works, and mining. During the Renaissance the clock was perfected, proved useful for astronomy and navigation, and quickly was taken as a model of the universe. Machines in general then led to theoretical mechanics and to the scientific study of motion and of change. Galileo Galilei (1565-1642) advanced the theory of hydrodynamics further by introducing the concepts of inertia and momentum and laid the foundation for the study of the dynamics of bodies[2]. Above all we owe to Galileo, more than to any other man of his period, the spirit of modern science based on the interplay of experiment and theory, with stress on the intensive use of mathematics. He knew that a rolling ball experiences both friction and especially air resistance which he tried to understand but incorrectly conjectured to be proportional to velocity. He did succeed in proving experimentally that a body floats, not because of it shape as Aristotle believed, but because of its density relative to the density of the fluid in which it is immersed. Galilei also showed that the effect of force was not to produce motion, but to change motion, that is to produce acceleration. But it was Sir Isaac Newton (1642-1727) who later gave this conclusion the mathematical form of his famous Second Law upon which all branches of dynamics including hydrodynamics could grow on a sound philosophical basis. Blaise Pascal (1623-1662) developed these ideas further into a coherent theory of hydrostatics by establishing the basic laws for water at rest. He proved that the pressure at any point within a fluid is the same in all directions and depends only on the depth. Earlier investigators all held the hypothesis of simple proportionality between resistance and velocity until Christian Huyghens (1629-1695) found the proportionality of the resistance to the square of the velocity to be more in accord with his experiments. Newton later derived this law, which Huyghens discovered experimentally, by deduction for certain special flow conditions. It is usually referred to as Newtonian theory.

1.2 The Enlightenment: the Age of Reason

Newton began his study by stating that resistance depends on three factors: the density of the fluid, its velocity and the shape of the body in motion. He observed also that different phenomena contribute to resistance. One is due to inertia and varies with the density of the fluid, but a second results from the viscosity of the fluid itself, and the third from the friction between the body and the fluid. Both these latter parts, Newton reasoned, could only be very small especially at high velocities and therefore could be neglected in the first analysis. But resistance coming from the inertia of matter must always be accounted for because it constitutes the essential mechanical property of matter. Thus began classical work on the understanding of ideal inviscid fluids, and as far as the description of drag is concerned led to the paradox of d'Alembert (1717-1783).

Mathematical productivity in the eighteenth century concentrated on the calculus and its applications to mechanics. The major figures in this activity were Leibniz, the Bernoulli brothers, Euler, Lagrange, and Laplace and others all closely connected with the philosophers of the Enlightenment. Scientific activity usually centered around academies, and not universities, of which those at Paris, Berlin, and St. Petersburg were outstanding. It was a period in which enlightened rulers like Louis XIV, Frederick the Great, and Catherine the Great surrounded themselves with learned men.

1.2.1 Leonhard Euler

Up to this time the development of fluid mechanics rested heavily on experimental observations. This changed with Leonhard Euler (1707-1783), the great mathematical architect and founder of fluid mechanics as a true analytical science. He is the key figure in 18th century mathematics and the dominant theoretical physicist of the century. He published more than 500 books and papers during his life-time, and it has been estimated that during his adult life he averaged writing about 800 pages a year[3].

In 1741 Frederick the Great invited him to join the Berlin Academy and it was there that he did his greatest work, including infinite processes of analysis, the function concept, infinite series, calculus of variations, differential equations, analytical and differential geometry, topology, number theory, astronomy, and mechanics. A flourishing branch of eighteenth century mathematics to which Euler made many contributions, and of special interest to us here, was that of differential equations. He investigated the existence of singular solutions of first-order equations, and stated the conditions for such equations to be exact. He also developed the technique of making a change of variable for solving second-order equations. Euler's work on problems of elasticity theory later led him to consider the general solution of a differential equation, of whatever order, with constant coefficients. He gave a complete treatment of this problem, first for homogeneous, and then for non-homogeneous equations. Euler also investigated systems of linked differential equations, with their obvious application to dynamical astronomy.

The study of partial differential equations really began, almost as a by-product, as part of the sustained eighteenth century attack on the problem of the vibrating string. Euler's most important paper on the subject, "On the Vibration of Strings", appeared in 1749. He extended his earlier concept of a function to include piecewise continuous curves. In 1748 he established that

$$y = \phi(ct + x) + \psi(ct - x),$$

where ϕ and ψ are arbitrary functions, is a solution of the wave equation

$$\frac{\partial^2 y}{\partial t^2} = c^2 \frac{\partial^2 y}{\partial x^2} .$$

He took the view that the only relevant part of the initial curve $y=f(x)$, representing the shape of the string at $t=0$, is that for which $0 \leq x < L$, where L is the length of the string. He proved that after a time t the transverse displacement of the string at the point x will be given by[4]:

$$y(t,x) = \frac{1}{2} f(x + ct) + \frac{1}{2} f(x - ct) .$$

He obtained a number of special solutions and advanced the idea of the superposition of principal modes of vibration. Several alternative treatments of the vibrating string problem were put forward by other mathematicians, notably by John and Daniel Bernoulli, d'Alembert, Lagrange and Laplace. The debate between them raged for many years. Euler also investigated various extensions of the wave equations, e.g. where c is variable, or where the equation contains additional terms or relates to more than one dimension.

In laying the analytical foundations of hydrodynamics Euler overcame a fundamental contradiction between the concepts of mathematics and mechanics. For example, a point is usually defined as an element of geometry which has position but no extension; a line is defined as a path traced out by a point in motion; and motion is defined as a change of position in space. But motion and matter cannot be divorced. A point that has no extension lacks volume and, consequently mass, and can have no momentum. Instead he introduced his historic "fluid particle" concept and thus gave fluid mechanics a powerful instrument of physical and mathematical analysis. A fluid particle is imagined as an infinitesimal body, small enough to be treated mathematically as a point, but large enough to possess such physical properties as volume, mass, density, inertia, etc. A fluid particle to Euler was not a mathematical construct, but a physical point possessing volume, weight, mass, and density.

Its mass is $dm=\rho dV$, whose integration

$$m = \iiint_V \rho \, dxdydz$$

represents, in fact, the law of conservation of mass in fluid flows. Moreover, he recognized that pressure was a point property that varied throughout a flow, and that differences in the pressure at two different points provided a mechanism to accelerate a fluid particle.

He treated incompressible flow in a paper in 1752; 3 years later he generalized his equations to embrace compressible flow. His approach was to consider the forces acting on a volume-element of the fluid of density ρ, subject to a pressure p and to external forces with components (X,Y,Z) per unit mass. He put these ideas in terms of an equation expressed, in the usual notation, as

$$\frac{Du}{Dt} \equiv \frac{\partial u}{\partial t} + u\,\frac{\partial u}{\partial x} + v\,\frac{\partial u}{\partial y} + w\,\frac{\partial u}{\partial z} = X - \frac{1}{\rho}\,\frac{\partial p}{\partial x}\,,$$

with two similar equations for the material derivatives

$$\frac{Dv}{Dt} \quad \text{and} \quad \frac{Dw}{Dt}\,,$$

which he derived for the momentum balance during his residence in St. Petersburg in 1748. He also generalized d'Alembert's equation of continuity to include compressible flow. Euler went further to explain that the force on an object moving in a fluid is due to the pressure distribution over the objects surface. But the derivation of the energy equation did not come about until the development of thermodynamics in the nineteenth century, beginning with B. de Saint-Venant who used a one-dimensional form of the energy equation to derive an expression for the exit velocity from a nozzle in terms of the pressure ratio across the nozzle.

These non-linear differential equations that result are very difficult to handle. As Euler put it in his 1755 paper: "If it is not permitted to us to penetrate to a complete knowledge concerning the motion of fluids, it is not to mechanics, or to the insufficiency of the known principles of motion, that we must attribute the cause. It is analysis itself which abandons us here, since all the theory of the motion of fluids has just been reduced to the solution of analytical formulae"[5]. Even so, Euler continued to work on the subject and was indeed engaged in writing a treatise on hydromechanics at the time of his death[3].

He tried various ways to simplify his equations, including the idea of a velocity potential. In "Principles of the Motion of Fluids" (1752) he found what we now call Laplace's equation for the velocity potential. In dealing with the components u,v, and w of the velocity of any point in a fluid, Euler had shown that udx+vdy+wdz must be an exact differential. He introduced the function S such that dS=udx+vdy+wdz.

Then $u = \frac{\partial S}{\partial x}\,, \quad v = \frac{\partial S}{\partial y}\,, \quad w = \frac{\partial S}{\partial z}\,.$

When the continuity of incompressible motion is taken into account, it follows that

$$\frac{\partial^2 S}{\partial x^2} + \frac{\partial^2 S}{\partial y^2} + \frac{\partial^2 S}{\partial z^2} = 0\,.$$

How to solve this equation generally, Euler says, is not known, so he considers just special cases where S is a polynomial in x,y, and z.

We should add that Euler did not use these names. The idea that a force can be derived from a potential function, and even the term, "potential function" were used by Daniel Bernoulli in Hydrodynamics (1738). The potential equation only became known as the Laplace equation after the publication of Laplace's five volume work, Celestial Mechanics, in 1825 for his work on the problem of gravitational attraction. The function S was later called the velocity potential by Helmholtz in 1868 in analogy with the term potential which Gauss introduced into mechanics in 1840.

1.3 The 19th Century: Mathematical Fluid Mechanics

Louis de Lagrange (1736-1813) also came to the conclusion that Euler's equations could be solved only for irrotational flows. For such flows he derived a general integral of Euler's equations, most usually, but mistakenly, ascribed to Daniel Bernoulli. The reduction of Euler's equations of motion to a single Laplace equation made it possible to carry out extremely complex mathematical operations. It became a cornerstone in the mathematical theory of fluid mechanics. But as this theory grew more refined from Newton to Euler and Lagrange, mathematicians began to recognize its shortcoming in the prediction of the drag of bodies. The first man to develop what we may call a rational theory of air resistance was d'Alembert, a great mathematician and one of the Encyclopaedists of France. He published his findings in a book called Essai d'une Nouvelle Théorie de la Resistance des Fluides. In spite of his important contributions to the mathematical theory of fluids, he got a negative result. He ends with the following conclusions:

"I do not see then, I admit, how one can explain the resistance of fluids by the theory in a satisfactory manner. It seems to me on the contrary that this theory, dealt with and studied with profound attention, gives, at least in most cases, resistance absolutely zero; a singular paradox which I leave to geometricians to explain".

This statement is what we call the paradox of d'Alembert. It means that purely mathematical theory leads to the conclusion that if we move a body through the air and neglect friction, the body does not encounter resistance. Evidently this was a result which could not be of much help to practical designers, and is far from what we experience in reality.

1.3.1 Vortex Discontinuities and Resistance

With Hermann Helmholtz (1821-1894) hydrodynamics made the most most notable progress since d'Alembert, Euler and Lagrange. Like them he also investigated irrotational motion. However, Helmholtz realized, as Euler had already observed in his paper "Principes Généraux du Mouvements des Fluides" of 1755 that there may be cases in which no velocity potential exists. Helmholtz went on to discover special cases of such motion not satisfying a velocity potential which he named "vortex motions", and thus opened a new field of research. Before Helmholtz it was generally held that flow without vorticity was a well-founded theoretical assumption. Doubts arose when large discrepancies with reality had been recognized, and even Euler had already pointed out that the assumption of potential flow is not always justified. But it was Helmholtz who openly rejected the potential assumption and, with his paper on vortical motions, opened the door to important new discoveries. This was the beginning of the classical theory of vorticity. His studies of jets in air led to the hypothesis that a surface of discontinuity forms in the velocity field. Up to this time such discontinuities in velocity had not been considered by the theory. He argued, however, that there is nothing in the theory that forbids two adjacent fluid layers from slipping past one another with a finite velocity, i.e. a tangential discontinuity, or vortex sheet. Helmholtz did retain the idea that viscous forces may be neglected in a fluid with vorticity. These subtle but extremely important differences among the concepts of potential flow, fluid with vorticity, and viscous flow are expressed in precise terms as follows: In potential flow vorticity is zero in the whole fluid (except for singular points), and viscous forces are neglected. In inviscid flow, a mathematical model of real flow, vorticity is present but viscous forces are still neglected (no diffusion of vorticity). No restricting assumptions at all on vorticity and viscous forces apply in real viscous flow. The idea of the Helmholtz surface of discontinuity prompted Lord Rayleigh (1842-1919) in 1876 to propose an inviscid wake behind a plate perpendicular to a stream which then experiences drag and thus solves the d'Alembert paradox. That Kelvin's theorem on the constancy of the circulation is not violated by vortex-sheets springing from the boundaries of immersed bodies, has been clearly explained by Prandtl. And so began another line of attack on the problem to understand resistance[6].

The first theoretical deduction of a formula for the drag was provided by the so-called theory of free stream-lines, developed for flow past a flat plate, by Kirchhoff and Rayleigh according to the methods used by Helmholtz for two-dimensional jets, and extended by Levi-Civita and others to the case of curved rigid boundaries. According to this theory there is, in two-dimensional flow past a flat plate, for example, a mass of fluid at rest behind the plate, separated from the stream by two stream-lines springing from the edges of the plate, as in Fig. 1.1.

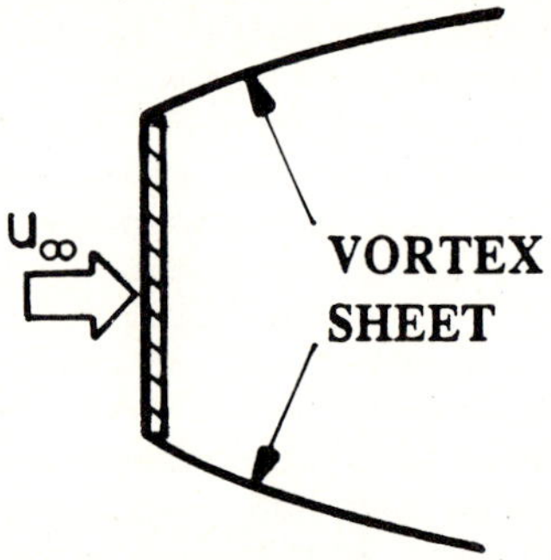

Fig. 1.1 Flow past flat plate separating in vortex sheets

The velocity is discontinuous across these stream-lines, which are therefore the traces of vortex-sheets. The velocity just outside the "free" stream-lines is constant, and equal to the velocity u_∞ of the undisturbed stream. The pressure in the stagnant fluid is constant and equal to p_∞, the pressure at infinity. This theory has considerable theoretical, but very little practical importance, its results being largely in disagreement with the results of observation in real fluids of small viscosity. Thus, for two-dimensional motion past a flat plate at right angles to the stream, the theoretical result for the drag coefficient, $D/(1/2\ \rho u_\infty{}^2 b)$, is 0.880, where b is the breadth of the plate: the measured value is nearly 2. The discrepancy arises largely from the fact that there is actually a defect of pressure, or suction, at the rear, the pressure being much less than p_∞. For the plate at right angles to the stream it is nearly constant and equal to $p_\infty - 0.7\rho u_\infty{}^2$ right across the rear, and similar features are present in other typical cases. In other important respects the theory is widely at variance with reality, since behind a bluff obstacle in a stream the observed motion either is an irregular, eddying one, or for two-dimensional motion at certain Reynolds numbers has, for some distance behind the obstacle, the appearance of a double trail of vortices with opposite rotations. Even if a motion like that in Fig. 1.1, or any similar one with vortex-sheets, is allowed to occur in an inviscid fluid, it would not persist, since it would be unstable. The notion that the stream leaves the plate at the edges is, however, valuable and in accordance with reality; and a vortex-sheet (more accurately, for real fluids, a thin vortex-layer) does begin to be formed from the edges of the plate. But this vortex-sheet or layer is not fully developed either in a real or an inviscid fluid; it curls round on itself, and something much more in the nature of concentrated vortices is formed.

To summarize what was the state of affairs around 1900, there was a mathematical theory of the mechanics of ideal, i.e. non-viscous, fluids. The first result of this theory was the paradox of d'Alembert, stating that the resistance of a body moving uniformly in a non-viscous fluid is zero if the fluid closes behind the body. If a "separation" of the flow from the body is assumed, as for example by Rayleigh, the theory leads to a value of the force quantitatively at variance with experimental facts[7].

One needs of course to bring viscosity into account in order to predict drag with quantitative accuracy. The use of vortex sheets then is one of modelling. If one has to start with the solution of the Euler equations, because the Navier-Stokes equations are too difficult to solve, one must introduce the discontinuity line artificially, that is, in an axiomatic way. This is certainly true for smooth bluff bodies. If the body has a streamline shape and a salient edge, the vortex-sheet theory comes much closer to reality, as evidenced by the numerical solution of the Euler equations for flow around airfoils with sharp trailing edges, because the sharp edge fixes the point of separation. In three dimensions reasonable predictions are given for flow past delta wings with sharp leading edges, presumably because, unlike the bluff body, the shed vortex-sheet here is stable due to the rollup process. But outside of these two regimes, one must be rather dubious about the realism of an Euler solution because of the uncertainty in the place of separation.

1.3.2 Shock Waves

The explanation of drag was also sought by investigating other phenomena, namely compressibility. Galileo was the first to point out that the denser the fluid, the greater is its resistance. Christian Huyghens, René Descartes (1596-1650) and others developed this obvious fact of nature into the broader concept that a given body moving in a homogeneous fluid can experience different levels of drag. But the conclusion lacked clarity and remained abstract for a long time. Trying to work out the significance of compressibility, Euler and especially Lagrange sought mathematical relationships between density and pressure, but made little progress. Rather it was Helmholtz's ideas on a surface of tangential discontinuity that led others to explore further and generalize the concept of discontinuities. A tangential discontinuity, or vortex sheet, is a stream surface and no fluid passes through it. Once Felix Savart (1791-1841) showed that sound waves propagate in water in the same way as in solids, the way opened to see discontinuities as propagating waves. Pierre Henry Hugoniot (1851-1887) put this together in his concept of an "acceleration wave". He wrote the mathematical equations that hold across such surfaces and proved that there are only two kinds of discontinuities possible in a non-viscous compressible fluid: 1) longitudinal ones which propagate as a wave with finite velocity, and 2) transverse ones which only move with the fluid particles. G.F.B. Riemann (1826-1866), however, was the first who tried to calculate the relations between the states of the gas before and behind the shock wave. He mistakenly thought that the change across a shock is isentropic, and his results, of course were wrong. They were corrected by W.J.M. Rankine (1820-1872), the Scottish engineer, in a paper "On the Thermodynamic Theory of Waves of Finite Longitudinal Disturbance" in the Philosophical Transactions of the Royal Society, 1870. By assuming that the internal structure of the shock wave was a region of dissipation, and not isentropic, he was able to derive the equations for the change across the shock. Hugoniot independently rediscovered these equations and published them in a paper in 1887 in a form much like those we use today.

Both Rankine and Hugoniot noted that rarefaction as well as compression shocks were possible mathematically, but the ambiguity was not resolved until 1910. First Lord Rayleigh (1842-1919) and then G.I. Taylor invoked the second law of thermodynamics to show that a compression shock is the only one physically possible. Thus the fundamental understanding of shock waves had evolved over the course of 40 years.

The phenomenon of shock waves was also being studied experimentally[2]. Ernst Mach (1838-1916), an Austrian physicist and ballistician used the schlieren technique to make shocks visible. Invented by A. Töpler (1836-1912), a German physicist who worked in the field of acoustics, this technique is an optical system that records density changes. In Berlin the ballistician C. Crantz studied experimentally the relationship between the speed of its flight and the drag of a bullet, the behaviour of the air in front of a fast moving body, and the influence of the shape of a body on the buildup of drag. He also conceived the idea of the shock tube. Later J. Schatte further refined the optical method of visualization of waves and flow patterns. His outstanding schlieren and interferometer pictures of supersonic flow and shock waves created by bullets received an enthusiastic reaction from many physicists, aerodynamicists, and military experts. Mach, however, was the first to show that compressibility effects in gas depend not simply on the flow velocity but rather on the ratio of velocity to the speed of sound. In honor of his work J. Ackeret named this ratio the "Mach number".

While Mach was studying the ballistics of supersonic projectiles, another field of progressing technology in compressible flow was the design of turbines, starting with steam turbines late last century, and extending to gas turbines early this century with the work of A. Stodola in Zürich.

1.4 The 20th Century: The Computational Era

In 1888 the Swedish engineer, Carl G.P. de Laval, constructed a single stage steam turbine whose blades were driven by a stream of high-pressure steam from a series of novel convergent-divergent nozzles to speeds previously unattainable, over 30,000 revolutions per minute. He was not entirely certain in 1888 that the flow actually reached supersonic speed in his "Laval nozzle". The possibility of supersonic flow in such nozzles had been established theoretically but it had not been verified experimentally, and therefore it was a matter of controversy. In 1903 Stodola measured the pressure distribution along the axis of a convergent-divergent nozzle. In his data he found a large increase in pressure near the exit which agreed with the shock equations of Rankine and Hugoniot. This was the first real quantitative verification of the shock-wave theory. The news of this discovery reached Ludwig Prandtl in Göttingen who went on to contribute greatly to the understanding of quasi-one-dimensional supersonic nozzle flow by producing outstanding schlieren photographs of shocks and Mach waves in the flow.

1.4.1 Early Methods

Some effort was then made to describe the flow pattern mathematically in the minimum cross section of the Laval nozzle and the shape of the plumes in the jet behind such nozzles by seeking solutions to the Euler equations. Quite independently B. Riemann in 1860, P. Molenbrock in 1890, and S.A. Chaplygin in 1902 all came upon the idea of introducing new independent variables to transform the Euler equations into equations in the hodograph plane where they were linearized. However, these attempts came before the boundary-layer theory established the practical value of ideal mathematical flow theory, and they did not attract the immediate interest of large groups of scientists and engineers[7].

During the period 1905 to 1908 Prandtl and Theodor Meyer extended normal shock wave theory to two dimensions and laid down the fundamentals of both oblique shock and expansion wave theory for supersonic flow. Meanwhile the growth of the young airplane industry added a practical interest to the advancement of compressible flow theory in the second decade of the century. Although the flight speeds of all airplanes at that time were certainly within the realm of incompressible flow, the tip speeds of the propellers regularly approached the speed of sound and focussed attention on the effects of compressibility on propeller airfoils. The basis for the theoretical approach here rested on the linearization of the governing equations followed by a search for an analytic solution. Prandtl, Jakob Ackeret, H. Glauert among others made major contributions. Mathematicians had been developing the theory of characteristics as a means to solve general systems of partial differential equations of first order. The French mathematician Jacques Salomon Hadamard in 1903 and the Italian Tullio Levi-Civita in 1932 made major contributions. But Prandtl and Adolf Busemann in 1929 were the first to apply this method to supersonic flow problems, and found exact non-linear solutions to the Euler equations for two-dimensional supersonic flow. Busemann went on to use this method to design supersonic nozzles and ultimately the first practical supersonic wind tunnel in the mid-1930s. But in a potential formulation of supersonic flow around a pointed cone G.I. Taylor and J.W. Maccoll solved the resulting ordinary differential equation for conical flow by numerical integration in 1933. It signified the use of numerical methods for compressible flow problems.

For the most part the development of the airplane at that time called for the analysis of low-speed subcritical flow. That meant that viscosity and vorticity were neglected, and the flow model was the Laplace equation. Based on the theory of complex variables, the approach taken was analytical, the superposition of elementary solutions[8]. As the aerodynamic shapes under investigation grew more complex, it matured in later decades into the computational singularity techniques called boundary integral, or more commonly, panel methods[9]. Viscosity was accounted for by solving the boundary-layer equations of Prandtl's theory using finite-difference methods and mechanical calculating machines[10]. Later there were also attempts, in an iterative fashion, to couple together the external potential flow with the boundary-layer solution. The relaxation method was also being applied to solve the Laplace equation by finite differences[11].

Meanwhile the mathematicians like Hadamard, Courant, and Friedrichs, were building the theory of hyperbolic partial differential equations, with the goal of understanding the fundamental issues like the well-posedness of the problem, the propagation of waves, the smoothness of the solution, and its uniqueness. It was in establishing a fundamental result on uniqueness that led Courant, Friedrichs, and Lewy to their famous stability condition, necessary for the analysis of any practical computing method[12].

1.4.2 Methods to Solve the Euler Equations: 1950-1970

During the 1940s, however, the two groups, the theoreticians and the practitioners, began to draw closer together. The advent of the jet plane, supersonic missiles, and high-energy blast waves brought demands for solutions to practical problems that went beyond the reach of methods based on the current theory of potential and linear hyperbolic equations. The heart of the difficulty was the numerical treatment of the non-linear occurence of shock waves. This instigated a large effort by von Neumann, Richtmyer, Lax, and others working closely with computing methods to establish a mathematical theory of non-linear hyperbolic conservation laws for the purpose of computing flows with shocks. (The book by Fox[13] reflects how far these efforts progressed during the 1950s.)

But because many of the transonic and hypersonic problems in aerodynamics are steady, the aeronautical community did not immediately embrace the newly emerging hyperbolic methods. Instead, as was commonplace during the earlier decades, special methods were sought to solve the specific non-linear steady problem. The so-called blunt-body problem is a good example[14]. When a blunt obstacle travels through air at a constant supersonic speed, a shock wave appears in the flow, termed a bow shock because it stands detached from and ahead of the body. If the goal is to predict the location of the bow shock and the flow properties between it and the body, then the appropriate model is the steady Euler equations. Except in a small region between the body and the shock, the speed of the flow is always supersonic. This subsonic pocket is what characterizes the problem and makes it difficult because the equations are of mixed type -- elliptic within the pocket and hyperbolic outside where the flow is supersonic. No general mathematical theory has been proposed to solve mixed-type equations, but a number of special methods were devised in the late 1950s to solve specifically the blunt-body problem. Among them were Van Dyke's inverse method[15] that first assumed a shape for the bow shock and performed an unstable but controllable numerical march from it inward to determine the corresponding body shape, and then adjusted the shock shape until the desired body was obtained. Another one was Dorodnitsyn's method of integral relations which reduced the problem to a set of coupled ordinary differential equations[16,17]. All of these specialized blunt-body methods, however, were restricted to flows at substantial supersonic speeds. The other important aerodynamic problem of mixed type, the case of subsonic but supercritical flow past an airfoil where now a supersonic pocket is embedded in a subsonic field, for example, could not be solved satisfactorily by these methods. The solution of the transonic airfoil problem was first obtained in 1970 by the relaxation procedure of Murman and Cole for the non-linear small-disturbance potential equation and was the initial use of an upwind scheme in aerodynamics. Oswatitsch and Rues[18] present a good survey of the methods being used up to 1975 to solve transonic aerodynamic problems. It is significant to point out that at the Symposium Transsonicum II in 1975 only one paper (by Rizzi) was given on the use of the hyperbolic time marching method to solve the Euler equations for transonic flow. And even at the Workshop on Numerical Method's for Transsonic Flow in 1979 in Stockholm the majority of the papers still treated the potential equations. It was not until the 1980s that the interest in solving the Euler equations for transonic aerodynamics blossomed out.

For truly time-varying flow problems, however, practitioners of Computational Fluid Dynamics (CFD) primarily in fields other than aerodynamics like meteorology, plasma physics, and geophysics, were beginning to apply the theory that the mathematicians had been laying down for hyperbolic evolutionary equations[12,13]. By now the development of the theory had advanced from purely linear problems to the understanding of weak solutions to conservation laws. During the 1960s news of the success with the general time-dependent hyperbolic approach in these other fields spread to the aerodynamics community where it was adapted for the solution of steady flows. The idea was to integrate the unsteady hyperbolic full potential and Euler equations forward in time, while maintaining steady boundary conditions so that, as all the transient fluctuations began to disperse, the steady state was reached asymptotically. Although it demands more arithmetic operations, the resulting time-asymptotic method proved to be both more effective and applicable to a wider class of problems than any of the other more specialized methods, e.g. the blunt-body procedures. This conclusion came about in part because of the broad latitude for algorithm modification afforded by the underlying hyperbolic theory. Another factor was the newly developed stability theory for difference approximations of time-dependent partial differential equations by Lax, Kreiss and others (see Ref. 19). Perhaps an even greater influence on the development came from the increasing computer power which became generally available at that time, making the additional computational work irrelevant. Supercomputer in its day, the Control Data 6600 appeared in 1964 with the power of 1 Mflops, (1 million floating point operations per second) and was followed by the 7600 in 1968 offering 4 Mflops. And if the user was willing to program with special assembly-language techniques, the performance of these two machines could be doubled to 2 and 8 Mflops respectively. Here then is another important and recurring theme of CFD. If there are computing machines readily available that can carry out the calculations in a reasonable period of time, it can be more feasible to use a more straight-forward method built from a general theory, even though it requires more computational work, than to use a more detailed method based on a narrower theory with limited application. The blunt-body problem with its various methods is a case in point. Rusanov[20-21] and Moretti[22] were two pioneers of the time-asymptotic approach for the blunt-body problem which now is used almost to the exclusion of all specialized methods in aerodynamics. The need to study the flow patterns around the space shuttle was one of the driving forces in the development of this technique, and led, for example, to the first application of the finite-volume method to the blunt-body problem by Rizzi and Inouye[23].

1.4.3 Methods to Solve the Euler Equations: 1970-1990

With the formulation of the problem decided, the debate in the 1970s then circled around the treatment of shock waves. Traditional thinking suggested that if a discontinuity exists in the flow, it should be treated as an internal boundary. Once the shock is located, appropriate boundary conditions can be prescribed across it, and then the regions of smooth flow on either side can be handled with well-established methods. The concept is one of tracking or fitting the shock by special features of the algorithm to the surrounding flow. All the blunt-body methods of the 1970s treated the bow wave in this way because it can be intense, and yet has a simple geometry that is easy to track. Based on the prior theoretical and computational work of von Neumann, Richtmyer, and Lax, the alternative concept was to disregard the shock as an internal boundary and rather compute the entire flow as an approximate weak solution to properly formulated conservation laws, often called the shock-capturing approach. Among others Rizzi and Bailey[24] took a hybrid approach in the study of hypersonic flow past the space shuttle at $M_\infty=20$ and $\alpha=40$ deg. Using the split MacCormack scheme in finite-volume form, they tracked the bow wave, but captured all of the other weaker shocks that develop around the canopy and wing leading edge at this high angle of attack. Although largely successful, this concept of capturing a shock does produce in practice small but unwanted side effects like anomalous oscillations in the solution near each side of the discontinuity. Such effects have in turn spurred the theoreticians to formulate entropy conditions in order to obtain the correct jumps in the solution, and to refine the details of the differencing scheme in order to avoid the undesirable oscillations. These theoretical endeavours to capture shock waves with more and more accuracy continue under lively development today.

Propelled by the current crop of supercomputers, of which the first was the CRAY 1, these hyperbolic methods for non-linear conservation laws are being applied now to simulate a host of non-smooth flow phenomena including fundamental macro-scale structures like shock waves, vortices, and wakes, as well as complex interactions involving shock waves and boundary layers, shock waves and vortices, thermal heating layers, base flows, vortical instabilities, and even transition and turbulence. As their predecessors did in the past, these demanding practical applications today are going beyond the frontiers of the existing theory and thus are helping to motivate further theoretical work to move them ahead. Examples are the recent attempts to capture vortex-sheets in numerical solutions and to compute unstable flows, i.e, to solve ill-posed mathematical problems.

The problems of capturing vortex sheets in Euler solutions also led to a new consideration of the vortex singularities of potential theory. The classical view is based completely on vortex laws. As is shown in Chapter X such singularities now are understood as high Reynolds number limits of boundary layers and wakes. Especially the wakes of finite lifting wings with their different kinematic properties are understood better now in their potential flow and Euler limit.

1.5 Brief Overview of Field

The previous section simply hinted at the evolution in the work during the past two decades. It is not really our intention here to present a comprehensive review of all the methods conceived by all the workers over the last twenty years, because of the sheer magnitude of the task. Indeed, a critical and analytical review of this body of literature might well run to book-length on its own.

Our goal is somewhat different in this book. Instead of a comprehensive presentation and discussion of all methods developed to solve the Euler equations, we focus, to a very large extent, on our own methods and the results which have been obtained with them. In this regard it is less a textbook where one can grasp an entire field, and more a monograph which sums up our own work over the last ten years. Readers searching for the former type of book are hereby forewarned that they may be disappointed.

However, even a monograph is not complete without a cursory review of other work in the field, and this is the purpose of this section. It is not intended to be exhaustive, we can only sketch the main line of development carried out by the main workers in our field of interest: numerical solution of the Euler equations applied to computational aerodynamics. The sketch begins by broadly outlining the major reference sources.

1.5.1 Secondary Reference Sources

With our limited ambition in this survey, it is not possible to cite all the papers in which the methods we wish to review are published for the first time, the so-called primary references. To a large degree, instead, we rely on textbooks and proceedings of workshops and conferences devoted to the theme of this book, i.e. secondary reference sources. The interested reader should be able to follow the references given in the secondary sources back to the primary papers.

As we saw in the preceeding sections, the methods in the 1960s were primarily limited to the solution of the blunt-body problem. It was not until 1970 that Magnus and Yoshihara[25] attempted to solve the Euler equations for transonic flow past an airfoil, a central problem of computational aerodynamics. The Symposium Transsonicum II[18] provides a good glimpse of the status of the field half-way through the decade of the 70s. This symposium marked a turning point from classical analytical methods to numerical methods. The overwhelming majority of the numerical methods were being used to solve the non-linear transonic potential equation. Very few papers at that symposium were concerned with solving the Euler equations. This situation changed rather rapidly over the next four years, as evidenced at the GAMM Workshop[26] in 1979 in Stockholm. There the mix of methods treating the potential equation or the Euler equations was in much better balance. It was here that workers began to realize the limits to the potential model (sufficiently weak shocks) and its deficiency (non-unique solutions). If the work in the 1960s and 1970s brought about the awakening of this field, it was the 1980s that saw the watershed in the number of methods and the number of workers contributing to the field. These years witnessed the full blossomming and maturing of the discipline. Good sources of reference for this active period are the AIAA papers that are proceedings of the fluid-mechanics-related conferences. In particular we can point out the bound proceedings of the CFD Conferences in 1981, 1983, and 1987, as well as the Tullahoma short courses[27]. Proceedings of the INRIA Workshop[28] provide a good overall summary of the status for the first third of the decade, but perhaps the most extensive comparison and analysis of the methods of this era is given by the AGARD Working Group 7[29]. The two-week long symposium of mathematicians in San Diego[30] presents a good overview of the fundamental schemes under construction at the time. Further developments and refinements came later and are reflected in the GAMM Workshop[31] and the Vortex Flow Experiment[32]. Here we see the growing ambition to model not only shock waves, but also vortices, to such an extent that the latter may be the dominant motivating factor for solving the Euler equations.

In the latter part of the decade Symposium Transsonicum III[33] is yet another landmark. All methods are numerical, but the mix now is not between potential and Euler, but rather between Euler and Navier Stokes. We see this not only in transonic aerodynamics, but also hypersonic aerothermodynamics[34], a field which 30 years ago pioneered the development of Euler methods to solve the blunt-body problem. In some sense the circle has closed.

1.5.2 Three Categories of Methods

The previous section presented the thread of references upon which we build our survey. Here we outline the main categories into which the structured-grid methods group themselves.

During the 1970s Lax-Wendroff-type of methods, like the MacCormack scheme, became very popular because of their effectiveness. During the 1980s two other classes, explicit centered Runge-Kutta schmes and upwind schemes, began to compete with the MacCormack scheme. To some extent the review Ref.35, but more thoroughly the textbook of Hirsch[36], presents a thorough development of these three different categories and provides many of the primary references to their formulations. In this book we are primarily concerned with the latter two classes: centered and upwind schemes. It should be pointed out, however, that within the Lax-Wendroff category Lerat and his co-workers continue to advance a very elegant implicit scheme (see Hirsch[36]).

At the 1979 Workshop, the majority of Euler methods used the MacCormack scheme, but the upwind method was represented by the Roe scheme. Pandolfi and his colleagues also presented some of the first results with the lambda scheme (see Ref.37 for a recent account). This scheme is unusual in that it is non-conservative and shocks have to be tracked explicitly.

Two years later Jameson introduced his centered Runge-Kutta scheme, and this quickly blossomed into the third category. All three categories were well represented by the contributions to the AGARD Working Group 7[29], but now the centered and upwind schemes were in the majority. By the time of the second Euler Workshop[31] these two classes were the two dominant methods. These two methods are presented in detail in later chapters where many more references are given.

The early 1980s also saw the emergence of finite-element methods being applied to the Euler equations. Many of the leading methods in this category were represented at the two Euler Workshops held at INRIA[28,31]. The reader interested in these methods is also referred to the short-course notes of Desideri and Dervieux[38], and Morgan and Peraire[39].

1.6 Outline of the Remaining Chapters

We aim to survey some of the current advancements in the development of evolutionary hyperbolic methods for non-linear conservation laws, particularly as they apply to the treatment of flows with non-smooth structures. This means prime attention is directed to the purely inviscid non-linear advection problem as embodied in the Euler equations. In keeping with the themes culled from past developments sketched above, we illustrate how in current work theory and practice go hand in hand, and how advanced computer architectures and improvements in the hardware may favor the application of one or another numerical algorithm. Our sketch of the origins of CFD indicates that practically all of the past developments were made in the context of the finite-difference method for numerical approximation. More recently of course other approximation schemes have come forth, notably the finite-element method, discrete vortex methods, and the spectral method, which in many applications do offer properties superior to finite differences. But for illustrating the treatment of non-smooth phenomena, the framework of finite differences, or closely related to it the finite-volume method, is still a good context since it is with these that much of the theory is being advanced.

The book blends together some very fundamental concepts upon which CFD methods are based and more specialized considerations adapted to specific techniques. In this way the material ranges widely from the elementary to the advanced.

For example a discussion of the properties of the Euler equations is presented in Chapter II, and some fundamentals of discrete solution methods like truncation error, stability, and consistency are given in Chapter III. That chapter also discusses some of the theory behind the continuous equations, e.g. characteristics, Riemann invariants, weak solutions, and shock waves, because these enter into some of the numerical schemes later. Chapter IV introduces the finite-volume concept including the handling of geometry and boundary conditions. Then centered differencing methods including the artificial viscosity model as well as the time integration scheme are presented in Chapter V. Upwind differencing follows in Chapter VI. These first six chapters together form the overall numerical approach to the problem.

Since so much interest lies in the solution to stationary problems, Chapter VII examines the issue of convergence to steady state, and some techniques that can improve the convergence. Chapter VIII discusses aspects of the use of vector and parallel supercomputers. Chapters IX and X together take up the historical question raised in Chapter I about the role of viscosity and how that is accounted for in an inviscid model. Also the modern questions of vorticity creation and entropy rise in discrete Euler solutions are answered.

In the last part of the book Chapter XI presents results computed by some of these methods and discusses their interpretation. The book closes with Chapter XII and some comments on the future prospects of new methods and interdisciplinary problems.

As the reader sees, the topic of the book is the numerical solution of the Euler equations. The historical perspective just presented, as well as Chapters IX and X, emphasize how important it is to account for the viscous effects. Why then write a book on the Euler equations, why not the Navier-Stokes equations? One answer is that because the Euler equations are a subset of the Navier-Stokes equations all of the techniques presented in this book carry over to the complete system. In other words you cannot build a good Navier-Stokes solver today without first creating a good Euler solver.

1.7 References

1. Giacomelli, R.: "The Aerodynamics of Leonardo da Vinci". Aero J. Vol. 24, No. 240, 1930, pp. 121-134.

2. Tokaty, G.A.: "A History and Philosophy of Fluid Mechanics". Foulis, London, 1971.

3. Bell, E.T.: "Men of Mathematics". Simon and Schuster, New York, 1937.

4. Kline, M.: "Mathematical Thought from Ancient to Modern Times". Oxford University Press, 1972.

5. Boyar, C.B.: "A History of Mathematics". Wiley & Sons, New York, 1968.

6. Giacomelli, R., and Pistolesi, E.: "Historical Sketch". In: Aerodynamic Theory, W.F. Durand (ed.), Vol. I, Springer, Berlin/ Heidelberg/New York, 1934.

7. von Kármán, T.: "Aerodynamics". Cornell Univ. Press, 1954.

8. Milne-Thomson, L.M.: "Theoretical Hydrodynamics". 5th ed., Macmillan Co, New York, 1968.

9. Hunt, B.: "The Mathematical Basis and Numerical Principles of the Boundary Integral Method for Incompressible Potential Flow Over 3D Aerodynamic Configurations". Numerical Methods in Fluid Dynamics, B. Hunt (ed.), Academic Press, New York, 1980, pp. 49-135.

10. Schlichting, H.: "Boundary Layer Theory". 6th ed., McGraw-Hill, New York. 1968, pp. 178-184.

11. Southwell, R.V.: "Relaxation Methods in Theoretical Physics". Oxford Univ. Press, 1946.

12. Courant, R., Friedrichs, K.O., Lewy, H.: "Uber die partiellen Differenzengleichungen der mathematischen Physik". Math. Ann. 100, 1928, p. 32.

13. Fox, L. (ed): "Numerical Solution of Ordinary and Partial Differential Equations". Pergamon Press, Oxford, 1962.

14. Rusanov, V.V.: "A Blunt Body in a Supersonic Stream". Ann. Rev. Fluid Mech., Vol. 8, 1976, pp. 377-404.

15. Van Dyke, M.D.: "The Supersonic Blunt Body Problem — Review and Extension." J. Aerospace Sci., Vol. 25, 1958, pp. 485-496.

16. Gilinskii, S.M., Telenin, G.F., Tinyakov, G.P.: "A Method for Computing Supersonic Flow Around Blunt Bodies". Izv. Adad. Nauk, SSSR Rekh. Mash. 4, 1964, 9 (translated as NASA TT F297).

17. Dorodnitsyn, A.A.: "A Contribution to the solution of Mixed Problems of Transonic Aerodynamics". Advances in Aeronautical Sciences, Vol. 2, Pergamon Press, New York, 1959.

18. Oswatitsch, K., Rues, D. (eds): "Transsonicum Symposium II". Springer, Berlin, 1976.

19. Richtmyer, R.D., Morton, K.W.: "Difference Methods for Initial-Value Problems". Interscience Publishers, New York, 1967.

20. Rusanov, V.V.: "The Calculation of the Interaction of Non-Stationary Shock Waves and Obstacles". Zh. Vych. Mat., Vol. 1, No. 2, 1961, pp. 267-279.

21. Rusanov, V.V.: "A Three-Dimensional Supersonic Gas Flow Past Smooth Blunt Bodies". In Proc. 11th Int. Congress Appl. Mech, H. Görtler (ed.), Springer, Berlin, 1966.

22. Moretti, G., Abbett, M.: "A Time Dependent Computational Method for Blunt Body Flows". AIAA J., Vol. 4, 1966, pp. 2136-2141.

23. Rizzi, A., Inouye, M.: "Time-Split Finite-Volume Method for Three-Dimensional Blunt-Body Flow". AIAA Journal, Vol. 11, No. 11, Nov 1973, pp. 1478-1485.

24. Rizzi, A., Bailey, H.E.: "Finite Volume Solution of the Euler Equations for Steady Three-Dimensional Transonic Flow". Proc. Fifth Internat. Conf. on Numer. Methods in Fluid Dynamics, A.I. van der Vooren and P.J. Zandbergen (eds.), Lecture Notes in Physics, 59, Springer, 1976, pp. 347-357.

25. Magnus, R., Yoshihara, H.: "Inviscid Transonic Flow Over Airfoils", AIAA J. Vol. 8, No. 12, 1970, pp. 2157-2162.

26. Rizzi, A., Viviand, H. (eds.): "Numerical Methods for the Computation of Inviscid Transonic Flows with Shock Waves". NNFM Vol. 3, Vieweg, Braunschweig/Wiesbaden, 1981.

27. Steinhoff, J.S., Reddy, K.C. (eds.): "Computational Fluid Dynamics in Aerospace Design". UTSI Publication, Tullahoma, 1985.

28. Angrand, F., Dervieux, A., Desideri, J.A., Glowinski, R. (eds.): "Numerical Methods for the Euler Equations of Fluid Dynamics". SIAM, Philadelphia, 1985.

29. Yoshihara, H., Sacher, P. (eds.): "Test Cases for Inviscid Flow Field Methods". AGARD-AR-211, Paris, 1985.

30. Engqvist, B., Osher, S., Sommerville, R. (eds.): "Large Scale Computations in Fluid Mechnics". Lectures in Applied Mathematics, Vol. 22, Ann Math. Soc., Providence, 1985.

31. Dervieux, A., Van Leer, B., Periaux, J., Rizzi, A. (eds.): "Numerical Simulation of Compressible Euler Flows". NNFM Vol. 26, Vieweg, Braunschweig/Wiesbaden, 1989.

32. Elsenaar, A., Eriksson, L.-E. (eds.): "International Vortex Flow Experiment on Euler Code Validation". FFA, Stockholm, 1987.

33. Zierep, J., Oertel, H. (eds.): "Symposium Transsonicum III". Springer, Berlin 1989.

34. Bertin, J.J., Glowinski, R., Periaux, J. (eds.): "Hypersonics". Birkhäuser, Boston, 1989.

35. Rizzi, A., Engqvist, B.: "Selected Topics in the Theory and Practice of Computational Fluid Dynamics - a Review". J. Comp. Phys, Vol. 72, No. 1, Sept 1987, pp. 1-69.

36. Hirsch, C.: "Computational Methods for Inviscid and Viscous Flows". John Wiley and Sons, Chichester, 1990.

34. Pandolfi, M.: "A Contribution to the Numerical Prediction of Unsteady Flows". AIAA J., Vol. 22, No. 5, 1984, pp. 1217-1225.

38. Desideri, J.A., Dervieux, A.: "Compressible Flow Solvers Using Unstructured Grids". VKI Lecture Series Notes 1988-05, Brussels, 1988.

39. Morgan, K., Peraire, J.: "Finite Element Methods for Compressible FLows". VKI Lecture Series Notes 1987-08, Brussels, 1987.

II THE EULER EQUATIONS

The Euler equations are an approximation of the Navier-Stokes equations with the viscous forces and the volume forces being neglected. The reason for doing so is a) because viscosity and heat conduction in a gas usually play a role only in a thin layer near solid surfaces, the thickness of which is much smaller than the characteristic length of the object being immersed in the fluid flow, b) because the volume forces are usually much smaller than the global forces generated by the dynamics of the flow.

This chapter is devoted to the discussion of certain properties of the Euler equations, as well as to the provision of transformations necessary for their numerical solution. The matter of describing vortex flows by means of the Euler equations will be discussed in Chapter X.

2.1 The Classical Euler Equations in Gas Dynamics

Since the derivation of the governing equations of fluid motion can be found in a number of textbooks, for instance References 1 and 2, this issue is not repeated in this book. In the following only the results of these derivations, the Euler equations in their classical form are repeated.

Continuity equation

$$\dot{\rho} + u\rho_x + u\rho_y + w\rho_z + \rho(u_x+v_y+w_z) = 0 \,, \tag{2.1}$$

Momentum equations

$$\begin{aligned} &\rho(\dot{u} + uu_x + vu_y + wu_z) + p_x = 0 \,, \\ &\rho(\dot{u} + uv_x + vv_y + wv_z) + p_y = 0 \,, \\ &\rho(\dot{w} + uw_x + vw_y + ww_z) + p_z = 0 \,, \end{aligned} \tag{2.2}$$

Energy equation (perfect gas)

$$\dot{p} + up_x + vp_y + wp_z + \gamma p(u_x + v_y + w_z) = 0 \,. \tag{2.3}$$

The conservative or divergence form can be found from Eq.(2.1) (2.2) and (2.3):

Continuity equation

$$\dot{\rho} + (\rho u)_x + (\rho v)_y + (\rho w)_z = 0 \,, \tag{2.4}$$

Momentum equations

$$\begin{aligned} (\rho u)^{\bullet} + (p+\rho u^2)_x + (\rho uv)_y + (\rho uw)_z &= 0 \,, \\ (\rho v)^{\bullet} + (\rho uv)_x + (p+\rho v^2)_y + (\rho vw)_z &= 0 \,, \\ (\rho w)^{\bullet} + (\rho uw)_x + (\rho vw)_y + (p+\rho w^2)_z &= 0 \,, \end{aligned} \tag{2.5}$$

Energy equation (perfect gas)

$$\dot{e} + [u(e+p)]_x + [v(e+p)]_y + [w(e+p)]_z = 0 \,. \tag{2.6}$$

This equation is supplied by the definition of the specific energy e:

$$e = \frac{p}{\gamma-1} + \rho \frac{q^2}{2} \,, \tag{2.7}$$

where $q^2 = u^2+v^2+w^2$.

2.2 Basic Results of the Non-Conservative Equations

2.2.1 Isentropic Flow

The energy equation (2.3) can be rewritten such that the differential relation

$$dp - \gamma \frac{p}{\rho} d\rho = 0 \tag{2.8}$$

holds along a streamline. Note that the Euler derivative is by definition the total derivative of a convected quantity along a streamline in space and time. That is why the last equation can be integrated along a streamline

$$p = A \rho^{\gamma} \,,$$

with A being at least piecewise constant along a streamline. Therefore the constancy of A can be used to check the accuracy of numerical Euler solutions along solid bodies which are enveloped by streamlines.

2.2.2 Homentropic Flow

If the differential equation (2.8) is true not only along streamlines but also along any arbitrary path at any location in the flowfield, then not only the pressure is a unique function of the density as in the last paragraph but also the flow is irrotational. As a consequence the energy equation can be integrated analytically in advance and also the momentum equations. The reason for this can be obtained from equation (2.8) by the following manipulation:

$$\gamma \frac{dp}{\rho} + \frac{dp}{\rho} - \gamma \frac{p}{\rho^2} \, d\rho = \gamma \frac{dp}{\rho} ,$$

where $\gamma(dp/\rho)$ has been added on both sides.

Rearrangement yields

$$dp = \frac{\gamma}{\gamma-1} \, \rho d(\frac{p}{\rho}) , \qquad (2.9)$$

such that the non-conservative momentum equations read now

$$\begin{aligned} \dot{u} + uu_x + vu_y + wu_z + \frac{\gamma}{\gamma-1} (\frac{p}{\rho})_x &= 0 , \\ \dot{v} + uv_x + vv_y + wv_z + \frac{\gamma}{\gamma-1} (\frac{p}{\rho})_y &= 0 , \qquad (2.10) \\ \dot{w} + uw_x + vw_y + ww_z + \frac{\gamma}{\gamma-1} (\frac{p}{\rho})_z &= 0 . \end{aligned}$$

If the first equation is differentiated by y and the second by x then the difference of both equations is

$$\begin{aligned} (u_y-v_x)^{\bullet} + u(u_y-v_x)_x + v(u_y-v_x)_y + w(u_y-v_x)_z + \\ + (u_x+v_y)(u_y-v_x) + u_z w_y - v_z w_x = 0 . \end{aligned}$$

A solution of this equation is

$$u_y = v_x , \qquad u_z = w_x , \qquad v_z = w_y . \qquad (2.11)$$

It can be checked easily that this relation of irrotationality also holds for the other two possible cross differentiations applied to the momentum equations. Introducing a potential by the definition

$$u = \phi_x , \qquad v = \phi_y , \qquad w = \phi_z , \qquad (2.12)$$

the kinematic conditions of irrotational flow are fulfilled automatically, and the scalar product of the momentum equations with an arbitrary displacement vector

$$d\underline{r} = \begin{pmatrix} dx \\ dy \\ dz \end{pmatrix}$$

leads to the total differential

$$d(\dot{\phi} + \frac{q^2}{2} + \frac{\gamma}{\gamma-1}\frac{p}{\rho}) = 0 \ , \tag{2.13}$$

being zero, or upon integration

$$\dot{\phi} + \frac{q^2}{2} + \frac{\gamma}{\gamma-1}\frac{p}{\rho} = B \ , \tag{2.14}$$

where B is the Bernoulli constant.

There is also another important quantity which is called total enthalpy, defined as

$$H = \frac{\gamma}{\gamma-1}\frac{p}{\rho} + \frac{q^2}{2} \ . \tag{2.15}$$

If the flow is steady, this quantity becomes a constant throughout the flowfield. This also can be used for checking the accuracy of Euler solutions if the flow is isenthalpic. Finally, with the previous equations we can eliminate the pressure and calculate the density as a function of the potential

$$\rho = [\frac{\gamma-1}{\gamma A}(B - \dot{\phi} - \frac{q^2}{2})]^{(1/\gamma-1)} \ . \tag{2.16}$$

Inserting this expression into the continuity equation, only a one-variable partial differential equation has to be solved. At this stage it is recalled that the wide field of potential theory has its limitations wherever in the flowfield the condition of irrotationality is violated. Although it is an excellent tool for aerodynamic design considerations, vortical flows, strong shock phenomena and all flows being known a priori to be rotational (say in helicopter aerodynamics, most turbo machinery flows, etc.) are excluded from the treatment by potential theory.

2.3 Basic Results of the Conservative Equations

2.3.1 Isenthalpic Flow

If the Euler equations written in form (2.4 - 2.6) are used for finding steady-state solutions only, then the energy equation may be integrated in advance under certain circumstances being specified later. Putting $\partial/\partial t=0$, the continuity equation (2.4) and the energy equation (2.6) read:

$$(\rho u)_x + (\rho v)_y + (\rho w)_z = 0 \ ,$$

$$[\rho u \frac{e+p}{\rho}]_x + [\rho v \frac{e+p}{\rho}]_y + [\rho w \frac{e+p}{\rho}]_z = \frac{e+p}{\rho}[(\rho u)_x + (\rho v)_y + (\rho w)_z] +$$

$$+ \rho[u(\frac{e+p}{\rho})_x + v(\frac{e+p}{\rho})_y + w(\frac{e+p}{\rho})_z] = 0 \ .$$

The first bracket on the right-hand side of the energy equation is zero because it is the continuity equation. The second bracket is nothing but the streamline (or Euler) derivative of

$$\frac{e+p}{\rho} = \frac{\gamma}{\gamma-1}\frac{p}{\rho} + \frac{q^2}{2} = H \ . \tag{2.17}$$

Therefore the total enthalpy H is constant along streamlines. If now all streamlines emanate from a uniform farfield state, which often is the case in aerodynamics, then each streamline carries the same total enthalpy, and the energy equation can be abandoned from the solution procedure, if only the steady state is of interest.

2.3.2 Shock Flow

There are flow situations where sudden changes of the flow variables may occur which will be analyzed now in more detail by the use of the Euler equations in the conservative form.

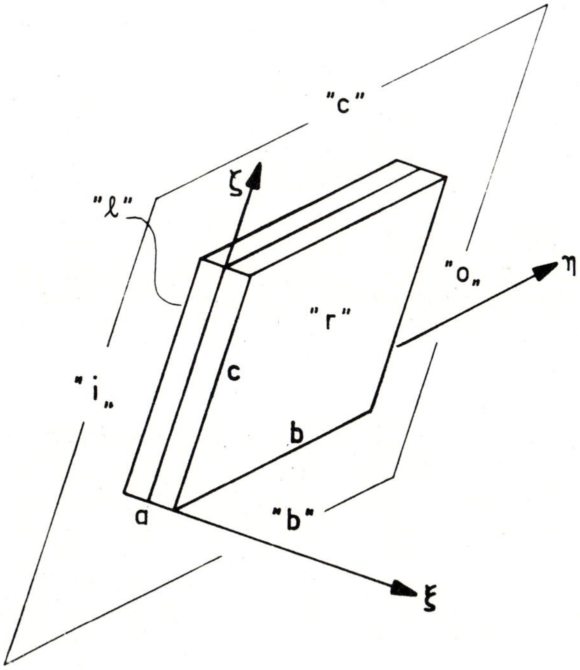

Fig. 2.1 Sketch of the control volume enclosing a surface of a discontinuity

For this purpose let us assume that there is a surface of discontinuity in the flowfield. We arrange a local Cartesian coordinate system ξ, η, ζ such that the plane spanned by the η- and ζ-axis is tangential to the surface of discontinutiy. The ξ-coordinate is normal to the latter. The normal speed of the discontinuity is simply $\dot{\xi}$. The velocity components u,v,w are understood now as being aligned with the ξ-, η- and ζ-axis respectively. For examining the flow through the layer of discontinuity a control volume is arranged as shown in Fig. 2.1.

The faces of the volume are labeled "left" (ℓ), "right" (r), "inboard" (i), "outboard" (o), "bottom" (b) and "ceiling" (c).

The Euler equations read in symbolic form

$$\dot{U} + E_\xi + F_\eta + G_\zeta = 0 \ . \tag{2.18}$$

The origin of the coordinate system moves together with the discontinuity with the speed

$$\dot{\underline{r}} = \begin{pmatrix} \dot{\xi} \\ \dot{\eta} \\ \dot{\zeta} \end{pmatrix} \ . \tag{2.19}$$

Therefore the partial time derivative has to be transformed to the moving coordinate system using a similar form as the Euler derivative applied to the displacement vector not of the fluid particle, but of the control volume center.

The Euler equations assume then the form

$$\overset{\circ}{U} - U_\xi \dot{\xi} - U_\eta \dot{\eta} - U_\zeta \dot{\zeta} + E_\xi + F_\eta + G_\zeta + U(\dot{\xi}_\xi + \dot{\eta}_\eta + \dot{\zeta}_\zeta) - U(\dot{\xi}_\xi + \dot{\eta}_\eta + \dot{\zeta}_\zeta) = 0, \tag{2.20}$$

whereby a dummy difference has been added. Upon rearrangement we arrive at

$$\overset{\circ}{U} + U(\dot{\xi}_\xi + \dot{\eta}_\eta + \dot{\zeta}_\zeta) + (E - U\dot{\xi})_\xi + (F - U\dot{\eta})_\eta + (G - U\dot{\zeta})_\zeta = 0 \ . \tag{2.21}$$

Now the space derivatives are replaced by divided differences (for the indices see above):

$$\overset{\circ}{U} + U\left(\frac{\dot{\xi}_r - \dot{\xi}_\ell}{a} + \frac{\dot{\eta}_o - \dot{\eta}_i}{b} + \frac{\dot{\zeta}_c - \dot{\zeta}_b}{c}\right) + \frac{(E-U\dot{\xi})_r - (E-U\dot{\xi})_\ell}{a} +$$

$$+ \frac{(F-U\dot{\eta})_o - (F-U\dot{\eta})_i}{b} + \frac{(G-U\dot{\zeta})_c - (G-U\dot{\zeta})_b}{c} = 0 \ .$$

Upon multiplication with the volume D=abc the following form is obtained:

$$\overset{\circ}{U}D + U\{bc(\dot{\xi}_r-\dot{\xi}_\ell)+a[c(\dot{\eta}_o-\dot{\eta}_i)+b(\dot{\zeta}_c-\dot{\zeta}_b)]\}+bc[(E-U\dot{\xi})_r-(E-U\dot{\xi})_\ell]$$

$$+ a\{c[(F-U\dot{\eta})_o-(F-U\dot{\eta})_u]-b[(G-U\dot{\zeta})_c-(G-U\dot{\zeta})_b]\} = 0. \qquad (2.22)$$

Shrinking the volume with $a \to 0$ leads to the result

$$\dot{\xi} = \dot{\xi}_\ell = \dot{\xi}_r \, , \qquad (2.23)$$

and

$$E_r - E_\ell - \dot{\xi}(U_r - U_\ell) = 0 \, . \qquad (2.24)$$

For convenience the subscript for the fluxes and flow quantities on the left side of the assumed discontinuity is replaced by ∞ and that of the right side is simply dropped. Identifying the flux E and the vector of the Euler equations, one finds with ease:

$$\rho u-(\rho u)_\infty - \dot{\xi}(\rho-\rho_\infty) = 0 \, , \qquad (2.25)$$

$$\begin{aligned} p+\rho u^2 - (p+\rho u^2)_\infty &- \dot{\xi}[\rho u-(\rho u)_\infty] = 0 \, , \\ \rho uv-(\rho uv)_\infty &- \dot{\xi}[\rho v-(\rho v)_\infty] = 0 \, , \\ \rho uw-(\rho uw)_\infty &- \dot{\xi}[\rho w-(\rho w)_\infty] = 0 \, , \end{aligned} \qquad (2.26)$$

$$u(e+p)-[u(e+p)]_\infty - \dot{\xi}(e-e_\infty) = 0 \, . \qquad (2.27)$$

The last equation reads with Eq.(2.7)

$$u(\frac{\gamma}{\gamma-1}\, p+\rho\, \frac{q^2}{2})-u_\infty(\frac{\gamma}{\gamma-1}\, p+\rho\, \frac{q^2}{2})_\infty - \dot{\xi}(\frac{p}{\gamma-1} +\rho\, \frac{q^2}{2} - \frac{p_\infty}{\gamma-1} - \rho_\infty \frac{q_\infty^2}{2}) = 0.$$

In order to simplify the formulas describing the discontinuous flow we use a Galilei transformation for the normal velocities

$$u = u_2 + a \, , \qquad u_\infty = u_1 + a \, , \qquad (2.28)$$

such that the continuity equation becomes

$$\rho(u_2 + a - \dot{\xi}) - \rho_\infty(u_1 + a - \dot{\xi}) = 0 \, . \qquad (2.29)$$

The choice $a=\dot{\xi}$ leads to

$$u = u_2 + \dot{\xi} \; , \quad u_\infty = u_1 + \dot{\xi} \; , \quad \rho u_2 - \rho_\infty u_1 = 0 \; . \tag{2.30}$$

The normal momentum equation becomes

$$p + \rho u_2(u_2+\dot{\xi}) - p_\infty - \rho_\infty u_1(u_1+\dot{\xi}) = 0 \; ,$$

or with the continuity equation inserted

$$p - p_\infty + \rho_\infty u_1(u_2 - u_1) = 0 \; . \tag{2.31}$$

The first tangential momentum equation becomes

$$\rho u_2 v - \rho_\infty u_1 v_\infty = 0 \; ,$$

or with the continuity equation inserted

$$v = v_\infty \; . \tag{2.32}$$

For the same reason we obtain:

$$w = w_\infty \; . \tag{2.33}$$

That means that the tangential velocity is not altered by the discontinuity.

The energy equation becomes

$$p\left(\frac{\gamma u_2}{\gamma-1} + \dot{\xi}\right) - p_\infty\left(\frac{\gamma u_1}{\gamma-1} + \dot{\xi}\right) + \rho u_2 \frac{(u_2+\dot{\xi})^2 + v_\infty^2 + w_\infty^2}{2} - \rho_\infty u_1 \frac{(u_1+\dot{\xi})^2 + v_\infty^2 + w_\infty^2}{2} = 0 \; ,$$

or with the continuity and the normal momentum equation inserted

$$(u_2-u_1)\left[\frac{\gamma \, p_\infty}{\gamma-1} + \frac{\rho_\infty u_1}{2}\left(u_1 - \frac{\gamma+1}{\gamma-1} u_2\right)\right] = 0 \; . \tag{2.34}$$

As expected one possible solution is the identity solution

$$u_2 = u_1 \; , \tag{2.35}$$

which holds if there is no discontinuity. The other solution is the well known Prandtl relation

$$u_2 u_1 = 2\,\frac{\gamma-1}{\gamma+1}\left(\frac{\gamma}{\gamma-1}\,\frac{p_\infty}{\rho_\infty} + \frac{u_1^2}{2}\right) = 2\,\frac{\gamma-1}{\gamma+1}\,H_1 \; . \tag{2.36}$$

From this solution all other quantities aft of the shock can be obtained by back substitution.

Introducing new normalized velocities gives us the possibility of a unique representation of discontinuous flow

$$u_1 = U_1 \left(2 \frac{\gamma - 1}{\gamma + 1} H_1\right)^{0.5} , \qquad u_2 = U_2 \left(2 \frac{\gamma - 1}{\gamma + 1} H_1\right)^{0.5} . \quad (2.37)$$

Regardless of the size of the pre-shock total enthalpy the shock polar is with these new dimensionless velocities always

$$U_1 U_2 = 1 . \quad (2.38)$$

There is obviously an upper limit for the dimensionless pre-shock velocity, which we find from putting $0.5u_1^2 = H_1$ to be

$$U_{1max} = \left(\frac{\gamma + 1}{\gamma - 1}\right)^{0.5} . \quad (2.39)$$

The flow situation is at best summarized in the following Fig. 2.2. The identity solution is the straight line, while

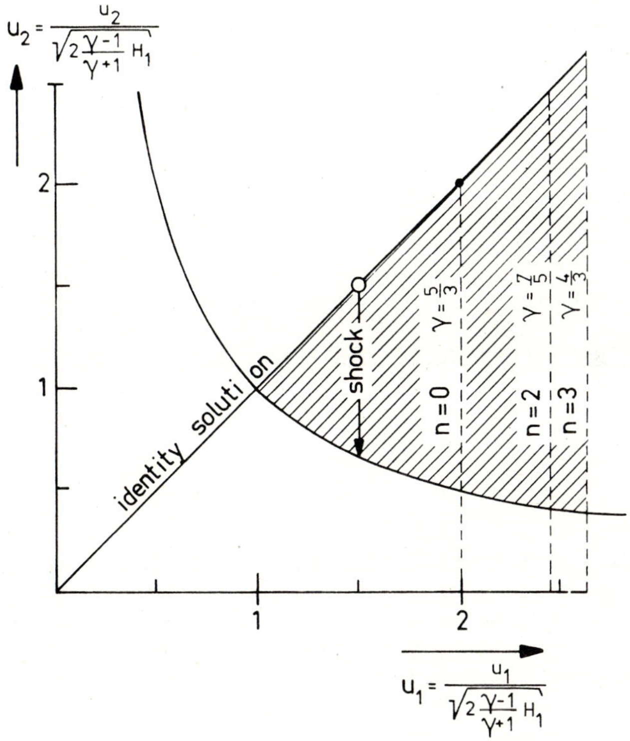

Fig. 2.2 Normalized shock polar. The number n indicates the number of rotational degrees of freedom of an ideal gas. The dashed lines indicate the maximum possible shock strength

the discountinuous solution is represented by the unit hyperbola. It is now evident that for $u_1<1$ there may be no discontinuous solution since this would mean that the fluid velocity would abruptly be accelerated to arbitrarily high values as u_1 approaches zero. On the other hand if $u_1>1$ abrupt decelerations may happen, which will be called shocks. Since there is an upper limit for u_1, the shock jump is always finite. So the feasible region for shock discountinuities can only be the shaded area of the graph.

2.3.3 Speed of Sound

There is a distinct lower limit of the flow velocity below which no shock solution may exist. This limit is called the speed of sound and is expressed following the last paragraph by

$$U_1^2 = 1 \ , \qquad (2.40)$$

or in terms of the non-normalized velocity

$$u_1^2 = \gamma \frac{p_\infty}{\rho_\infty} \ . \qquad (2.41)$$

Dropping subscripts for generalization we find the speed of sound:

$$s = (\gamma\ p/\rho)^{1/2} \ .$$

2.3.4 Eigenvalues of Pressure Waves

Going back to the original definition of the speed $u_1 = u_\infty - \dot{\xi}$, we find from the last paragraph

$$\dot{\xi} = u \pm \sqrt{\gamma\ p/\rho} \equiv u \pm s \ . \qquad (2.42)$$

These are two velocities of the wave fronts, which are generated by small pressure discontinuities of vanishing strength. In the development of numerical up-wind schemes these velocities, which also may be called eigenvalues or characteristic slopes, play an important role.

2.3.5 Homogeneous Property of the Euler Equations

Finally an important property of the Euler equations written in conservative form is considered, the homogeneous property. It is expressed by the formula

$$E_U U \equiv E \; . \tag{2.43}$$

This means that the partial vector derivative of a flux of the Euler equations with respect to the conservative flow variables multiplied scalarly by the conservative flow variable vector recovers the original flux. The verification of the homogeneous property is based on the fact that the polynomials of the velocity components occuring in the individual fluxes are multiplied only by the density:

continuity equation ρu ,

momentum equation ρu^2 ,

energy equation ρu^3 .

If the velocity components are replaced by the conservative mass flux components we obtain the following flux components:

continuity equation ℓ ,

momentum equation $\frac{\ell^2}{\rho}$,

energy equation $\frac{\ell^3}{\rho^2}$.

The general form of the flux components is therefore

$$\ell^i \rho^{1-i} \qquad (i = 1,2,3) \; .$$

The homogeneous property is verified by the following observation:

$$\begin{aligned} \ell^i \rho^{1-i} &= (\ell^i \rho^{1-i})_\ell \ell + (\ell^i \rho^{1-i})_\rho \rho = \\ &= i \ell^{i-1} \rho^{1-i} \ell + \ell^i \rho^{-i} (1-i) \rho = \\ &= i \ell^i \rho^{1-i} + (1-i) \ell^i \rho^{1-i} = \\ &= \ell^i \rho^{1-i} \; . \end{aligned}$$

The homogeneous property of the conservative Euler equations has an important impact on the accuracy of numerical flux differences:

$$\Delta E = E_r - E_\ell = (E_U U)_r - (E_U U)_\ell \, , \tag{2.44}$$

where r means right and ℓ now means left. A Taylor expansion about the centerpoint gives

$$(E_U U)_r = E + (E_U U)_u \, (U_r - U) =$$

$$= E + (E_{UU} U + E_U) \, (U_r - U) \, ,$$

$$(E_U U)_\ell = E + (E_{UU} U + E_U) \, (U_\ell - U) \, .$$

The difference is

$$\Delta E = (E_{UU} U + E_U) \, (U_r - U_\ell) \, . \tag{2.45}$$

Because of the homogeneous property we have

$$E_U = (E_U U)_U = E_{UU} U + E_U \, ,$$

and therefore

$$E_{UU} U \equiv 0 \, . \tag{2.46}$$

Thus the flux difference becomes

$$\Delta E = E_r - E_\ell = E_U (U_r - U_\ell) \, . \tag{2.47}$$

If other variables but the conservative are applied then

$$E^{**}_{UU} \overset{*}{U} \neq 0 \, ,$$

and the chain rule of differentiation would enter an a priori error into the finite difference analogue.

2.4 Coordinate Transformations

When we derived the shock relations, we used a form of the Euler equations which reads in Cartesian coordinates (see Eq. 2.21)

$$\overset{\circ}{U} + U(\dot{x}_x + \dot{y}_y + \dot{z}_z) + (E - U\dot{x})_x + (F - U\dot{y})_y + (G - U\dot{z})_z = 0 \, .$$

The second term is nothing but the relative volume dilatation multiplied by the flow-variable vector:

$$U(\dot{x}_x + \dot{y}_y + \dot{z}_z) = \frac{\overset{\circ}{D}}{D}\, U \,, \qquad (2.48)$$

where D is the control volume, which should not be mixed up with the particle volume V.

If it is desired to carry the flow quantities together with the instantaneous location of the moving coordinate system, we can use the Euler equations in the form

$$(UD)^{\circ} + D\left[(E - U\dot{x})_x + (F - U\dot{y})_y + (G - U\dot{z})_z\right] = 0 \,, \qquad (2.49)$$

where the dot o means the time derivative in the moving coordinate system, and the vector

$$\dot{\underline{r}} = \begin{pmatrix} \dot{x} \\ \dot{y} \\ \dot{z} \end{pmatrix}$$

is the local speed of the coordinate system.

Unfortunately this form of the flow equations is no more in divergence form because of the coefficient D. Furthermore, the coordinate system is still Cartesian and thus not well suited for an evaluation in curvilinear coordinates. With the short hand writing

$$\overset{*}{E} = E - U\dot{x} \,, \quad \overset{*}{F} = F - U\dot{y} \,, \quad \overset{*}{G} = G - U\dot{z} \,, \qquad (2.50)$$

we arrive after chain ruling of the flow equations for an arbitrary coordinate system ξ, η, ζ at

$$\begin{aligned}(UD)^{\circ} + D(&\overset{*}{E}_{\xi}\bar{\xi}_x + \overset{*}{E}_{\eta}\bar{\eta}_x + \overset{*}{E}_{\zeta}\bar{\zeta}_x + \\ &+ \overset{*}{F}_{\xi}\bar{\xi}_y + \overset{*}{F}_{\eta}\bar{\eta}_y + \overset{*}{F}_{\zeta}\bar{\zeta}_y + \\ &+ \overset{*}{G}_{\xi}\bar{\xi}_z + \overset{*}{G}_{\eta}\bar{\eta}_z + \overset{*}{G}_{\zeta}\bar{\zeta}_z) = 0 \,. \end{aligned} \qquad (2.51)$$

This is even farer away from the divergence form than the previous formulation. Therefore, we try to analyze the metrics $\bar{\xi}_{x,y,z}$ etc. in more detail, and use the trivial statement that the Cartesian coordinates are independent from each other:

$$\begin{aligned} x_x &= 1 \,, & x_y &= 0 \,, & x_z &= 0 \,, \\ y_x &= 0 \,, & y_y &= 1 \,, & y_z &= 0 \,, \\ z_x &= 0 \,, & z_y &= 0 \,, & z_z &= 1 \,. \end{aligned} \qquad (2.52)$$

Chain ruling of the first column of the equations gives

$$x_\xi \bar{\xi}_x + x_\eta \bar{\eta}_x + x_\zeta \bar{\zeta}_x = 1 ,$$
$$y_\xi \bar{\xi}_x + y_\eta \bar{\eta}_x + y_\zeta \bar{\zeta}_x = 0 , \qquad (2.53)$$
$$z_\xi \bar{\xi}_x + z_\eta \bar{\eta}_x + z_\zeta \bar{\zeta}_x = 0 ,$$

the inversion of which is

$$\bar{\xi}_x = \frac{y_\eta z_\zeta - z_\eta y_\zeta}{D} , \quad \bar{\eta}_x = \frac{y_\zeta z_\xi - y_\xi z_\zeta}{D} , \quad \bar{\zeta}_x = \frac{y_\xi z_\eta - y_\eta z_\xi}{D} ,$$

with

$$D = x_\xi(y_\eta z_\zeta - z_\eta y_\zeta) + y_\xi(x_\zeta z_\eta - x_\eta z_\zeta) + z_\xi(x_\xi y_\eta - x_\eta y_\xi) . \qquad (2.54)$$

In a similar way we find from chain ruling of the second and the third column

$$\bar{\xi}_y = \frac{x_\zeta z_\eta - x_\eta z_\zeta}{D} , \quad \bar{\eta}_y = \frac{x_\xi z_\zeta - z_\xi x_\zeta}{D} , \quad \bar{\zeta}_y = \frac{x_\eta z_\xi - z_\eta x_\xi}{D} ,$$

$$\bar{\xi}_z = \frac{x_\eta y_\zeta - x_\zeta y_\eta}{D} , \quad \bar{\eta}_z = \frac{x_\zeta y_\xi - x_\xi y_\zeta}{D} , \quad \bar{\zeta}_z = \frac{x_\xi y_\eta - x_\eta y_\xi}{D} .$$

Upon that at best short hand writing is introduced

$$\xi_x = y_\eta z_\zeta - z_\eta y_\zeta = D\bar{\xi}_x , \quad \eta_x = y_\zeta z_\xi - y_\xi z_\zeta = D\bar{\eta}_x , \quad \zeta_x = y_\xi z_\eta - y_\eta z_\xi = D\bar{\zeta}_x ,$$

$$\xi_y = x_\zeta z_\eta - x_\eta z_\zeta = D\bar{\xi}_y , \quad \eta_y = x_\xi z_\zeta - z_\xi x_\zeta = D\bar{\eta}_y , \quad \zeta_y = x_\eta z_\xi - z_\eta x_\xi = D\bar{\zeta}_y ,$$

$$\xi_z = x_\eta y_\zeta - x_\zeta y_\eta = D\bar{\xi}_z , \quad \eta_z = x_\zeta y_\xi - x_\xi y_\zeta = D\bar{\eta}_z , \quad \zeta_z = x_\xi y_\eta - x_\eta y_\xi = D\bar{\zeta}_z . \qquad (2.55)$$

Now, the quantities $\xi_{x,y,z}$, $\eta_{x,y,z}$, $\zeta_{x,y,z}$ are cell-face area projections in the Cartesian coordinate directions x,y,z spanned by the coordinate surfaces ξ=const, η=const. and ζ=const. It is also easy to see that the quantity D is the volume of an element $d\xi, d\eta, d\zeta$. Thus the flow equations assume the form

$$(UD)^{o} + \overset{*}{E}_\xi \xi_x + \overset{*}{E}_\eta \eta_x + \overset{*}{E}_\zeta \zeta_x + \overset{*}{F}_\xi \xi_y + \overset{*}{F}_\eta \eta_y + \overset{*}{F}_\zeta \zeta_y + \overset{*}{G}_\xi \xi_z + \overset{*}{G}_\eta \eta_z + \overset{*}{G}_\zeta \zeta_z +$$
$$+ \overset{*}{E}(\xi_{x\xi} + \eta_{x\eta} + \zeta_{x\zeta}) + \overset{*}{F}(\xi_{y\xi} + \eta_{y\eta} + \zeta_{y\zeta}) + \overset{*}{G}(\xi_{z\xi} + \eta_{z\eta} + \zeta_{z\zeta}) = 0 ,$$

where the three last brackets are dummy terms which are zero, which can be found by verification. Upon rearrangement we finally obtain:

$$(UD)^{o} + (\overset{*}{E}\xi_x + \overset{*}{F}\xi_y + \overset{*}{G}\xi_z)_\xi + (\overset{*}{E}\eta_x + \overset{*}{F}\eta_y + \overset{*}{G}\eta_z)_\eta + (\overset{*}{E}\zeta_x + \overset{*}{F}\zeta_y + \overset{*}{G}\zeta_z) = 0 \, . \tag{2.56}$$

In this way the divergence form is obtained for the flow equations with respect to a moving grid, the coordinates of which need not necessarily be orthogonal. By the definition of the new fluxes indicated by the star even coordinate distortions with time are allowed. This transformation is ideally suited for the numerical evaluation on structured computational grids in which the mapping from the Cartesian coordinates to the curvilinear grid coordinates can be calculated numerically.

With Eqs. (2.4-2.6) and Eq. (2.50) the fluxes with the star symbol as well as the flow quantities can be identified. For short hand notation the following abbreviations are introduced:

$$\dot{\xi}_o = u\xi_x + v\xi_y + w\xi_y \; , \quad \dot{\eta}_o = u\eta_x + v\eta_y + w\eta_z \; , \quad \dot{\zeta}_o = u\zeta_x + v\zeta_y + w\zeta_z \; ,$$
$$\dot{\xi} = \xi_x X + \xi_y Y + \xi_z Z \; , \quad \dot{\eta} = \eta_x X + \eta_y Y + \eta_z Z \; , \quad \dot{\zeta} = \zeta_z X + \zeta_y Y + \zeta_z Z \; ,$$
$$X = \rho(\dot{\xi}_o - \dot{\xi}) \; , \quad Y = \rho(\dot{\eta}_o - \dot{\eta}) \; , \quad Z = \rho(\dot{\zeta}_o - \dot{\zeta}) \; . \tag{2.57}$$

The transformed Euler equations are then:

$$(D\rho)^{o} + X_\xi + Y_\eta + Z_\zeta = 0 \; , \tag{2.58a}$$

$$(D\rho u)^{o} + (uX + p\xi_x)_\xi + (uY + p\eta_x)_\eta + (uZ + p\zeta_x)_\zeta = 0 \; , \tag{2.58b}$$

$$(D\rho v)^{o} + (vX + p\xi_y)_\xi + (vY + p\eta_y)_\eta + (wZ + p\zeta_y)_\zeta = 0 \; , \tag{2.58c}$$

$$(D\rho w)^{o} + (wX + p\xi_z)_\xi + (wY + p\eta_z)_\eta + (wZ + p\zeta_z)_\zeta = 0 \; , \tag{2.58d}$$

$$(De)^{o} + [(e+p)\dot{\xi}_o - e\dot{\xi}]_\xi + [(e+p)\dot{\eta}_o - e\dot{\eta}]_\eta + [(e+p)\dot{\zeta}_o - e\dot{\zeta}]_\zeta = 0 \; . \tag{2.58e}$$

In the steady state the dot becomes a partial derivative indicated by a point (•) and the grid velocities $\dot{\xi}, \dot{\eta}, \dot{\zeta}$ become zero.

2.5 Stokes' Integral

If the geometric quantites $\xi_{x,y,z}$, $\eta_{x,y,z}$, $\zeta_{x,y,z}$ in Eq. (2.58) are thought of being numerically calculated cell-face area projections, then Eq. (2.58) is already a particular form of Stoke's integral written for an arbitrary but structured coordinate system. In order to find a form free of a chosen coordinate system the individual terms are reinterpreted for the case $d\xi=d\eta=d\zeta=1$. In this case the partial differential operators $\partial/\partial\xi$, $\partial/\partial\eta$, $\partial/\partial\zeta$ convert simply to differences of the function under consideration with respect to the coordinate directions ξ,η,ζ. Upon integration with respect to $d\xi, d\eta, d\zeta$ within the interval 0 and 1 and interchanging differentiation and integration we arrive at

$$\left(\iiint_D \rho dD\right)^{\circ} + \iint_S \rho(\underline{q_n}-\dot{\underline{n}})\, d\underline{S} = 0 \;, \tag{2.59a}$$

$$\left(\iiint_D \rho u dD\right)^{\circ} + \iint_S [\rho u(\underline{q_n}-\dot{\underline{n}})d\underline{S} + p n_x dS] = 0 \;, \tag{2.59b}$$

$$\left(\iiint_D \rho v dD\right)^{\circ} + \iint_S [\rho v(\underline{q_n}-\dot{\underline{n}})d\underline{S} + p n_y dS] = 0 \;, \tag{2.59c}$$

$$\left(\iiint_D \rho w dD\right)^{\circ} + \iint_S [\rho w(\underline{q_n}-\dot{\underline{n}})d\underline{S} + p n_z dS] = 0 \;, \tag{2.59d}$$

$$\left(\iiint_D e dD\right)^{\circ} + \iint_S [(e+p)\underline{q_n}-e\dot{\underline{n}}]d\underline{S} = 0 \;, \tag{2.59e}$$

where D is the volume under consideration and S is its surface. q_n is the fluid velocity normal to the surface. $\dot{\underline{n}}$ is the normal geometric surface velocity of S. n_x, n_y, n_z are the components of the normal vector in x,y,x-direction. At this stage we recognize at best the conservative nature of the Euler equations: The total change of the fluid properties with respect to time depends only on the surface flux integral but not on internal sources. Later this property will be postulated also for the construction of numerical methods solving the discrete analogue of the Euler equations. The last form of the latter (Stokes integral) is particularly well suited for methods working with unstructured grids, say, forms of finite element discretizations such as triangles.

2.6 Physical Boundary Conditions

Let us consider the solution of a flow problem by a numerical method. Then it is obvious that the physical space we can approximate is confined by more or less artificial boundaries. If we know the flow variables there it is clear that we can prescribe them there and the answer of our solution is that which fulfills these boundary conditions. How these are implemented in the numerical algorithm and whether the algorithm under consideration tolerates the prescribed values without numerical contradiction is the question of formulating numerical boundary conditions which will be considered later.

A particular boundary which often occurs is the solid body boundary condition. It simply says that a solid body cannot be penetrated by the fluid surrounding it. Mathematically this fact is expressed by the vanishing of the fluid velocity normal to the solid body surface:

$$\underline{q}_n - \dot{\underline{n}} = 0 \; . \tag{2.60}$$

If the numerical scheme uses the form of Eq.(2.58) of the Euler equations, the solid body boundary condition is expressed by $\dot{\xi}_o - \dot{\xi} = 0$, if the solid body surface is a ξ=const surface, $\dot{\eta}_o - \dot{\eta} = 0$, if the solid body surface is a η=const surface, $\dot{\zeta}_o - \dot{\zeta} = 0$, if the solid body surface is a ζ=const surface. All these equations express the fact that fluid particles may not enter a solid surface, while the tangential flow of fluid particles is unhampered. This is in contrast to the Navier-Stokes equations which call for zero total velocity at body surfaces.

2.7 Other Forms of the Euler Equations

Of course the non-conservative form of the Euler equations can be recast in an arbitrary coordinate system ξ, η, ζ by chain ruling the equations written for a Cartesian coordinate system. With the abbreviation

$$\frac{1}{V}\frac{dV}{dt} = u_x + v_y + w_z = (u_\xi \xi_x + u_\eta \eta_x + u_\zeta \zeta_x + v_\xi \xi_y + v_\eta \eta_y + v_\zeta \zeta_y + \\ + w_\xi \xi_z + w_\eta \eta_z + w_\zeta \zeta_z)/D \; , \tag{2.61}$$

and Eq.(2.1 to 2.3) the following form is obtained

$$D\dot{p} + \rho \frac{dV}{V} + \rho_\xi \dot{\xi}_o + \rho_\eta \dot{\eta}_o + \rho_\zeta \dot{\zeta}_o = 0 \,, \tag{2.62a}$$

$$\rho(D\dot{u} + u_\xi \dot{\xi}_o + u_\eta \dot{\eta}_o + u_\zeta \dot{\zeta}_o) + p_\xi \xi_x + p_\eta \eta_x + p_\zeta \zeta_x = 0 \,, \tag{2.62b}$$

$$\rho(D\dot{v} + v_\xi \dot{\xi}_o + v_\eta \dot{\eta}_o + v_\zeta \dot{\zeta}_o) + p_\xi \xi_y + p_\eta \eta_y + p_\zeta \zeta_y = 0 \,, \tag{2.62c}$$

$$\rho(D\dot{w} + w_\xi \dot{\xi}_o + w_\eta \dot{\eta}_o + w_\zeta \dot{\zeta}_o) + p_\xi \xi_z + p_\eta \eta_z + p_\zeta \zeta_z = 0 \,, \tag{2.62d}$$

$$D\dot{p} + p_\xi \xi_o + p_\eta \eta_o + p_\zeta \zeta_o + \gamma p \frac{dV}{V} = 0 \,. \tag{2.62e}$$

This form of the Euler equations will be used for the derivation of diagonalized matrix forms and is therefore the most important form besides the conservation form.

There are of course numerous other non-conservative forms which depend on the choice of variables. An example is the transformation to the conservative variables as follows

$$u = \frac{\ell}{\rho} \,, \quad v = \frac{m}{\rho} \,, \quad w = \frac{n}{\rho} \,, \quad p = (\gamma-1)(e - \frac{\ell^2+m^2+n^2}{2\rho}) \,. \tag{2.63}$$

The differentials are

$$du = \frac{d\ell - u d\rho}{\rho} \,, \quad dv = \frac{dm - v d\rho}{\rho} \,, \quad dw = \frac{dn - w d\rho}{\rho} \,,$$

$$dp = (\gamma-1)(de - \frac{q^2}{2} d\rho - u d\ell - v dm - w dn) \,, \tag{2.64}$$

from which the quasi-conservative form can be obtained by inserting the differentials into the non-conservative equations. Since the formulae are lengthy the reader is asked to do this exercise on his own.

Another transform can be introduced via the definition of the speed of sound s and of the entropy S:

$$p = \rho^\gamma e^S \quad ; \quad p = \frac{\rho s^2}{\gamma} \,.$$

From the first equation we get

$$dp = \rho s^2 (\frac{d\rho}{\rho} + \frac{dS}{\gamma}) \,,$$

and from the second equation we get

$$dp = \frac{\rho s^2}{\gamma} (\frac{d\rho}{\rho} + 2 \frac{ds}{s}) \,.$$

Upon subtraction the differential of the density is obtained

$$d\rho = \frac{\rho}{\gamma-1} (2 \frac{ds}{s} - dS) \,. \tag{2.66}$$

The differential of the pressure is therefore

$$dp = \frac{\rho s^2}{\gamma-1} \left(2 \frac{ds}{s} - \frac{1}{\gamma} dS\right) . \qquad (2.67)$$

Upon inserting both differentials into the energy equation

$$dp - s^2 d\rho = 0 ,$$

we obtain for the energy equation

$$dS = 0 , \quad \text{or}$$

$$\dot{S} + uS_x + vS_y + wS_z = 0 , \qquad (2.68a)$$

which holds piecewise along streamlines.

The continuity equation becomes

$$\dot{s} + us_x + vs_y + ws_z + \frac{\gamma-1}{2} s(u_x + v_y + w_z) = 0 . \qquad (2.68b)$$

The momentum equations are

$$\dot{u} + uu_x + vu_y + wu_z + \frac{s^2}{\gamma-1}\left(2 \frac{s_x}{s} - \frac{1}{\gamma} S_x\right) = 0 , \qquad (2.68c)$$

$$\dot{v} + uv_x + vv_y + wv_z + \frac{s^2}{\gamma-1}\left(2 \frac{s_y}{s} - \frac{1}{\gamma} S_y\right) = 0 , \qquad (2.68d)$$

$$\dot{w} + uw_x + vw_y + ww_z + \frac{s^2}{\gamma-1}\left(2 \frac{s_z}{s} - \frac{1}{\gamma} S_z\right) = 0 . \qquad (2.68e)$$

2.8 References

1. Liepmann, H.W., Roshko, A.: "Elements of Gasdynamics". J. Wiley & Sons, New York/London/Sydney, 1957.

2. Bird, R.B., Stewart, W.E., Lightfoot, E.N.: "Transport Phenomena". J. Wiley & Sons, New York/London/Sydney, 1960.

III FUNDAMENTALS OF DISCRETE SOLUTION METHODS

Wave propagation is an inherent feature of the solutions to the Euler equations. The first part of this chapter surveys the basic theory of wave propagation encountered in hyperbolic differential equations including the concept of characteristics,[1] discontinuities, and weak solutions.[2] Boundary conditions and how they affect stability are also discussed.[3] The second part introduces basic ideas on discrete methods, their accuracy and consistency, numerical stability and numerical boundary conditions.

Before looking closer at some of these ideas, let us first consider what is meant by a wave. There are, by tradition, two quite different intuitive notions of waves. One of these involves motion, and can be used to describe physical phenomena in which an identifiable characteristic propagates, like the changing profile of a water wave. The other is that which involves periodic behaviour, and it need not necessarily describe motion. For example, the undular pattern left on the sand of a beach as the tide recedes involves periodicity of pattern, as does the optical diffraction pattern caused by a slit.

A wave propagating as a disturbance which varies with time throughout some region in space, has two important properties. Firstly energy is transmitted to distant points, and, secondly the disturbance travels through the medium without giving the medium as a whole any permanent displacement.

3.1 Hyperbolic Equations and Waves

Solutions to hyperbolic partial differential equations exhibit the familiar properties of waves, namely that signals propagate at a finite speed and are localized within a finite region so that, at least initially, there can be a start and finish to them. A simple example demonstrates this point for the quasilinear system of equations

$$U_t + Aq_x + Bq_y + Cq_z = 0 \ . \tag{3.1}$$

Consider what happens to a signal of the form

$$U = U_o e^{st} e^{-i(\alpha x+\beta y+\varepsilon z)} \tag{3.2}$$

obeying Eq.(3.1). It must satisfy

$$[sI - i(\alpha A + \beta B + \varepsilon C)]U = 0$$

and we see that the time coefficient is

$$s = i\,\lambda_D \,,$$

where λ_D are the eigenvalues of the matrix $D=\alpha A+\beta B+\varepsilon C$ which is the linear combination of the coefficient matrices.

If all λ_D are real, s is imaginary and we find that the signal (3.2) is a traveling plane wave moving with velocity λ_D in the direction with normal $\underline{\nu}=\alpha\underline{e}_x+\beta\underline{e}_y+\varepsilon\underline{e}_z$. This notion leads us to the formal definition of a hyperbolic system of equations.

Definition

The quasi-linear system (3.1) is called hyperbolic at the point (t,x,y,z,U) if

i) for all real values of α,β,ε the eigenvalues λ_i of the matrix $D=\alpha A+\beta B+\varepsilon C$ are all real, and

ii) there exists a non-singular matrix $T\,(\alpha,\beta,\varepsilon)$ that diagonalizes D such that

$$T^{-1}DT = \text{diag}\,\lambda_i \,, \tag{3.3}$$

where the norms of T and T^{-1} are uniformly bounded for all α,β,ε:

$$\|T\|,\ \|T^{-1}\| \leq \text{const.}$$

Example

Consider the Euler equations in non-conservative quasilinear form:

$$q = \begin{vmatrix} \rho \\ u \\ v \\ w \\ p \end{vmatrix} , \qquad A = \begin{vmatrix} u & \rho & 0 & 0 & 0 \\ 0 & u & 0 & 0 & 1/\rho \\ 0 & 0 & u & 0 & 0 \\ 0 & 0 & 0 & u & 0 \\ 0 & \rho c^2 & 0 & 0 & u \end{vmatrix} ,$$

$$B = \begin{vmatrix} v & 0 & \rho & 0 & 0 \\ 0 & v & 0 & 0 & 0 \\ 0 & 0 & v & 0 & 1/\rho \\ 0 & 0 & 0 & v & 0 \\ 0 & 0 & \rho c^2 & 0 & v \end{vmatrix} , \quad C = \begin{vmatrix} w & 0 & 0 & \rho & 0 \\ 0 & w & 0 & 0 & 0 \\ 0 & 0 & w & 0 & 0 \\ 0 & 0 & 0 & w & 1/\rho \\ 0 & 0 & 0 & \rho c^2 & w \end{vmatrix} .$$

The eigenvalues λ_i of the linear combination $D=\alpha A+\beta B+\varepsilon C$ are

$$\lambda_1 = \lambda_2 = \lambda_3 = \alpha u + \beta v + \varepsilon w\ , \qquad \lambda_{4,5} = \alpha u + \beta v + \varepsilon w \pm c\ ,$$

where we have assumed $\alpha^2+\beta^2+\varepsilon^2=1$, i.e. $\underline{v}$ is the unit normal vector, and c is the speed of sound.

The diagonalizing matrices work out to

$$T = \begin{vmatrix} \alpha & \beta & \varepsilon & \rho/c\sqrt{2} & \rho/c\sqrt{2} \\ 0 & -\varepsilon & \beta & \alpha/\sqrt{2} & -\alpha/\sqrt{2} \\ \varepsilon & 0 & -\alpha & \beta/\sqrt{2} & -\beta/\sqrt{2} \\ -\beta & \alpha & 0 & \varepsilon/\sqrt{2} & -\varepsilon/\sqrt{2} \\ 0 & 0 & 0 & \rho c/\sqrt{2} & \rho c/\sqrt{2} \end{vmatrix}\ ,$$

where the jth column of T is a right eigenvector of D corresponding to the eigenvalue λ_j. The jth row of T^{-1} is a left eigenvector of D corresponding to the eigenvalue λ_j

$$T^{-1} = \begin{vmatrix} \alpha & 0 & \varepsilon & -\beta & -\alpha/c^2 \\ \beta & -\varepsilon & 0 & \alpha & -\beta/c^2 \\ \varepsilon & \beta & -\alpha & 0 & -\varepsilon/c^2 \\ 0 & \alpha/\sqrt{2} & \beta/\sqrt{2} & \varepsilon/\sqrt{2} & 1/\rho c\sqrt{2} \\ 0 & -\alpha/\sqrt{2} & -\beta/\sqrt{2} & -\varepsilon/\sqrt{2} & 1/\rho c\sqrt{2} \end{vmatrix}\ .$$

The norms of T and T^{-1} can be shown to be uniformly bounded for all $\alpha, \beta, \varepsilon$, and thus according to the definition the Euler equations are hyperbolic.

3.2 Characteristics[1]

We now introduce the concept of characteristics for the one-dimensional system of equations that serves as a model to system (3.1)

$$\frac{\partial \underline{U}}{\partial t} + A\frac{\partial \underline{U}}{\partial x} = 0\ , \tag{3.4}$$

in which $\underline{U}$ are n element column vectors and A is an n×n matrix with elements a_{ij}. Here, and in the next two sections, the presentation closely follows that of Ref. 1.

Although x,t are the natural variables to use when deriving systems of equations describing motion in space and time, they are not necessarily the most appropriate ones from the mathematical point of view. So, as we are interested in the way a solution evolves with time, let us leave the time variable unchanged, but replace x by some arbitrary curvilinear coordinate ξ and then try to choose ξ in a manner which is convenient for our mathematical arguments. Accordingly, our starting point will be to change from (x,t) to the arbitrary semi-curvilinear coordinates

$$\xi = \xi(x,t) , \qquad t' = t .$$

If the Jacobian of the transformation is non-vanishing we may thus transform (3.4) to

$$\frac{\partial \underline{U}}{\partial t'} + \left(\frac{\partial \xi}{\partial t} I + \frac{\partial \xi}{\partial x} A\right) \frac{\partial \underline{U}}{\partial \xi} = 0 . \tag{3.5}$$

This equation is an algebraic relationship connecting the matrix vector derivatives $\partial \underline{U}/\partial t'$ and $\partial \underline{U}/\partial \xi$. It may only be used to determine $\partial \underline{U}/\partial \xi$ if the determinant of the coefficient matrix of $\partial \underline{U}/\partial \xi$ is non-vanishing. This condition obviously depends on the nature of the curvilinear coordinate lines $\xi(x,t)$=const. Suppose now that for some particular choice $\xi \equiv \phi$ the determinant does vanish, giving the condition

$$\left|\frac{\partial \phi}{\partial t} I + \frac{\partial \phi}{\partial x} A\right| = 0 . \tag{3.6}$$

Then because of this the derivative $\partial \underline{U}/\partial \phi$ will be indeterminate on the family of lines ϕ=const. Consequently, across such lines $\phi(x,t)$=const., $\partial \underline{U}/\partial \phi$ may actually be discontinuous.

Let us confine our attention to solutions $\underline{U}$ which are everywhere continuous but for which the derivative $\partial \underline{U}/\partial \phi$ is discontinuous across the line ϕ. Because of the continuity of U, and the continuity of the elements a_{ij} of A, the matrix A will experience no discontinuity across ϕ. In Eq.(3.5) there is no indeterminacy of $\partial \underline{U}/\partial t'$ across the lines ϕ=const., and as $\partial/\partial t'$ denotes differentiation along these lines it must follow that $\partial \underline{U}/\partial t'$ is everywhere continuous across the line ϕ. Taking these facts into account and differencing Eq.(3.5) across the line $\xi \equiv \phi$ gives

$$\left(\frac{\partial \phi}{\partial t} I + \frac{\partial \phi}{\partial x} A\right) \left[\frac{\partial \underline{U}}{\partial \phi}\right] = 0 , \tag{3.7}$$

where $[\alpha]\equiv\alpha_- - \alpha_+$ signifies the discontinuous jump in the quantity α across the line ϕ. It expresses compatibility conditions to be satisfied by the component of the derivative of U on either side of and normal to these curves in the (x,t)-plane. Now there will only be a non-trivial solution to (3.7) if

$$\left|\frac{\partial\phi}{\partial t} I + \frac{\partial\phi}{\partial x} A\right| = 0 \ . \tag{3.8}$$

However, along the lines ϕ=const. we have, by differentiation,

$$\frac{\partial\phi}{\partial t} + \frac{\partial\phi}{\partial x}\frac{dx}{dt} = 0 \ ,$$

so that these lines have the gradient

$$\frac{dx}{dt} = -\frac{\partial\phi}{\partial t}\Big/\frac{\partial\phi}{\partial x} \equiv \lambda(\text{say}) \ . \tag{3.9}$$

Combining (3.8) and (3.9) we deduce that λ must be such that

$$\left|A-\lambda I\right| = 0 \ . \tag{3.10}$$

Consequently the λ in (3.9) can only be one of the eigenvalues of A, and since (3.7) can be re-written

$$(A-\lambda I)\left|\frac{\partial \underline{U}}{\partial\phi}\right| = 0 \ , \tag{3.11}$$

the column vector $\left|\partial U/\partial\phi\right|$ must be proportional to the corresponding right eigenvector of A:

$$\left|\frac{\partial \underline{U}}{\partial\phi}\right| = k\,\underline{r} \ . \tag{3.12}$$

As A is a n×n matrix, it will have n eigenvalues. If these are real and distinct, integration of equations (3.9) will give rise to n distinct families of real curves $C^{(1)},C^{(2)},\ldots,C^{(n)}$ in the (x,t) plane:

$$C^{(i)}:\ \frac{dx}{dt} = \lambda^{(i)}, \qquad i=1,2,..,n \ . \tag{3.13}$$

Any one of these families of curves $C^{(i)}$ may be taken for our curvilinear coordinate lines ϕ=const. The $\lambda^{(i)}$ associated with each family will then be the speed of propagation of the vector $\left|\partial \underline{U}/\partial\phi\right|$ along the curves $C^{(i)}$ belonging to that family.

When the eigenvalues $\lambda^{(i)}$ of A are all real and distinct, so that the propagation speeds are also all real and distinct, and there are n distinct linearly independent right eigenvectors $\underline{r}^{(i)}$ of A satisfying the defining relation

$$A\underline{r}^{(i)} = \lambda^{(i)}\underline{r}^{(i)} , \qquad \text{for } i=1,2,..,n , \tag{3.14}$$

the system of equations (3.5) will be said to be totally hyperbolic. We may, if we desire, replace the words right eigenvector by left eigenvector in this definition, where the left eigenvectors $\underline{\ell}$ of A satisfy the defining relation

$$\underline{\ell}^{(i)}A = \lambda^{(i)}\underline{\ell}^{(i)} , \qquad \text{for } i=1,2,..,n . \tag{3.15}$$

This follows because simple linear algebra arguments establish that when n linearly independent vectors $\underline{r}^{(i)}$ exist, then so also do n linearly independent vectors $\underline{\ell}^{(i)}$

Hereafter our concern will be with such systems, since they characterize the type of wave propagation that has been the object of our study so far. The families of curves $C^{(i)}$ defined by integration of equations (3.13) are called the families of characteristic curves of system (3.5). A totally hyperbolic system (3.5) is thus one in which there are n distinct real speeds of propagation of a disturbance, each of which when characterized by the appropriate right eigenvector is different. The precise nature of these differences will be examined shortly.

The relationship between characteristic curves and the solution vector $\underline{U}$ to system (3.5) is illustrated in Fig. 3.1 in the case of a typical element u_i of $\underline{U}$. Here it has been assumed that initial conditions have been specified for system (3.4) in the form

$$\underline{U}(x,0) = \Psi(x) ,$$

where the ith element u_i of $\underline{U}$ has for its initial condition $u_i(x,0)=\psi_i(x)$.

The line PQ in the solution surface S is the one across which $\partial u/\partial\phi$ is discontinuous, and its projection onto the (x,t)-plane is the characteristic which has equation $\phi(x,t)=k$. Since such a line marks the boundary between the different solutions to the left and right of it, it is natural to think of the solution to the left of $\phi=k$ as a propagating

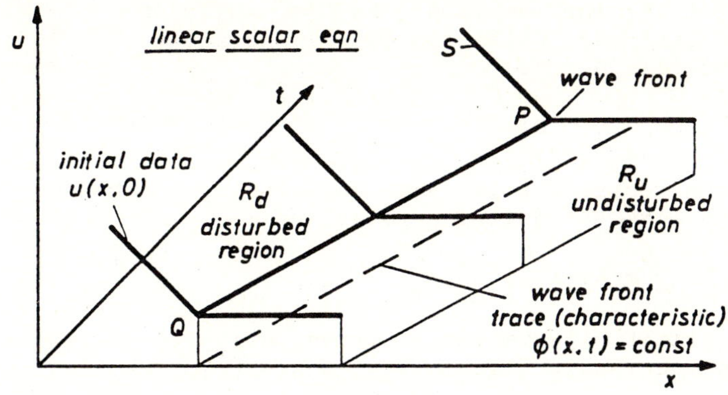

Fig. 3.1 Propagation of wave front Q in space and time

disturbance wave, and the solution to the right as the solution in a region not yet reached by the disturbance. With these ideas in mind we shall call the line PQ in S the solution surface wavefront, its projection onto the (x,t) plane, which forms the characteristic curve $\phi=k$, the disturbed region, and the region R_u to the right of $\phi=k$ the undisturbed region. At any time t_1 the physical wavefront is at the intersection of the wavefront trace and the line $t=t_1$.

3.3 Wavefronts Bounding a Constant State[1]

In physical problems the solution vector U describes the "state" of the system governed by equations (3.4). Thus we refer to a region in which $\underline{U}$ is non-constant as as a disturbed state, and a region in which $\underline{U}$ is constant as a constant state. Our purpose here will be to examine the simplification that results in equation (3.12) when a wavefront bounds a constant state.

First, as the element a_{ij} of A are continuous functions of their arguments, it follows directly that the eigenvalues $\lambda^{(i)}$ of A are continuous functions of a_{ij}, and hence of the elements $u_1, u_2, \ldots, u_n$ of $\underline{U}$. Since $\underline{U}$ is itself continuous across a wavefront we conclude that $\lambda^{(i)}=\lambda_o^{(i)}=\text{const.}$, on a wavefront bounding the constant state $\underline{U}=\underline{U}_o$, where $\lambda_o^{(i)}=\lambda^{(i)}(\underline{U}_o)$. From equations (3.13) we thus see that if a characteristic curve from the ith family $C^{(i)}$ bounds a constant state, then it must be a straight line.

If such a straight-line characteristic $C_o^{(i)}$ belonging to the ith family $C^{(i)}$ bounds a constant state $\underline{U}=\underline{U}_o$ that lies to its right (say), then because $(\partial\underline{U}/\partial\phi)_+=\partial\underline{U}_o/\partial\phi\equiv 0$ we have

$$\left[\frac{\partial u_j}{\partial\phi}\right]\equiv\left(\frac{\partial u_j}{\partial\phi}\right)_- -\left(\frac{\partial u_j}{\partial\phi}\right)_+ =\left(\frac{\partial u_j}{\partial\phi}\right)_- \quad \text{for } j=1,2,\ldots,n \,. \qquad (3.16)$$

Now $\partial \underline{U}/\partial t'$ is continuous across $C_o^{(i)}$ while $\partial \underline{U}_o/\partial t' \equiv 0$. Thus in the disturbed region immediately adjacent to $C_o^{(i)}$ the total differential $d\underline{U}$ reduces to

$$d\underline{U} = \underline{r}^{(i)} d\phi \ , \qquad (3.17)$$

where the constant k is taken to be unity. Combining this last equation with the defining relationship

$$A\underline{r} = \lambda \underline{r}$$

for the right eigenvector $\underline{r}$ corresponding to the eigenvalue λ gives the result immediately adjacent to the constant state $\underline{U}=\underline{U}_o$

$$(A_o - \lambda_o I) d\underline{U} = 0 \ , \qquad (3.18)$$

where $A_o = A(\underline{U}_o)$ and $\lambda_o = \lambda(\underline{U}_o)$, and $d\phi$ is chosen as the unit increment.

3.4 Riemann Invariants[1]

In this section we offer a brief discussion of an important technique that can lead directly to a solution when certain calculations can be performed, and which in any case provides a valuable insight into the nature of solutions for wave propagation. A system of two dependent variables u_1 and u_2 illustrate the concepts first. The system is hyperbolic provided the two eigenvalues $\lambda^{(i)}$, i=1,2 of

$$|A - \lambda I| = 0$$

are real and A has linearly independent eigenvectors. We make use of the corresponding left eigenvectors $\underline{\ell}$ defined by

$$\underline{\ell}^{(i)} A = \lambda^{(i)} \underline{\ell}^{(i)}, \qquad \text{for } i=1,2 \ . \qquad (3.19)$$

If, now, we pre-multiply (3.4) by $\underline{\ell}^{(i)}$, we obtain the result

$$\underline{\ell}^{(i)} \left(\frac{\partial \underline{U}}{\partial t} + \lambda^{(i)} \frac{\partial \underline{U}}{\partial x} \right) = 0 \qquad \text{for } i=1,2 \ . \qquad (3.20)$$

The bracketed expression will be recognised as the directional derivative of $\underline{U}$ with respect to time along the family of characteristics $C^{(i)}$. Denoting differentiation with respect to time along members of the $C^{(1)}$ family of characteristics by $d/d\alpha$ and differentiation with respect to time along members of the $C^{(2)}$ family of characteristics by $d/d\beta$ enables us to replace (3.20) by the following pair of ordinary differential equations which are defined along the $C^{(1)}$ characteristics by $\underline{\ell}^{(1)} \frac{d\underline{U}}{d\alpha} = 0$, (3.21)

and along the $C^{(2)}$ characteristics by $\underline{\ell}^{(2)} \frac{d\underline{U}}{d\beta} = 0$. (3.22)

Hence β=constant, along $C^{(1)}$ characteristics and α=constant, along $C^{(2)}$ characteristics as indicated in Fig. 3.2.

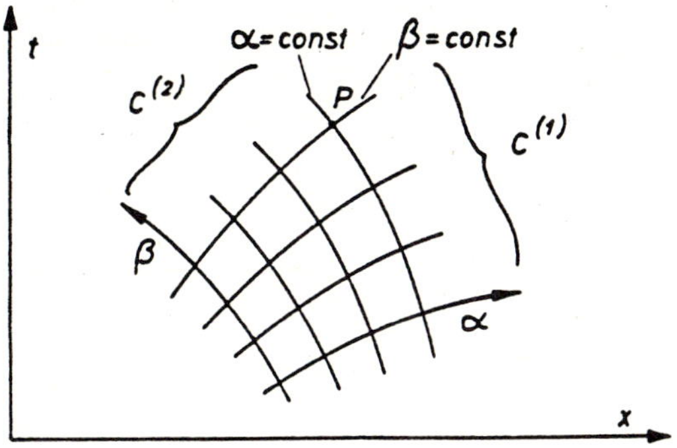

Fig. 3.2 The two characteristic families for the 2×2 system

Since, by supposition, A depends only on u_1 and u_2, so also will the coefficients $\ell_j^{(i)}$ of the left eigenvectors $\underline{\ell}^{(1)}$, $\underline{\ell}^{(2)}$. Consequently, both (3.21) and (3.22) will always be integrable along their respective characteristics, though they may first require multiplication by a suitable integrating factor μ.

Integrating them with respect to α along the $C^{(1)}$ characteristics, and with respect to β along the $C^{(2)}$ characteristics gives:

along $C^{(1)}$ characteristics $\int \mu \ell_1^{(1)} du_1 + \int \mu \ell_2^{(1)} du_2 = r(\beta)$, (3.23)

and

along $C^{(2)}$ characteristics $\int \mu \ell_1^{(2)} du_1 + \int \mu \ell_2^{(2)} du_2 = s(\alpha)$, (3.24)

where r,s are arbitrary functions of their respective arguments β and α. The two families of characteristic curves are themselves given by integration of the equations

$$C^{(i)}: \frac{dx}{dt} = \lambda^{(i)}, \qquad \text{for} \quad i=1,2 . \qquad (3.25)$$

The functions $r(\beta)$ and $s(\alpha)$ are called Riemann invariants and, by virtue of their manner of derivation, r and s are constant along their respective families of characteristics. To be more precise, $r(\beta)$ is constant along any $C^{(1)}$ characteristic, though as it is a function of β (which in turn identifies the characteristics) it will, in general, be different for different characteristics. Correspondingly, $s(\alpha)$ is constant along any $C^{(2)}$ characteristic, though here again the constant will be different for different characteristics depending on the value of α associated with each characteristic.

Equations (3.23) and (3.24) enable u_1 and u_2 to be expressed in terms of r and s, the values of which are determined at points of the initial line t=0 by the initial data. Suppose $r(\beta)$ is denoted by $R(u_1,u_2)$ and $s(\alpha)$ is denoted by $S(u_1,u_2)$. Then along the $C^{(i)}$ characteristic issuing out from the point $(x_0,0)$ of the initial line $r(\beta)$

$$R(u_1,u_2) = R\left(u_1(x_0),\; u_2(x_0)\right) . \tag{3.26}$$

Similarly, along the $C^{(2)}$ characteristic issuing out from the point $(x_1,0)$ of the initial line

$$S(u_1,u_2) = S\left(u_1(x_1),\; u_2(x_1)\right) . \tag{3.27}$$

Solving these two implicit equations for u_1 and u_2 then determines the solution at the point P in Fig. 3.2, which is the point of intersection of the $C^{(1)}$ and $C^{(2)}$ characteristics along which the respective constant values R and S are transported. In principle the initial value problem is now solved. However, in any particular case, the task of solving the two implicit relationships and of finding the characteristic curves in order to determine their point of intersection P is usually difficult. Nevertheless, this method of characteristics can often be used to solve problems and it is, in any case, of considerable theoretical importance.

The preceding discussion was for the special case of two dependent variables. In general we have the following definition:

A Riemann invariant is a function of $\underline{U}$ that is constant along a characteristic associated with the eigenvalue λ.

The Riemann invariant $\underline{f}(\underline{U})$ satisfies the condition:

$$A^T \frac{\partial \underline{f}}{\partial \underline{U}} = \lambda \frac{\partial \underline{f}}{\partial \underline{U}} ,$$

i.e. $\partial \underline{f}/\partial \underline{U}$ is the matrix of left eigenvectors of A.

This can be demonstrated by differentating $\underline{f}$ along a characteristic curve

$$\frac{d\underline{f}}{d\phi} = \frac{\partial \underline{f}}{\partial \underline{U}} \frac{d\underline{U}}{d\phi} = \frac{\partial \underline{f}}{\partial \underline{U}} \left(\frac{\partial \underline{U}}{\partial t} \frac{dt}{d\phi} + \frac{\partial \underline{U}}{\partial x} \frac{d\lambda}{d\phi}\right) ,$$

which by (3.4)

$$\frac{\partial \underline{f}}{\partial \phi} = \frac{\partial \underline{f}}{\partial \underline{U}} \left(-A^T \frac{\partial \underline{U}}{\partial x} + \lambda \frac{\partial \underline{U}}{\partial x}\right) = \left(-A^T \frac{\partial \underline{f}}{\partial \underline{U}} \frac{\partial \underline{U}}{\partial x} + \lambda \frac{\partial \underline{f}}{\partial \underline{U}} \frac{\partial \underline{U}}{\partial x}\right) = 0 .$$

Example 1

Consider the 1-D Euler equations for isentropic flow with $p=p(\rho)$ and $dp/d\rho=c^2$, then

$$A = \begin{vmatrix} u & \rho \\ c^2/\rho & u \end{vmatrix} , \qquad \underline{U} = \begin{vmatrix} \rho \\ u \end{vmatrix} .$$

The eigenvalues are $\lambda^{(1)}=u+c$ and $\lambda^{(2)}u-c$. A simple calculation shows that the left eigenvector $\underline{\ell}^{(1)}$ corresponding to $\lambda^{(1)}$ is

$$\underline{\ell}^{(1)} = [1, \rho/c] ,$$

and the left eigenvector corresponding to $\lambda^{(2)}=u-c$ is

$$\underline{\ell}^{(2)} = [1,-\rho/c] .$$

With an integrating factor $\mu=c/\rho$ Eqs.(3.23) and (3.24) become

$$r(\beta) = u + \int_{\rho_0}^{\rho} \frac{c}{\rho} d\rho \quad \text{along } C^{(1)} \text{ characteristics} ,$$

and

$$s(\alpha) = u - \int_{\rho_0}^{\rho} \frac{c}{\rho} d\rho \quad \text{along } c^{(2)} \text{ characteristics} .$$

For a perfect gas where $p=\kappa\rho^{\gamma}$, the speed of sound is

$c = (\gamma\ p/\rho)^{1/2}$ and the last two equations become

$$\binom{r}{s} = u \pm \frac{2}{\gamma-1} c . \qquad (3.28)$$

It is possible to express the dependent variables in terms of the Riemann invariants r and s, but the determination of the characteristics by integrating

$$C^{(1)}: \frac{dx}{dt} = u + c \quad \text{and} \quad C^{(2)}: \frac{dx}{dt} = u - c \tag{3.29}$$

is only possible in special cases.

Application to Illustrate the Cauchy Problem

Consider the 1-D isentropic flow of the perfect gas of Example 1. The initial data are given on a non-characteristic curve Γ of infinite extent that is transverse to the characteristics (3.28). Determining the dependent variables c and u in the field is called the Cauchy problem.

It is solved by picking two points 1 and 2 on Γ (see Fig. 3.3), constructing the left and right-going characteristics

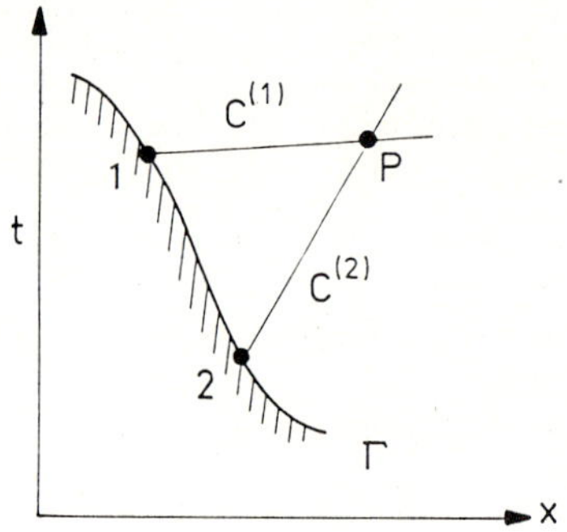

Fig. 3.3 Solving the Cauchy Problem by the method of characteristics

through these two points respectively, and then finding the point P where they intersect. Since the Riemann invariants (3.28) are constant along the characteristics, we obtain the linear system

$$u + \frac{2}{\gamma-1}\, c = u_1 + \frac{2}{\gamma-1}\, c_1 \,, \qquad u - \frac{2}{\gamma-1}\, c = u_2 - \frac{2}{\gamma-1}\, c_2$$

with solutions

$$u = \tfrac{1}{2}(u_1+u_2) + \frac{1}{\gamma-1}(c_1-c_2), \quad \frac{2}{\gamma-1}\, c = \tfrac{1}{2}(u_1-u_2) + \frac{1}{\gamma-1}(c_1+c_2) \,. \tag{3.30}$$

The characteristics have to be constructed to find P. The only inaccuracy of the method comes in by the assumption that the characteristics from 1 to P and from 2 to P are straight lines with the slopes

$$\left.\frac{dx}{dt}\right|_1 = \frac{1}{2}(u_1+c_1+u+c) , \quad \left.\frac{dx}{dt}\right|_2 = \frac{1}{2}(u_2-c_2+u-c) .$$

If we assume that the initial data are analytic on the analytic curve C, and that the coefficient matrix A is an analytic function of x,t and $\underline{U}$, and that $|\partial\underline{U}/\partial x|$ is bounded in some region of x,t around P, then by the Cauchy-Kovaleski theorem the Cauchy problem has an unique solution in the neighborhood of P.

3.5 Well-Posed and Unique Solutions[2]

The notion of a well-posed problem finds its origins in the early work of Hadamard[4]. He considered a problem concerning a system of partial differential equations to be well-posed, or correctly set, if the data imposed on the system to identify a particular solution is such that:

i) a solution exists,
ii) the solution so determined is unique, and
iii) the solution depends continuously on the data.

In the event that these three conditions are not all satisfied the problem is said to be improperly-posed, or ill-posed.

The ideas underlying these requirements for a well-posed problem are simply that in the physical world a solution is usually expected to exist to a real problem and, furthermore, the solution is expected to be unique. However, still more is expected of the solution to a physical problem, since when the data used to specify a particular solution is changed slightly then it normally is anticipated that the solution will only exhibit a correspondingly small change. Naturally, when expressed in mathematical terms, this last requirement necessitates the specification of the class to which the solution belongs and the criterion by which the continuous dependence of the solution on the data is to be judged. Examples of simple improperly-posed problems belonging to each of these three categories have been constructed by Jeffreys[2] and are presented below, the first two of which concern scalar first-order equations and the third a scalar second-order equation. For more information see the works by John[5] and Lavrentiev[6].

(i) Non-Existence of Solution

Consider the linear first-order scalar equation (the so-called Kreiss equation)

$$\frac{\partial u}{\partial t} + \frac{\partial u}{\partial x} = 0 ,$$

subject to the initial condition

a) $u(x,0) = x$,

and the boundary condition

b) $u(0,t) = f(t)$ with $f(0) = 0$.

Then the solution satisfying both the equation and the data in a) is easily seen to be $u(x,t)=x-t$. At the boundary thus becomes $u(0,t)=-t$ which shows that the data represented by a) and b) is over-prescribed, so that no solution will exist, unless $f(t)\equiv -t$.

ii) Non-Uniqueness of Solution

Consider the Kreiss equation as in i) above but with conditions a) and b) replaced by $u_x(x,0)=1$ on the initial line $t=0$. Then the data implies $u(x,0)=x+d$, with d an arbitrary constant of integration. The solution is thus $u(x,t)=x-t+d$ which is not unique because of the arbitrariness of d.

(iii) Non-Continuous Dependence of Solution on Data

The best known example in this category is the following one due to Hadamard and concerns Laplace's equation. Consider the equation

$$\frac{\partial^2 u}{\partial x^2} + \frac{\partial^2 u}{\partial y^2} = 0 ,$$

subject to the Cauchy data

$$u(x,0) = 0, \qquad u_y(x,0) = \frac{1}{n} \sin nx.$$

Then the solution is easily found to be

$$u(x,y) = \frac{1}{n^2} \sin nx \sinh ny.$$

However, as $n\to\infty$ the Cauchy data approaches zero uniformly while the solution oscillates unboundedly for $y\neq 0$. This behaviour of the solution is unexpected since the solution to the equation with identically zero Cauchy data is $u(x,t) \equiv 0$. This example serves to illustrate that Cauchy data is inappropriate for Laplace's equation. The reason for this behaviour is, of course, that Laplace's equation is an elliptic equation. Furthermore, Cauchy data for this elliptic equation is being imposed on an open region.

In the general quasilinear case Friedrichs[7] and Lax[8] have proved the following theorem. Let the quasilinear system of equations (3.4) be such that

i) it is hyperbolic in the t-direction throughout all the x,t plane,

ii) the coefficient matrix $A(u,x,t)$ has bounded continuous partial derivatives,

iii) it satisfies the initial condition $u(x,0)=u_o(x)$ $a<x<b$ for which du_o/dx is bounded continuous.

Then a unique solution exists in the neighborhood of the interval $a<x<b$ of the initial line $t=0$, and furthermore, within this neighborhood the solution has bounded continuous partial derivatives.

This theorem only ensures the existence of a unique solution for a finite elapsed time beyond $t=0$. To determine precisely how the solution behaves if non-uniqueness occurs, and whether or not it can be extended beyond this time in any meaningful way, requires further analysis, as we shall see.

3.6 Initial Boundary-Value Problems[3]

Up to now the discussion has focused primarily on the initial value problem in the absence of boundaries. New considerations enter when boundaries appear. In order to introduce the problem by way of examples consider the scalar equation

$$\frac{\partial u}{\partial t} = a\,\frac{\partial u}{\partial x} \quad \text{for } 0<x<\infty \text{ and } t>0\ . \qquad (3.31)$$

Let initial values be given by

$$u(x,0) = f(x), \quad 0<x<\infty\ . \qquad (3.32)$$

The solution is given by

$$u(x,t)=f(x+at)\ , \qquad (3.33)$$

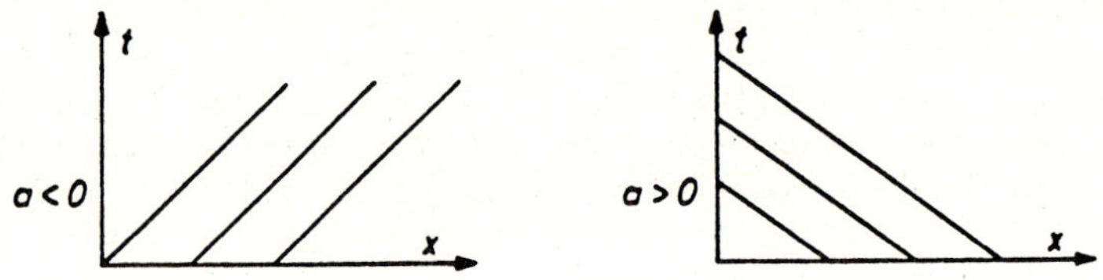

Fig. 3.4 The characteristics for the initial boundary value problem 3.31

which is constant on the characteristic lines $x+at=\text{constant}$. If $a>0$, then $u(x,t)$ is uniquely determined by (3.33) and it is not appropriate to specify a boundary condition at $x=0$. If $a<0$, then $u(x,t)$ is only determined in the triangular region (Fig. 3.4) $x+at\leq 0$. In this case a boundary condition

$$u(o,t) = g(t), \quad t\geq 0 \tag{3.34}$$

is required to determine the solution for $x+at>0$.

The solution $u(x,t)$ is continuous in a neighborhood of $x+at=0$ if and only if f and g are continuous and satisfy the compatibility condition

$$f(0) = g(0) \; .$$

More generally we are interested in the system of equations

$$\frac{\partial \underline{w}}{\partial t} = A \frac{\partial \underline{w}}{\partial x} \quad \text{for} \quad 0\leq x\leq 1, \qquad t\geq 0 \; , \tag{3.35}$$

with initital values

$$\underline{w}(x,0) = \underline{f}(x) \; , \tag{3.36}$$

where A can be transformed to real diagonal form by a nonsingular transformation T, i.e.

$$T^{-1}AT = \begin{vmatrix} -\Lambda_1 & 0 \\ 0 & \Lambda_2 \end{vmatrix} \; .$$

Let $\underline{u}=T^{-1}\underline{w}$. Then the system (3.35) becomes

$$\partial\underline{u}/\partial t = \begin{vmatrix} -\Lambda_1 & 0 \\ 0 & \Lambda_2 \end{vmatrix} \partial\underline{u}/\partial x, \quad x\geq 0, \quad t\geq 0 \; , \tag{3.37}$$
$$\underline{u}(x,o)= \underline{f}(x), \quad x\geq 0 \; ,$$

where

$$\Lambda_1 = \begin{vmatrix} \lambda_1 & & & \\ & \lambda_2 & & \\ & & \ddots & \\ & & & \lambda_r \end{vmatrix} > 0 \ , \quad \Lambda_2 = \begin{vmatrix} \lambda_{r+1} & & & \\ & \lambda_{r+2} & & \\ & & \ddots & \\ & & & \lambda_n \end{vmatrix} > 0$$

are positive matrices and

$$\underline{u} = \begin{vmatrix} u^I \\ u^{II} \end{vmatrix} , \quad u^I = \begin{vmatrix} u_1 \\ \vdots \\ u_r \end{vmatrix} , \quad u^{II} = \begin{vmatrix} u_{r+1} \\ \vdots \\ u_n \end{vmatrix} .$$

Then Eq.(3.37) can be expressed as two partitioned systems

$$\frac{\partial u^I}{\partial t} = -\Lambda_1 \frac{\partial u^I}{\partial x} , \tag{3.38}$$

$$\frac{\partial u^{II}}{\partial t} = \Lambda_2 \frac{\partial u^{II}}{\partial x} . \tag{3.39}$$

And we reason that the number of boundary conditions at $x=0$ must equal the number of negative eigenvalues of A and the number of boundary conditions at $x=1$ must correspond to the number of positive eigenvalues. This is true because the solution of (3.39) is determined by the initial values $f^{II}(x)$ as is the solution of equation (3.31) for $a>0$. Correspondingly, u^I has to be specified on the boundary $x=0$. We require

$$u^I(0,t) = S^I u^{II}(0,t) + g^I(t), \quad t>0 , \tag{3.40}$$

and assume its compatibility with the initial values. Here S^I is a $r\mathrm{x}(n-r)$ matrix and the term $S^I u^{II}(0,t)$ represents the dependence of u^I on u^{II}. Because of the different directions of the characteristics of u^I and u^{II}, u^{II} is considered an outgoing variable and u^I an ingoing variable on the line $x=0$. S^I then represents a generalized reflection of u^{II}.

Boundary conditions must also be specified at $x=1$. Proceeding in the same way as before we require

$$u^{II}(1,t) = S^{II} u^I(1,t) + g^{II}(t) , \tag{3.41}$$

where S^{II} is an $(n-r)\mathrm{x}r$ matrix representing a generalized reflection at $x=1$.

If $a=0$ in (3.31), then $\underline{u}(x,t)=\underline{f}(x)$ and we do not need to specify any boundary conditions. Similarly, if Λ_1 or Λ_2 are not positive definite, then the components $u_j(x,t)$ corresponding to a $\lambda_j=0$ can be considered as outgoing variables and will be included in u^{II} for x=0 and in u^I for x=1, i.e., we always assume that $\Lambda_1>0$ for x=0 and $\Lambda_2>0$ for x=1.

So the boundary conditions are clear enough in the 1-D constant coefficient problem. When the coefficients are not constant but vary smoothly, the same result holds but the analysis is more complicated. In the two-dimensional scalar case, the additional space dimension does not alter the result, and the same conditions hold.

This is also true for the system of equations in two space dimensions, but unfortunately for an arbitrary boundary not all boundary conditions yield well-posed problems. (See Higdon[9]).

3.7 Weak Solutions and Shocks[2]

In the discussion of the uniqueness of a solution u(x,t) the existence of bounded derivatives of u implies that u(x,t) is bounded continuous of order 1 with respect to x. In general, this suggests that to determine when a solution ceases to be unique we should consider it to belong to the class of bounded continuous solutions and seek to determine when it ceases to be bounded continuous. In terms of a general wavefront evolution, this approach may be interpreted as seeking the position on the wavefront trace at which the slope of the disturbance wave surface immediately behind the wavefront becomes infinite.

This is the case of a flow containing a shock wave where piecewise differentiable C^1 solutions are separated by shock discontinuities across which both u and its derivatives are discontinuous.

Mathematicians have found it desirable to unify these two types of solution by generalizing the whole concept of a "solution" to system (3.4) in such a way that strict differentiability and continuity are no longer required. This is precisely the motivation underlying the notion of a weak solution which has been developed by Lax[10,11], Gel'fand[12], Oleinik[13], and others. Jeffreys presents a thorough account of this development. The following, a summary from Refs. 2, 1, gives a brief introduction to the way in which such an extension may be achieved and the difficulties that ensue.

An example of the piston problem is useful to illustrate these ideas. It involves the determination of the one-dimensional flow of a polytropic gas in a semi-infinite tube, one end of which is sealed by a piston. There is thus only one boundary condition of a space-like type at the piston wall. In the simplest problem of this type the gas, which is assumed initially to be in equilibrium and at rest, is set in motion by moving the piston.

If the origin of the x-axis is taken at the piston wall with the positive x-axis being directed into the gas, the initial conditions in the gas will be (the origin is fixed in space)

$$\rho(x,0) = \rho_0(\text{const}) \text{ and } u(x,0) = 0 \quad \text{for } x>0 \ .$$

Since the gas in contact with the piston wall moves with the wall, the boundary condition must be that the gas in contact with the piston at time t has the piston velocity at that time. Suppose the piston path, as a function of time, is given by $x=\sigma(t)$. Then the piston speed at time t will be $d\sigma/dt$. The boundary condition on the moving piston wall then becomes

$$u(\sigma(t),t) = \frac{d\sigma}{dt} \ . \tag{3.42}$$

Because the piston starts from rest, we must require of $\sigma(t)$ that $d\sigma/dt=0$ when $t=0$. Since initially $u=0$ the characteristic curves through the origin determined by equations (3.29) will have slopes $\pm c_0$, where c_0 is the speed of sound in the equilibrium state determined by the initial conditions and the polytropic gas law involved.

The fact that the piston wall is, at least initially, space-like then follows from the fact that the tangent to the piston path at time $t=0$ (it is vertical when drawn in the (x,t)-plane) lies between the two characteristic curves drawn through the origin. This, then, is an example of a mixed initial and boundary value problem in which the space-like boundary (the piston) is moving. If the piston accelerates smoothly the $C^{(1)}$ characteristics originating from points on the piston path will converge and a shock forms.

The starting point for the concept of a weak solution is the non-linear conservation equation

$$\frac{\partial u}{\partial t} + \frac{\partial F}{\partial x} = 0 \ , \tag{3.43}$$

subject to the initial condition $u(x,0)=g(x)$.

We restrict ourselves to the half-plane $t>0$ and recall that in general a unique solution to (3.43) will only exist for a finite time because conservation equations possess discontinuous solutions, or shocks. Accordingly, and with reference now only to a general function F and initial condition g, let us consider some strip $0<t<T$ in which the classical unique C^1 solution exists everywhere except on certain shock lines across which the solution is bounded.

Then the bounded function u defined in the half plane $t>0$ will be called a weak solution of (3.43) if in this half plane it satisfies the condition

$$\iint \left(\frac{\partial w}{\partial t} u + \frac{\partial w}{\partial x} F(u)\right) dxdt = 0 \tag{3.44}$$

for every twice continuously differentiable function $w(x,t)$ that vanishes outside some finite region in the half plane $t>0$. Such functions w are called test functions and the closure of the region in which they are non-zero is then known as the support of the test functions. Furthermore, a piecewise C^1 weak solution satisfies the generalized Rankine-Hugoniot condition across a shock.

Consider the region R bounded by the closed arc ∂R and traversed by the line L across which a shock occurs. Denote the two sub-regions so defined by R_- and R_+ and their boundaries by ∂R_- and ∂R_+, and let the directed arcs along adjacent sides of L be ∂L_- and ∂L_+, as in Fig. 3.5.

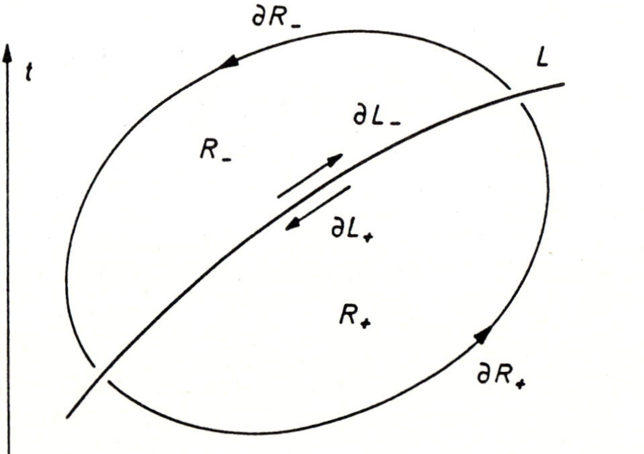

Fig. 3.5 Integration path over the region R cut by the shock L (from Ref. 2)

Then R is the union of R_- and R_+ and ∂R is the union of ∂R_- and ∂R_+. The test functions w in (3.44) will be assumed to have their support in R so that the test functions w will vanish on ∂R. Thus (3.44) may be written

$$\iint_R \left(\frac{\partial w}{\partial t} u + \frac{\partial w}{\partial x} F(u)\right) dxdt = 0 \,. \tag{3.45}$$

Now multiply (3.43) by w and integrate over R_- to obtain

$$\iint_{R_-} \left(w \frac{\partial u}{\partial t} + w \frac{\partial F}{\partial x}\right) dxdt = 0 \,,$$

which may also be written in the form

$$\iint_{R_-} \left\{\frac{\partial (wu)}{\partial t} + \frac{\partial (wF)}{\partial x}\right\} dxdt - \iint_{R_-} \left\{\frac{\partial w}{\partial t} u + \frac{\partial w}{\partial x} F\right\} dxdt = 0 \,. \tag{3.46}$$

Applying Green's theorem to the first terms in this result then transforms (3.46) to

$$\int_{\partial R_- + \partial L_-} \{-wFdt + wudx\} - \iint_{R_-} \left\{\frac{\partial w}{\partial t} u + \frac{\partial w}{\partial x} F\right\} dxdt = 0 \,. \tag{3.47}$$

However, as the support of the functions w lie in R, w will be zero on ∂R_- so that (3.47) reduces to

$$\int_{\partial L_-} \{-wF(u_-)dt + w\, u_- dx\} - \iint_{R_-} \left\{\frac{\partial w}{\partial t} u + \frac{\partial w}{\partial x} F\right\} dxdt = 0 \,. \tag{3.48}$$

A similar result applies with respect to R_+, where we find

$$\int_{\partial L_+} \{-wF(u_+)dt + w\, u_+ dx\} - \iint_{R_+} \left\{\frac{\partial w}{\partial t} u + \frac{\partial w}{\partial x} F\right\} dxdt = 0 \tag{3.49}$$

the integration along ∂L_- and ∂L_+ being oppositely directed, as indicated in Fig. 3.5.

If (3.48) and (3.49) are now added, the sign of the line integral in (3.48) is reversed with a corresponding replacement of ∂L_- by ∂L_+, and result (3.45) is used, we find

$$\int_{\partial L_+} w\left\{(u_+ - u_-) \frac{dx}{dt} - \left(F(u_+) + F(u_-)\right)\right\} dt = 0 \,, \tag{3.50}$$

where, as the point (x,t) is now constrained to lie on ∂L_+, the term (dx/dt) represents the speed of propagation $\tilde{\lambda}$ of the shock along L. As w is arbitrary, (3.50) can only be true if

$$\tilde{\lambda}(u_+ - u_-) = \left(F(u_+) - F(u_-)\right) \,, \tag{3.51}$$

which is the one-dimensional form of the generalized Rankine-Hugoniot condition. This holds degenerately when u is continuous across L.

If, now, the support of w is allowed to be arbitrary, the same form of argument proves that piecewise C^1 solutions of (3.43) satisfying (3.51) across a shock will also be a weak solution of (3.43). All this leads to the following definition and theorem.

Definition (Weak Solution)

The function u will be called a weak solution of

$$\frac{\partial u}{\partial t} + \frac{\partial F(u)}{\partial x} = 0 \; ,$$

if for all twice continuously differentiable test functions w with support in $t>0$ the function u is such that

$$\iint \left\{ \frac{\partial w}{\partial t} u + \frac{\partial w}{\partial x} F(u) \right\} dx dt = 0 \; ,$$

the integration being extended over the upper half plane $t>0$.

Theorem (Properties of Weak Solutions)

Let u be a weak solution of

$$\frac{\partial u}{\partial t} + \frac{\partial F(u)}{\partial x} = 0 \; . \qquad (3.52)$$

The following results are then true:

a) If u is piecewise C^1 in addition to being a weak solution, it is also a piecewise C^1 classical solution of (3.52);

b) a piecewise C^1 weak solution of (3.52) satisfies the generalized Rankine-Hugoniot condition

$$\tilde{\lambda}(u_+ - u_-) = F(u_+) - F(u_-) \qquad (3.53)$$

across a discontinuity moving with speed $\tilde{\lambda}$;

c) a necessary and sufficient condition for a piecewice C^1 classical solution of (3.52) to be a weak solution is that across a discontinuity moving with speed $\tilde{\lambda}$ it satisfies the generalized Rankine-Hugoniot condition (3.53).

The general objective when introducing a weak solution was to lift the requirements of strict continuity and differentiability that need to be imposed on classical solutions, since it is not usually known a priori how long they will remain C^1. In this respect the notion of a weak solution is successful and, furthermore, because of its method of definition the class of weak solutions is even wider than the class of piecewise C^1 functions so that considerable generality has been achieved. But the attempt to overcome analytical difficulties caused by the loss of differentiability and the occurrence of shocks, by the introduction of weak solutions[11], is only partly successful, since weak solutions are not uniquely defined by the initial data so that an admissibility problem still remains. Thus, this generality has been obtained at the cost of the uniqueness of a weak solution. More precisely, unlike a strict classical C^1 solution, a weak solution to (3.43) is not determined uniquely by the initial data $\rho(x,0)$. This is so because the n non-linear algebraic equations (3.53) imply that even if n+1 quantities are specified between either side of the discontinuity the solution for the remaining n quantities need not necessarily be unique.

Thus additional admissibility or entropy conditions are needed to extend in a unique physical manner a differentiable solution of (3.4) to a solution comprising a piecewise differentiable solution joined by finite shocks. These conditions derive their name from the fact that when the non-uniqueness of shock solutions was encountered in gas dynamics the non-physical, but mathematically possible, rarefaction shock was first rejected in favour of the physically observed mathematically possible compression shock on the basis of the entropy increase required by the second law of thermodynamics.
Since then the selection problem has arisen in far more general situations where entropy is not involved, though the name entropy conditions is still usually used. Such admissibility conditions are now based on a great many different ideas which may involve convexity, singular perturbation, layering and smoothing arguments, and still more abstract approaches which are difficult to motivate in simple physical terms. Two examples hint at the ideas of artificial viscosity and total variation diminishing that are used in practical applications.

Example with viscosity

The solution to the 1-D Euler equations

$$\frac{\partial u}{\partial t} + \frac{\partial F}{\partial x} = 0 \, , \tag{3.54}$$

where $u = (\rho, \rho u, e)$, $F = (\rho u, \rho u^2+p, (e+p)u)$,

can be determined uniquely by solving

$$\frac{\partial u}{\partial t} + \frac{\partial F}{\partial x} = \nu \frac{\partial^2 u}{\partial x^2} \tag{3.55}$$

with ν=const.>0 in the limit of $\nu \to 0$.

When the viscosity ν is small but not zero, the thickness of a region with large gradients where deviations of the solutions to (3.46) and (3.47) can be expected, is so small that it generally cannot be described by a finite difference mesh. Thus, the numerical viscous solution to (3.47) behaves like the weak solution to (3.46) satisfying the entropy condition.

Example Burgers Equation

For the scalar conservation law

$$\frac{\partial u}{\partial t} + \frac{\partial}{\partial x} (u^2/2) = 0 \tag{3.56}$$

the entropy condition has the consequence that the "energy" $\int u^2(x,t)dx$ cannot increase. To make this plausible we consider the variation of the solution u(x,t) of (3.48)

$$\mathrm{var}[u(x,t)] = \sup_{\{x_n\}} \sum_{m=1}^{n} |u(x_m,t) - u(x_{m-1},t)| , \tag{3.57}$$

where $\{x_n\}$ denotes all possible partitions of the x-axis, i.e. $-\infty = x_0<x_1<x_2<..<x_{n-1}<x_n=\infty$. We assume $\mathrm{var}[u(x,0)]<\infty$.

Since a smooth solution is propagated along characteristics from the initial data, we have $\mathrm{var}[u(x,t)]=\mathrm{var}[u(x,0)]$ for smooth solutions. However, if u(x,t) has a shock satisfying the entropy condition, u(x,t) looses a part of its initial variation at any time the shock is present.

The property

$$\mathrm{var}[u(x,t+\Delta t)]\leq \mathrm{var}[u(x,t)] \tag{3.58}$$

of the scalar conservation law (3.48) with the entropy condition satisfied was used by Harten[14] (1983) to design TVD (Total Variation Diminishing) schemes for scalar conservation laws and systems like the 1-D Euler conservation equations.

3.8 Discrete Solution Methods

The preceding sections have established the basic properties and character of hyperbolic partial differential equations. The remaining sections will discuss and show how we construct discrete methods for solution of these equations so that they reflect these basic properties.

The overall goal of a discrete method is: Given a flow problem specified by particular boundary conditions and sometimes also initial conditions, describe the solution as a finite set of numbers distributed throughout the domain of the flow and obeying some functional relationship among them based on some approximation derived from the continuum equations chosen to govern the problem at hand. It is arrived at by first projecting the continuum problem of the differential equations to some finite-dimensional space for the dependent and independent variables and then by solving the resulting discrete equations for the final set of numbers. When solving the partial differential equations cast in the Eulerian formulation with reference to some coordinate system, the first step in the projection process is to discretize the domain of the flow by laying out a network of points situated at a finite number of different locations of the independent variables, i.e. to create a grid. The simplest one is the regular Cartesian grid. The grid points (x_j, y_k, z_l) are given by $x_j = x_o + j\Delta x$, $y = y_o + k\Delta y$, $x = z_o + l\Delta z$, and the approximation to the dependent variables U at these grid points are here denoted by u_{jkl}. Let u^n_{jkl} denote the time-dependent approximation ($u^n_{jkl} \approx u(x_j, y_k, z_l, t_n)$, $t_n = t_o + n\Delta t$). The extension to variable step size is simply ($x_j = x_o + \Sigma_j \Delta x_j$, etc.) Other extensions are discussed below based on transformation of the independent variables. The location of the spatial grid points may be time dependent as, e.g., in adaptive grid methods and Lagrangian grids. The grid is called unstructured when it does not have the Cartesian form. These grids are best used with finite element methods.

The computational algorithms are either explicit,

$$u^{n+1} = G(u^n) \ , \tag{3.59}$$

or implicit,

$$G(u^{n+1}, u^n) = 0 \ . \tag{3.60}$$

The vector u^n consists here of all unknowns (u_{jkl}) at the time level n. (The index n may also be the iteration number in a steady state computation.) The accuracy depends on the smoothness of the functions being represented and the density of grid. It is usually analyzed on the basis of a Taylor series expansion from one grid point to another. Thus it becomes a question of the resolution of scales in the fluid phenomena in relation to the distance between grid points, i.e., the mesh length. The scales of the physical features in the problem may only be very broad, on the order of the overall size of the domain, if the flow is inviscid; but they also may range from these broad ones down to the dissipation length scales if there are boundary layers or instabilities in the flow and turbulence occurs.

3.9 Classical Finite-Difference Approximations to Derivatives

Consider the function u(x) of a single independent variable. Expanding this in a Taylor series about the grid point x_j yields the classical central difference denoted by D_o

$$D_o u_j = \frac{u_{j+1}-u_{j-1}}{2\Delta x} = \frac{\partial u}{\partial x}(j) + \frac{\Delta x^2}{3!}\frac{\partial^3 u}{\partial x^3}(j) + \frac{\Delta x^4}{5!}\frac{\partial^5 u}{\partial x^5}(j) + \ldots, \quad (3.61)$$

which is an expression for the first derivative accurate to second order. The forward difference D_+ is

$$D_+ u_j = \frac{u_{j+1}-u_j}{\Delta x} = \frac{\partial u}{\partial x}(j) + \frac{\Delta x}{2!}\frac{\partial^2 u}{\partial x^2}(j) + \frac{\Delta x^2}{3!}\frac{\partial^3 u}{\partial x^3}(j) + \ldots, \quad (3.62)$$

and the backward difference D_-

$$D_- u_j = \frac{u_j-u_{j-1}}{\Delta x} = \frac{\partial u}{\partial x}(j) - \frac{\Delta x}{2!}\frac{\partial^2 u}{\partial x^2}(j) + \frac{\Delta x^2}{3!}\frac{\partial^3 u}{\partial x^3}(j) - \ldots. \quad (3.63)$$

Second derivative approximations can then be formed

$$D_+ D_- u_j = \frac{u_{j+1}-2u_j+u_{j-1}}{\Delta x^2} = \frac{\partial^2 u}{\partial x^2}(j) + \frac{2\Delta x^2}{4!}\frac{\partial^4 u}{\partial x^4}(j) + \ldots. \quad (3.64)$$

3.10 Computational Grid and Accuracy

One way to increase the accuracy of these approximations, i.e. to reduce the truncation error, is to raise the complexity of the functional relationship between the dependent variables evaluated at the grid points and thus to improve the information content being passed from one point to another, that is to use a higher-order numerical method. Provided that the function over the grid points is smooth, this approach does offer better resolution for a given number of grid points. But it also means a more complicated algorithm that may demand more arithmetic operations per step and more effort to implement on advanced computer architectures, and it therefore may be less universal in its applications.

The other way to improve the resolution of fine scales with a given algorithm is to increase the number or optimize the location of grid points in the independent variables over which the dependent variables are evaluated. The length scale of the phenomena resolved is thereby reduced in proportion locally to the mesh-length, raised to the order of the approximating scheme for the differential equations. This approach is particularly appropriate if the function being approximated is not smooth. It involves no changes in the algorithm itself. Instead, its drawback lies on the hardware side because the computational problem grows in size, it demands more memory, and the execution time increases. As partial relief to this, one tries to optimize the distribution of grid points locally in order to maximize the resolution in special regions of the flow. A long-standing example is the mesh aligned with, or fitted, to a solid-wall boundary. Here rapid gradients in boundary layers are known to occur, and the region can be identified in advance. With a boundary-fitted mesh the distribution of points is regular and it can be graded from a small size going outward from the wall to match smoothly with a larger mesh size in a region away from the wall. It requires no extra interpolation to set the boundary conditions and thus enhances the efficiency of the computation. A further enhancement comes by setting additional grid points into the mesh in a preselected local region. This local refinement establishes a second grid distribution identified by interior boundaries with its parent distribution across which the pattern of points may be smooth and regular or irregular. Different interior boundary conditions may be needed in each case.

When the region for grading or refinement cannot be identified in advance, some form of intelligent decision-making has to be built into the algorithm in order to sense the appropriate regions and then automatically to grade or refine the mesh. Adding complexity to the algorithm, this adaptive strategy affords better efficiency. However, no effective methods are known up to now for the treatment of general three-dimensional flow problems.

3.11 Local Truncation Error

We can also discuss the overall local error of the difference approximation. Let $F_j^n(u)=0$ represent the difference equation approximating the partial differential equation at the mesh point j, and the time level n with exact solution u. If u is replaced by U at the mesh points of the difference equation, where U is the exact solution of the partial differential equation, the value of $F_j^n(U)$ is called the local truncation error T_j^n at the j,n mesh point. $F_j^n(U)$ clearly measures the amount by which the exact solution values of the partial differential equation at the mesh points of the difference equation do not satisfy the difference equation at the point $(j\Delta x,\ n\Delta t)$.

Using Taylor expansion, it is easy to express T_j^n in terms of powers of j and n and partial derivatives of U at $(j\Delta x,\ n\Delta t)$. Although U and its derivatives are generally unknown, the analysis is worthwhile because it provides a method for comparing the local accuracies of different schemes approximating the partial differential equation.

A simple example illustrates the notion of local truncation error for the scalar equation

$$\frac{\partial U}{\partial t} + a\frac{\partial U}{\partial x} = 0 , \tag{3.65}$$

where a is a positive constant.

By Taylor's expansion

$$U_j^{n+1} = U(x_j, t_n+\Delta t) = U_j^n + \Delta t\left(\frac{\partial U}{\partial t}\right)_j^n + \frac{1}{2}\Delta t^2\left(\frac{\partial^2 U}{\partial t^2}\right)_j^n + \ldots .$$

The differential equation can now be used to eliminate the t-derivatives by replacing

$$\frac{\partial}{\partial t} = -a\frac{\partial}{\partial x} , \quad \text{so that}$$

$$U_j^{n+1} = U_j^n - \Delta t\ a\left(\frac{\partial U}{\partial x}\right)_j^n + \frac{1}{2}\Delta t^2 a^2\left(\frac{\partial^2 U}{\partial x^2}\right)_j^n + \ldots .$$

Then the replacement of the x-derivatives by central-difference approximations gives to second order the explicit difference equation

$$u_j^{n+1} = u_j^n - \frac{a\Delta t}{2\Delta x}(u_{j+1}^n - u_{j-1}^n) + \frac{1}{2}\left(\frac{a\Delta t}{\Delta x}\right)^2(u_{j-1}^n - 2u_j^n + u_{j+1}^n) , \tag{3.66}$$

which can also be written in predictor-corrector form as

$$\tilde{u}_j^{n+1} = u_j^n - \frac{a\Delta t}{2\Delta x}(u_{j+1}^n - u_j^n) , \tag{3.67}$$

$$u_j^{n+1} = u_j^n - \frac{\Delta\Delta t}{2\Delta x}(\tilde{u}_j^n - \tilde{u}_{j-1}^n) .$$

Its local truncation error T_j^n, found by substituting the exact solution U into the difference scheme and expanding each term by Taylor series about the point j,n, becomes

$$T_j^n = \left[\frac{1}{6}\,\Delta t^2\,\frac{\partial^3 U}{\partial t^3} + \frac{1}{6}\,a\,\Delta x^2\,\frac{\partial^3 U}{\partial x^3}\right]_j^n + \ldots\,, \tag{3.68}$$

which is accurate to second order in both space and time. The truncation error gives an indication of the error resulting from the replacement of the differential equation by F_j^n.

3.12 Consistency

It is sometimes possible to approximate a hyperbolic equation by a finite-difference scheme that is stable, (i.e. limits the amplification of all the components of the initial conditions) but which has a solution that converges to the solution of a different differential equation as the mesh lengths tend to zero. Such a difference scheme is said to be inconsistent or incompatible with the partial differential equation. The more precise definition is as follows.

Definition

Let $L(U)=0$ represent the partial differential equation in the independent variables x and t, with exact solution U. Let $F(u)=0$ represent the approximating finite-difference equation with exact solution u. Then the local truncation error is $T_j^n(U)=F_j^n(U)$. The difference equation is said to be consistent if the limiting value of $T_j^n(U)$ goes to zero as $\Delta x \to 0$ and $\Delta t \to 0$.

3.13 Convergence and Stability

The following sections are concerned with the conditions that must be satisfied if the solution of the finite-difference equations is to be a reasonably accurate approximation to the solution of the corresponding partial differential equation.

These conditions are associated with two different but inter-related problems. The first concerns the convergence of the exact solution of the approximating difference equations to the solution of the differential equation; the second concerns the unbounded growth, or controlled decay, or boundedness of the exact solution of the finite-difference equation, and therefore of all rounding errors introduced during the computation because the errors and the exact solution are processed by the same arithmetic operations. This is what we call the stability problem. (What we compute is the exact solution degraded by rounding errors.)

3.14 Notion of Convergence[15]

Let U represent the exact solution of a partial differential equation with independent variables x and t, and u the exact solution of the difference equations used to approximate the partial differential equation. Then the finite-difference equation is said to be convergent when u tends to U at a fixed point or along a fixed t-level as Δx and Δt both tend to zero.

Although the conditions under which u converges to U have been established for linear elliptic, parabolic and hyperbolic second-order partial differential equations with solutions satisfying fairly general boundary and initial conditions, they are not yet known for non-linear equations except in a few particular cases.

The difference (U-u) is called the discretization error. Some texts call it the truncation error but here the latter term means the difference between the differential equation and its approximating difference equation. The magnitude of the discretization error at any mesh point depends on the finite-sizes of the mesh lengths, Δx and Δt, i.e. on the distances between consecutive, discrete grid-points, and on the number of terms in the truncated series of differences used to approximate the derivatives.

The discretization error can usually be diminished by decreasing Δx and Δt, subject invariably to some relationship between them, but as this leads to an increase in the number of equations to be solved, this method of improvement is limited by such factors as cost of computation and computer storage requirements, etc.

The problem of convergence of a finite difference scheme for solving a partial differential equation (elliptic, hyperbolic or parabolic) consists of finding the criteria under which the local discretization error U-u at a fixed mesh point, tends to zero uniformly as the mesh is refined. For the first-order hyperbolic equation such as (3.65), this refinement means that $\Delta x, \Delta t \to 0$ and $j, n \to \infty$. In carrying out the convergence analysis, it may be convenient to assume that $\Delta x, \Delta t$ do not tend to zero independently but according to a relationship of the form $\Delta x = r(\Delta t)^{\alpha}$, where r is a constant and $\alpha > 1$ is some parameter.

In general, the problem of convergence is a difficult one to investigate usefully because the final expression for the discretization error is usually in terms of unknown derivatives for which no bounds can be estimated. Fortunately, however, the convergence of difference equations approximating linear parabolic and hyperbolic differential equations can be investigated in terms of stability and consistency, which are easier to deal with.

The real importance of the concept of consistency lies in a theorem by Lax (see Ref. 16) which states that if a linear finite-difference equation is consistent with a properly posed linear initial-value problem, then stability guarantees convergence of u to U as the mesh lengths tend to zero.

3.15 Notion of Stability[15]

The essential idea behind stability is that a numerical process, when applied exactly, should limit the amplification of all components of the initial conditions, including rounding errors. The error, $u_j^n - U_j^n$, at the mesh point $(j\Delta x, n\Delta t)$ in using a finite difference scheme to solve the differential equation must be bounded as $n \to \infty$ for fixed $\Delta x, \Delta t$ and as $\Delta x, \Delta t \to 0$ for a fixed value of $n\Delta t$. In both cases the number of applications of the difference scheme becomes infinite in the limit and there is a possibility of unbounded amplification of errors. The concept of stability is concerned with the boundedness of the solution of the finite-difference equations and this is examined by finding conditions under which the error

$$z_j^n = u_j^n - U_j^n \tag{3.69}$$

remains bounded as n increases for fixed $\Delta x, \Delta t$. Here U_j^n is, as before, the theoretical solution of the finite difference scheme and u_j^n is the solution of the scheme which is actually obtained, so that u_j^n contains rounding errors. The analysis considers the growth of perturbations in initial data or the growth of errors introduced at mesh points at a given time level.

There are three common methods of investigating stability: the von Neumann or Fourier method, the matrix method, and the energy method. Before presenting these methods, however, it is necessary to introduce a bound for the spectral radius.

3.15.1 A Bound for the Spectral Radius[17]

Let λ_i be an eigenvalue of the $n \times n$ matrix A and $\underline{x}_i$ the corresponding eigenvector. Hence

$$A\underline{x}_i = \lambda_i \underline{x}_i$$

and

$$\|A\underline{x}_i\| = \|\lambda_i \underline{x}_i\| = |\lambda_i| \, \|\underline{x}_i\| \, .$$

For all compatible matrix and vector norms it follows that

$$|\lambda_i| \, \|\underline{x}_i\| = \|A\underline{x}_i\| \leq \|A\| \, \|\underline{x}_i\| \, .$$

Therefore,

$$|\lambda_i| \leq \|A\| \, , \qquad i=1(1)n \, .$$

Hence, the spectral radius ρ

$$\rho(A) \leq \|A\| \, .$$

3.16 Von Neumann Method[15]

The Fourier method or von Neumann method, developed by J. von Neumann in the early 1940s, expresses the signal at grid points at a given time level in terms of a harmonic decomposition, i.e. the discrete equivalent to Eq.(3.2). This amounts to a separation of variables concept identical to that commonly used for solving partial differential equations. Then it determines the criterion governing the growth of the signal.

If the finite difference method to be analyzed is linear, so that separate solutions will be additive, it is only necessary to consider the propagation of the signal due to a single, typical Fourier mode. Consider the term $e^{i\beta j\Delta x}$ associated with the typical frequency $\beta_m = m\pi/J\Delta x$. The Fourier coefficient is constant and can be factored out of the analysis.

To examine the propagation of this single, typical mode as $t\to\infty$, the solution of the finite-difference scheme must be found which reduces to $\exp(i\beta j\Delta x)$ when $t=0$. Let such a solution be

$$e^{\alpha t}e^{i\beta x} = e^{\alpha n\Delta t}e^{i\beta j\Delta x} \, ,$$

where $\alpha=\alpha(\beta)$ is complex. The original Fourier component $e^{i\beta j\Delta x}$ will not grow with time (as n increases, that is) if

$$|e^{\alpha\Delta t}| \leq 1$$

for all α. This is von Neumann's criterion for stability. Often the notation $\xi=e^{\alpha\Delta t}$ is introduced, where ξ is known as the amplification factor.

If the exact solution of the difference equations does not increase exponentially with time, then a necessary and sufficient condition for stability is that

$$|\xi| \leq 1 \, .$$

If, however, u_j^n does increase with t, then the necessary and sufficient condition for stability is,

$$|\xi| \leq 1+K\Delta t = 1+O(\Delta t) \ ,$$

where the positive number K is independent of $\Delta x, \Delta t$ and β.

The following important points should be noted concerning the von Neumann method of examining stability[15].

(i) Because the method is based on Fourier series, it applies only if the coefficients of the linear difference equation are constant. If the difference equation has variable coefficients, the method can still be applied locally and it might be expected that a method will be stable if the von Neumann condition, derived as though the coefficients were constant, is satisfied at every point of the field. There is much numerical evidence to support this contention.

(ii) For two-level difference schemes with one dependent variable and any number of independent variables, the von Neumann condition is sufficient as well as necessary for stability. Otherwise, the condition is necessary only.

(iii) Boundary conditions are neglected by the von Neumann method which applies in theory only to pure initial value problems with periodic initial data. It does, however, provide necessary conditions for stability of constant-coefficient problems regardless of the type of boundary condition.

3.17 Matrix Method[17]

For linear initial-boundary-value problems, stability can be related to convergence via Lax's Equivalence Theorem by defining stability, in effect, in terms of the boundedness of the solution of the finite-difference equations at a fixed time-level T as $\Delta x \to 0$, i.e. as $J \to \infty$, it being assumed that Δx is related to Δt in such a way that $\Delta x \to 0$ as $\Delta t \to 0$ (see Refs. 16, 18, 19).

Assume that the vector of solution values $u^{n+1} = [u_1^{n+1}, u_2^{n+1}, \ldots, u_J^{n+1}]^T$ of the finite-difference equations at the (n+1)th time-level is related to the vector of solution values at the nth time-level by the equation

$$u^{n+1} = Au^n + b^n \ , \qquad (3.70)$$

where b^n is a column vector of known boundary-values and zeros, and matrix A an $(J) \times (J)$ matrix of known elements. Then it will be shown that the practical consequence of this definition of stability is that a norm of matrix A compatible with a norm of u must satisfy

$$\|A\| \leq 1 + O(\Delta t) ,$$

when the solution of the partial differential equation increases as t increases.

3.18 The Energy Method

The energy method is a powerful tool in dealing with particular equations or particular classes of equations. Its application can, unfortunately, become rather messy and each problem to which it is applied requires a different treatment. The successful application of the method will be due in no small part to the ingenuity of the user. The strength of the method lies in its ability to deal effectively with boundary conditions, variable coefficients and non-linear problems. Besides proving the stability of a finite-difference scheme, the energy method can indicate the correct choice of method. However, the method provides only sufficient conditions for stability which may be far removed from what is necessary in certain initial-boundary value problems.

The method calculates the sum of the squares of the errors $z_j^n(j=1,..,J;\ n=1,2,..)$ at time level $n\Delta t$. This sum of squared errors is called the energy from which the method gets its name. It must be noted, however, that the conserved quantity (the energy) is not the physical energy of the system, a point which can cause confusion. The energy method, it should be pointed out, is very difficult to use with three-level schemes.

This method of analyzing stability may be applied in principle to problems with variable coefficients and to non-linear equations. Each problem requires a different treatment but we can illustrate the general philosophy by means of an example (see also Richtmyer and Morton[16]).

Example

Consider the problem of linear wave propagation

$$\frac{\partial u}{\partial t} = a\frac{\partial u}{\partial x} , \quad a<0 \quad \text{in } (0\leq x\leq 1)x(t\geq 0) \text{ with initial data}$$
$$u(x,0)=f(x) . \qquad (3.71)$$

The solution propagates along the characteristics as shown in Fig. 3.6,

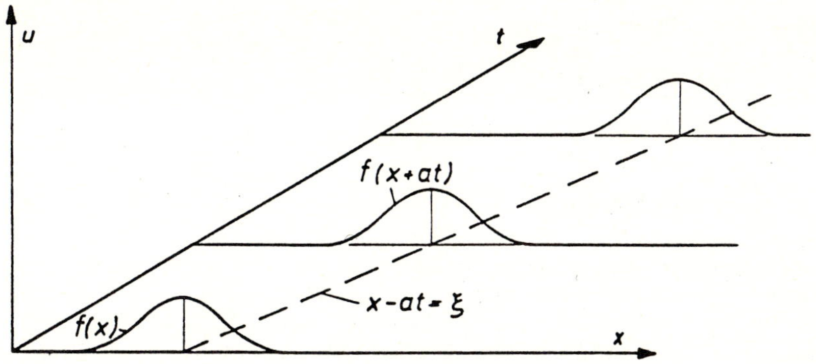

Fig. 3.6 Propagation of waveform f(x) along the characteristic of Eq.(3.71)

and it is clear that u must be specified along the boundary x=0. The solution u is then determined in the whole region and thus no boundary condition can be given at x=1. The appropriate homogeneous boundary condition thus is $u(0,t)=0$ $(t>0)$.

Multiplying the equation by u and integrating with respect to x, we obtain

$$\int_0^1 u\,\frac{\partial u}{\partial t}\,dx = a\int_0^1 u\,\frac{\partial u}{\partial x}\,dx \ ,$$

which on integration by parts on the right hand side gives

$$\frac{\partial}{\partial t}\int_0^1 u^2 dx = a\,u^2\Big|_0^1 = a\left[u^2(1,t)-u^2(0,t)\right] = a\,u^2(1,t) < 0 \qquad (3.72)$$

through the boundary conditions and the fact that $a<0$.

We deduce from this, that

$$\int_0^1 u^2(x,t)dx < \int_0^1 u^2(x,0)dx = \int_0^1 f^2(x)dx \ ,$$

and therefore the quantity $\int_0^1 u^2(x,t)dx$ remains bounded as $t\to\infty$.

If we define the scalar product for continuous variables by

$$(u,v) = \int_0^1 u\,v\,dx \ , \qquad (3.73)$$

and the L_2 norm by

$$\|u\| = (u,u)^{1/2} ,$$

we see that the L_2 norm decreases with time

$$\|u(t)\|^2 \leq \|u(0)\|^2 . \qquad (3.74)$$

The quantity $\|u(t)\|^2$ is called the energy from which the method gets its name, but as was already said, the successful application of this method to more general problems relies heavily on the ingenuity of the user to identify a suitable norm or energy quantity E_n. A comprehensive discussion, along with several examples, may be found in Richtmyer and Morton[16] (Chapter 6).

In the above example it was very simple to show well-posedness for the differential equation. Unfortunately it becomes more difficult to show this for the difference equations, but when successfully applied, the energy method offers the means to analyze the stability of boundary conditions. An example illustrates this point.

Example

Consider the Crank-Nicholson scheme applied to the linear wave equation (3.65)

$$(I - Q)u^{n+1} = (I + Q)u^{n} , \qquad (3.75)$$

where $Qu=1/2\ \Delta x\ a\ D_0 u$. The left boundary condition must be specified at j=0 and it is natural to set $u_0^n=0$ for all n. At the right boundary x=1, no condition is needed for the differential equation, as we saw above, but one is required for the difference equations in order to carry out the scheme near the boundary. Such conditions are called numerical or extra boundary conditions. For our example let us try setting the derivative zero: $u_x=0$ at the right boundary, i.e. $u_J^n=u_{J-1}^n$.

In analogy with (3.73) we define the discrete scalar product, equivalent to the differential one,

$$(u,v) = \sum_{j=1}^{J-1} u_j v_j\ \Delta x , \quad \text{where } u_j=u(j\Delta x) .$$

It follows then that the quadratic form is

$$(u,Qu) = \frac{a\ \Delta x}{2}\ (u,D_0 u) = \frac{a\ \Delta x}{4} \sum_{j=1}^{J-1} u_j(u_{j+1}-u_{j-1}) , \qquad (3.76)$$

and after writing out the sums

$$= \frac{a\ \Delta x}{4}\ [u_1(u_2-u_0)+u_2(u_3-u_1)+\ldots+u_{J-1}(u_J-u_{J-2})] .$$

Then cancellation and the boundary condition $u_0=0$ leave

$$(u,Qu) = \frac{a\ \Delta x}{4}\ u_{J-1}u_J = \frac{a\ \Delta x}{4}\ u_{J-1}^2 ,$$

where we have used the right boundary condition $u_{J-1}=u_J$.

Thus we see that $(u,Qu) \leq 0$ for all u which satisfy the boundary conditions. We must now show how this relates to the discrete energy. Rewrite Eq.(3.18.5) in the form

$$u^{n+1} - u^n = Q(u^{n+1} + u^n) .$$

Doing the equivalent of multiplying the differential equation by u and integrating over the interval, we scalar multiply with $u^{n+1} + u^n$, which yields

$$\|u^{n+1}\|^2 - \|u^n\|^2 = \left(u^{n+1} + u^n, Q(u^{n+1} + u^n)\right) \leq 0 ,$$

and thus

$\|u^n\| \leq \|u^0\|$, i.e. the energy decreases, exactly the same result we found for the differential equation.

3.19 Schemes for Non-Linear Equations

Lax and Wendroff[20] (1960) have shown that discontinuous solutions to

$$\frac{\partial u}{\partial t} + \frac{\partial F(u)}{\partial x} = 0 \qquad (3.77)$$

can be found by solving

$$\frac{\partial u}{\partial t} + \frac{\partial F(u)}{\partial x} = \mu \frac{\partial^2 u}{\partial x^2} , \qquad \mu=\text{const}>0 \qquad (3.78)$$

for $\mu \to 0$. When the viscosity μ is small but not zero, the thickness of a region with large gradients, where deviations between the solutions (3.77) and (3.78) can be expected, is so small that it generally cannot be described by a finite-difference mesh. Thus the numerical solution to the viscous problem (3.78) behaves like the weak solution to (3.77), satisfying the entropy condition. Solutions can thus be computed without special treatment of the discontinuity, if the differential equation is solved in conservation law form and if the scheme is conservative.

Definition

A two-level explicit scheme in time is said to be conservative if it is of the general form

$$u_j^{n+1} = u_j^n - \frac{\Delta t}{\Delta x}[H(u_{j+m}, ..u_{j-m+1}) - H(u_{j+m-1}, ..u_{j-m})] , \quad (3.79)$$

where H is a function of 2m arguments which must satisfy for purposes of consistency

$$H(u,..u) = F(u) \text{ for any } u \text{ where } u_j^n = u(j\Delta x, n\Delta t) .$$

The 1950s saw the development of first-order accurate methods to solve the Euler equations. These schemes were robust, but not very accurate. To increase the accuracy, second-order methods were sought in the 1960s. One of the first was proposed by Lax and Wendroff[20]. It is based on the third-order Taylor series expansion with respect to time:

$$u(t+\Delta t, x) = u(t,x) + \frac{\partial u}{\partial t} \Delta t + \frac{1}{2} \frac{\partial^2 u}{\partial t^2} \Delta t^2 + O(\Delta t^3). \quad (3.80)$$

The derivatives are calculated from the partial differential equation

$$\frac{\partial u}{\partial t} = - \frac{\partial F}{\partial x} ,$$

$$\frac{\partial^2 u}{\partial t^2} = - \frac{\partial^2 F}{\partial t \partial x} = - \frac{\partial^2 F}{\partial x \partial t} = - \frac{\partial}{\partial x} (\frac{\partial F}{\partial u} \frac{\partial u}{\partial t}) = \frac{\partial}{\partial x} (A(u)\frac{\partial F}{\partial x}),$$

where $A = \partial F/\partial u$ is the Jacobian matrix.

These expansions are introduced into the Taylor expansion (3.80) and the derivatives are approximated by second-order central differences

$$\frac{\partial F}{\partial x} \Big|_j = \frac{1}{2\Delta x} D_o F + O(\Delta x^2) ,$$

$$\frac{\partial}{\partial x} [(A(u) \frac{\partial F}{\partial x})] = \frac{1}{\Delta x^2} D_+[A(\mu_x u_j) D_- F]_j + O(\Delta x^2) ,$$

where $\mu_x u_j = 1/2(u_{j+1/2} + u_{j-1/2})$ is the classical averaging operator.

Thus we obtain the Lax-Wendroff scheme:

$$u_j^{n+1} = u_j^n - \frac{\Delta t}{2\Delta x}(F_{j+1}^n - F_{j-1}^n) + \\ + \frac{\Delta t^2}{2\Delta x^2}[A_{j+1/2}^n(F_{j+1}^n - F_j^n) - A_{j-1/2}^n(F_j^n - F_{j-1}^n)] \ . \tag{3.81}$$

The Lax-Wendroff scheme is second-order accurate in space and time. Furthermore (3.81) is conservative, because it is of the form (3.79) with m=1 and (L for left and R for right face)

$$H(L,R) = \frac{1}{2}[F(L) + F(R) - \frac{\Delta t}{2\Delta\lambda} A(\frac{L+R}{2}) \ [F(L)-F(R)] \ .$$

The stability of such a scheme is studied by assuming $A(u)=A(u_o) = \text{const}$.

Since the equations are hyperbolic, the system can be diagonalized and the resulting decoupled scalar equations can be studied for the "worst-case" eigenvalue λ

$$\frac{\partial u}{\partial t} + \lambda \frac{\partial u}{\partial x} = 0 \ . \tag{3.82}$$

Thus the von Newmann stability analysis is applied to (3.81) with F=A and $A=\lambda$. We check, whether the Fourier modes $v^n(k)$ $e^{ikj\Delta x}$ are damped by the finite difference scheme. Here v^n denotes the amplitude at $t=n\ \Delta t$, k the wave number and $i=\sqrt{-1}$. The amplification factor g defined by $v^{n+1}(k)=g(k)v^n(k)$ becomes

$$g(k)= 1 - \frac{\lambda\ \Delta t}{\Delta\lambda} \sin(k\Delta x)\ i + \frac{\lambda^2\ \Delta t^2}{\Delta x^2} [\cos(k\Delta x)-1] \ .$$

The stability condition $|g(k)|\leq 1$ for all $k>0$ is satisfied for the Courant-Friedrichs-Lewy condition

$$CN = |\lambda| \frac{\Delta t}{\Delta x} \leq 1 \ , \tag{3.83}$$

because for the Courant number CN, $0<CN<1$, the function $f(\cos(k\Delta x))=|g(k)|^2$ has its maximum at $\cos(k\Delta\lambda)=1$ with $f(1)=1$.

Since the stability condition (3.83) must hold for all eigenvalues of A, we obtain the stability condition for (3.81) from λ_{max}

$$(|u|+c) \frac{\Delta t}{\Delta x} \leq 1 \ . \tag{3.84}$$

Thus, the numerical domain of dependences of the scheme (3.81) contains the physical domain of dependence of the differential equation (3.77).

In order to avoid the evaluation of the Jacobian matrices $A_{j\pm 1/2}$, Richtmyer[18] (1962) proposed a two-step version of the Lax-Wendroff scheme, the <u>two-step Lax-Wendroff scheme</u>

$$\begin{aligned} &\underline{\text{predictor}}: \quad \tilde{u}_j^n = \frac{1}{2}(u_j^n + u_{j+1}^n) - \frac{1}{2}\frac{\Delta t}{\Delta x}(F_{j+1}^n - F_j^n) \ , \\ &\underline{\text{corrector}}: \quad u_j^{n+1} = u_j^n - \frac{\Delta t}{\Delta x}(\tilde{F}_j^n - \tilde{F}_{j-1}^n) \ , \quad \text{where } F_j^n = F(u_j^n) \ . \end{aligned} \qquad (3.85)$$

For A=const. (3.81) and (3.85) are identical. Thus, the linear stability condition (3.84) also applies to (3.85).

In Chapters V and VI we will study other classes of schemes which are not based on the Taylor expansion (3.80) in time. But first we must consider ways to treat higher space dimensions.

3.20 References

1. Coulson, C.A., Jeffrey, A.: "Waves". Longman, 2nd ed., London, 1977.

2. Jeffrey, A.: "Quasilinear Hyperbolic Systems and Waves", Res. Notes Math., Vol 5, Pitman, London, 1976.

3. Kreiss, H., Oliger, J.: "Methods for the Approximate Solution of Time Dependent Problems". GARD Pub. No.10, Geneva, 1973.

4. Hadamard, J.: "Lectures on Cauchys' Problem in Linear Partial Differential Equations". Dover, New York, 1952.

5. John, F.: "Continuous Dependence on Data for Solutions of Partial Differential Equations with a Prescribed Bound". Comm. Pure Appl. Math., Vol. 13, 1960, pp. 551-585.

6. Lavrentiev, M.M.: "Some Improperly Posed Problems of Mathematical Physics" (trans.). Springer, Berlin 1967.

7. Friedrichs, K.O.: "Nonlinear Hyperbolic Differential Equations for Functions of Two Independent Variables". Am. J. Math., Vol. 70, 1948, pp. 555-588.

8. Lax, P.D.: "Nonlinear Hyperbolic Equations". Comm. Pure Appl. Math., Vol 6, 1953, pp. 231-258.

9. Higdon, R.L.: "Initial-Boundary Value Problems for Linear Hyperbolic Systems". SIAM Rev, Vol 28, 1986, pp. 177-217.

10. Lax, P.D.: "Hyperbolic Systems of Conservation Laws II". Comm. Pure Appl. Math., Vol.10, 1957, pp. 537-566.

11. Lax, P.D.: "Weak Solutions of Nonlinear Hyperbolic Equations and their Numerical Computation". Comm. Pure Appl. Math., Vol.7, 1954, pp. 159-193.

12. Gel'fand, I.M.: "Some Problems in the Theory of Quasi-linear Equations". Am. Math. Soc. Trans., Series 2, Vol.29, 1959, pp. 295-381.

13. Oleinik, O.: "On Discontinuous Solutions of Nonlinear Differential Equations". Am. Math. Soc. Trans., Series 2, Vol.26, 1959, pp. 95-192.

14. Harten, A.: "High Resolution Schemes for Hyperbolic Conservation Laws". J. Comp. Phys., Vol.49, 1983, pp. 357-393.

15. Mitchell, A.R., Griffiths, D.F.: "The Finite Difference Method in Partial Differential Equations". John Wiley, Chichester, 1980.

16. Richtmyer, R.D., Morton, K.W.: "Difference Methods for Initial-Value Problems". Interscience, New York, 1967.

17. Smith, G.D.: "Numerical Solution of Partial Differential Equations". 3rd edition, Oxford Univ. Press, New York, 1985.

18. Richtmyer, R.D.: "A Survey of Difference Methods for Non-Steady Gas Dynamics". NCAR Tech. Note 63-2 Boulder, CO, 1962.

19. Morton, K.W.: "Stability of Finite-Difference Approximations to a Diffusion-Convection Equation". Intl. J. Num. Meth. Eng., Vol.12, 1980, pp. 899-916.

20. Lax, P.D., Wendroff, B.: "Systems of Conservation Laws". Comm. Pure Appl. Math., Vol.13, 1960, pp. 217-237.

IV THE FINITE VOLUME CONCEPT

The previous chapter presented fundamental concepts for the difference solution to hyperbolic problems in one space dimension. The situation becomes more complicated when the number of space dimensions increases, especially if the computational region does not conform naturally to a Cartesian mesh.

4.1 Coordinate Transformations

If the boundaries do not conform to the Cartesian mesh, then the boundary points occur at irregular locations in the mesh and special routines must be established to treat the boundary conditions. In practice this has been found to be tedious, and the results generally have been unsatisfactory. The alternative is to adopt a curvilinear coordinate system that conforms to the irregular boundary. This results in a mesh system that is aligned with the boundary, a so-called boundary-aligned mesh.

The equations then have to be transformed to these curvilinear coordinates. There are a number of ways of carrying this out. The first is the most conventional and stems from the concept of conformal mapping where the operator form of Laplace's equation remains invariant under all such conformal transformations. This approach offers the advantage of the invariant Laplace operator, useful in classical methods for solving the potential equation, but it implies a transformation not only of the independent variables x,y,z but the dependent vector variables as well, notably the velocity $\underline{V}=u\underline{e}_x+v\underline{e}_y+w\underline{e}_z$. The usual cylindrical or spherical coordinate systems are simple examples. This is unfortunate because the velocity vector must be differentiated and so it brings in derivatives of the curvilinear base vectors belonging to the curvilinear system. If the curvilinear system is orthogonal, the number of additional new terms grows but still remains manageable. If the system is non-orthogonal, general tensor analysis applies and the number of terms, including Christoffel symbols and their derivatives, quickly grows out of hand[1]. Nevertheless, during the 1950s and 1960s conformal mappings were the most commonly used transformations to obtain boundary-aligned meshes.

4.1.1 The Differential Approach

Non-orthogonal meshes, however, still presented a dilemma. One remedy sought in the 1970s was to break from the conformal mapping tradition and transform only the independent variables (even including time):

$$\tau=t \ , \quad \xi=\xi(x,y,z,t) \ , \quad \eta=\eta(x,y,z,t) \ , \quad \zeta=\zeta(x,y,z,t) \ .$$

Viviand[2] took this approach and worked out the chain rule for the transformed partial derivatives. When the velocity vector remains written in the Cartesian base, no terms result from the differentiation of the velocity and thus the simple form is retained

$$\frac{\partial}{\partial t}\begin{bmatrix}\rho \\ \rho u \\ \rho v \\ \rho w \\ e\end{bmatrix} + \frac{\partial}{\partial \xi}\begin{bmatrix}\rho U \\ \rho u U + \xi_x p \\ \rho v U + \xi_y p \\ \rho w U + \xi_z p \\ (e+p)U\end{bmatrix} + \frac{\partial}{\partial \eta}\begin{bmatrix}\rho V \\ \rho u V + \eta_x p \\ \rho v V + \eta_y p \\ \rho w V + \eta_z p \\ (e+p)V\end{bmatrix} + \frac{\partial}{\partial \zeta}\begin{bmatrix}\rho W \\ \rho u W + \zeta_x p \\ \rho v W + \zeta_y p \\ \rho w W + \zeta_z p \\ (e+p)W\end{bmatrix} = 0, \tag{4.1}$$

where the contravariant velocity components are

$$U=u\xi_x+v\xi_y+w\xi_z \ , \quad V=u\eta_x+v\eta_y+w\eta_z \ , \text{ and } \quad W=u\zeta_x+v\zeta_y+w\zeta_z \ .$$

In this way arbitrary regions of space can be treated in a generalized Cartesian manner.

4.2 The Finite-Volume Approach

An alternative to transforming to a global curvilinear coordinate system is to write the equations in integral form.

4.2.1 Continuum Equations

Representing the conservation of mass, momentum, and energy in any arbitrary volume Ω of space, the Euler equations of motion in integral form are

$$\frac{\partial}{\partial t}\int_{\Omega} q \, dvol + \iint_{\partial\Omega} \underline{H}\cdot\underline{n} \, ds = 0 \ , \tag{4.2}$$

where the dependent variables $q=[\rho, \rho u, \rho v, \rho w, e]$ comprise a column vector containing as elements the density and rectangular components of momentum and total energy per unit volume referred to a Cartesian system (x,y,z) fixed in space. The velocity of the fluid is

$$\underline{V} = u\underline{e}_x + v\underline{e}_y + w\underline{e}_z \ .$$

The quantity $\underline{H}(q)\cdot\underline{n}=[q\underline{V}+(0,\underline{e}_x,\underline{e}_y,\underline{e}_z)p]\cdot\underline{n}$ represents the net flux of q transported across, plus the pressure p acting on, the closed surface $\partial\Omega$ that bounds the volume Ω with unit normal $\underline{n}$. We only treat here flows of a perfect gas, so the equation of state $p=\kappa\rho(2e-u^2-v^2-w^2)$, where $\kappa=(\gamma-1)/2$, completes system (4.2).

Dividing by Ω and then shrinking Ω to a point leads to the differential conservation law valid at that point if the partial derivatives are continuous there. Conceptually, however, we find it more appealing to discretize the finite-domain integral system (4.2) directly, the so-called finite-volume approach[3], since the integral law formally does not exclude discontinuities from the interior of Ω. Our method therefore is a cell concept rather than a grid-point concept. The integral approach may be important for the correct capturing of discontinuities in the flow. It also lends itself to an obvious geometrical interpretation between the dependent and independent variables in the physical space and their counterparts in the computational space which makes the use of any arbitrary coordinate system more readily comprehensible.

4.2.2 Coordinate Geometry

Perhaps the most attractive feature of the integral approach is its readiness to accommodate any type of coordinate system. Any convenient grid-generation technique can be used simply to pack a grid of cells in an orderly fashion so that they discretize the entire flow field. Although any arbitrary mesh can be used, the one we found most practical is hexahedral cells orchestrated by a three-dimensional coordinate system for which the body surface is aligned with one of the three coordinate surfaces. Such an arrangement facilitates the enforcement of the numerical boundary conditions on solid walls. A variety of body-aligned coordinate topologies can be formulated, but we prefer for illustration purposes here the so-called O-O mapping (a generalization of the classical conical system) because it focuses grid points along all edges of the wing. Figure 4.1 illustrates this type of mesh and shows our placement of the singular lines which are unavoidable in a three-dimensional body-aligned mapping. We work with the transfinite interpolation procedure to construct our O-O type mesh based on the curvilinear coordinates

$$X_I = X_I(x,y,z)\ , \qquad X_J = X_J(x,y,z)\ , \qquad X_K = X_K(x,y,z)\ ,$$

where the surface X_J=constant aligns with the wing. The non-orthogonal coordinates (X_I,X_J,X_K) define in physical space the edges of the mesh cells (Fig. 4.1a), the integers I,J,K are the corresponding directions in the computational space, and in the physical space the unit vectors $\underline{n}_I,\underline{n}_J,\underline{n}_K$ are normal respectively to the cell surfaces X_I,X_J,X_K = constants. Complete details on the construction of such a mesh as well as a discussion of the relative economy of resolution afforded by the O-O mapping is found in Eriksson[4].

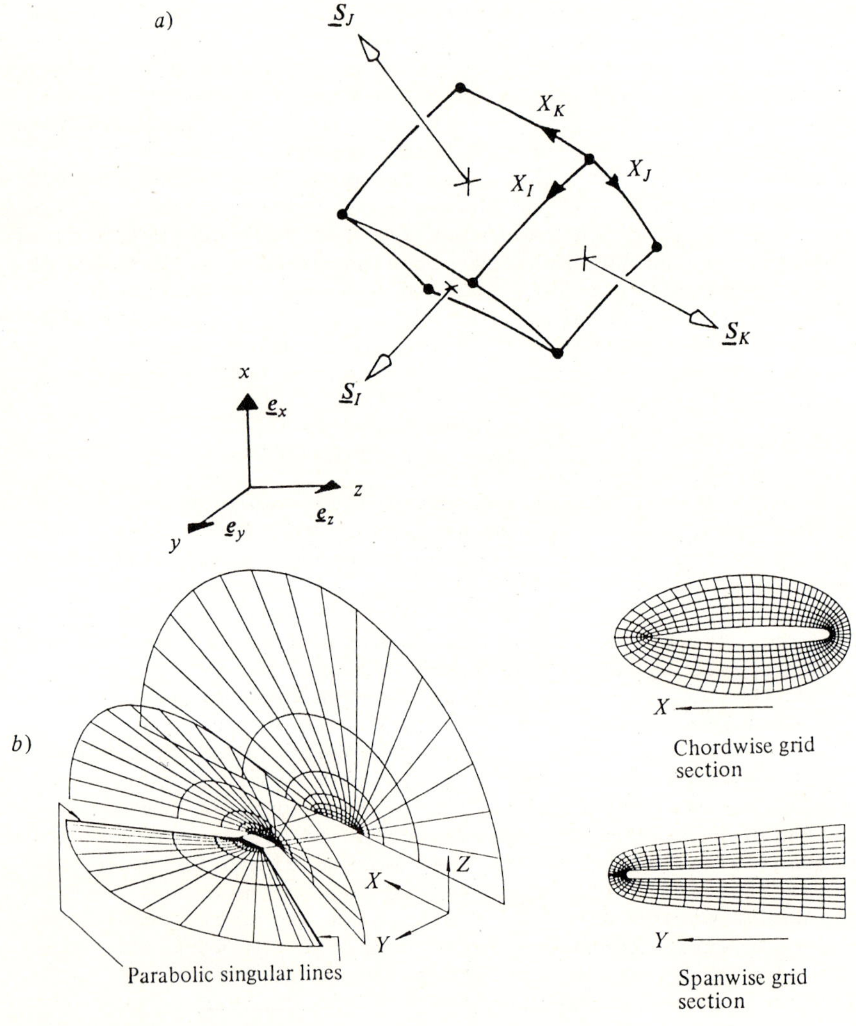

Fig. 4.1 The O-O mesh topology wraps cells around all the edges of a large-aspect-ratio wing and offers good resolution near the wing. a) The hexahedral cells of the mesh are defined by their eight vertices expressed in Cartesian coordinates (x,y,z). b) Basic features of the O-O mesh: oval surfaces encircling all edges and two parabolic singular lines starting at the tip

For the finite-volume method no global coordinate transformation needs to be specified. In fact the only details about the mesh that we transmit to the method are the three Cartesian coordinates of the eight vertices of every cell in the mesh. With this information it is not even necessary, as it is for a grid-point method, to formulate the local curvilinear coordinate system (X_I, X_J, X_K) in order to calculate the metric coefficients of the coordinate transformation. Instead the equivalent of these terms can be determined strictly by the principles of geometry. For example, altogether ten metric quantities are needed - the three components of each of the three surface areas $\underline{S}_I, \underline{S}_J, \underline{S}_K$ of a cell together with its volume Ω. If the four vertices defining a surface are coplanar, its area is given exactly by one half the cross product of its diagonal line segments $S=1/2\,\underline{\ell}_{31}\times\underline{\ell}_{42}$, and this is a good approximation even if it is non-planar.

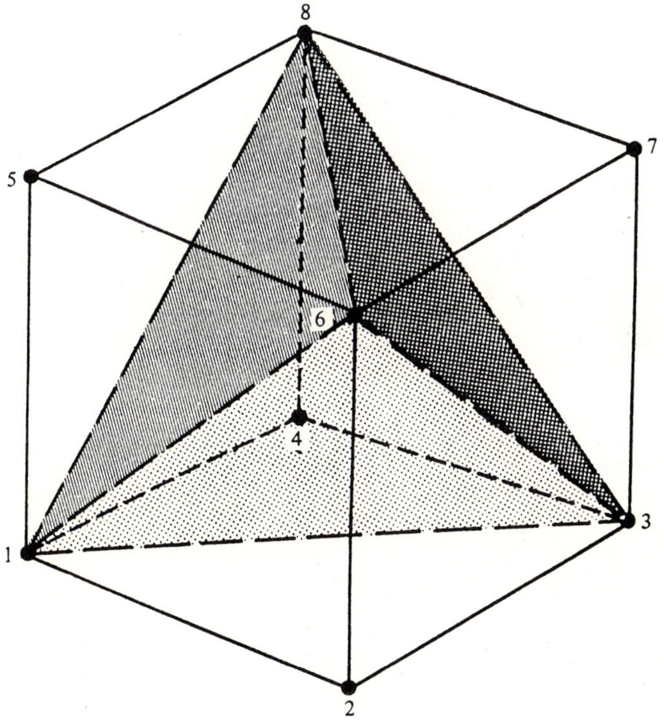

Fig. 4.2 The volume of a hexahedral cell is the sum of the five constituent tetrahedra:
$T_{1236}+T_{3867}+T_{3816}+T_{1685}+T_{1348}$

The volume Ω is computed in the following way. Without restriction, a general hexahedron is composed of five tetrahedra (Fig. 4.2), each of whose volume is determined exactly by

$$T_{1236} = 1/6 \begin{bmatrix} x_1 & y_1 & z_1 & 1 \\ x_2 & y_2 & z_2 & 1 \\ x_3 & y_3 & z_3 & 1 \\ x_6 & y_6 & z_6 & 1 \end{bmatrix} ,$$

where the integer subscripts on T_{1236} refer to the four vertices that define the tetrahedron. The volume of the hexahedron is then the sum of the volumes of these five tetrahedra. It is identical with the geometric Jacobian.

4.2.3 Spatial Finite-Volume Discretization

Since Eq. (4.2) is valid for any arbitrary volume it also holds locally for each individual cell ijk in the mesh, where the bounding surface $\partial\Omega_{ijk}$ now consists of the family of the three coordinate surfaces $\underline{S}=\{\underline{S}_I,\underline{S}_J,\underline{S}_K\}$ that delineate the hexahedral mesh cell (Fig. 4.1). In order to solve this continuum equation we must evaluate the integrals by some discrete approximation which then characterizes the class of the cell method. Here we consider only a single-point evaluation per cell for the dependent variables q, so that by the mean-value theorem Eq.(4.2) becomes

$$\Omega_{ijk} \frac{dq_{ijk}}{dt} + \delta[\underline{H}(q)\cdot\underline{S}]_{ijk} = 0 , \qquad (4.3)$$

where q_{ijk} is now interpreted as a volumetric average located at the centre of the cell, and $\underline{H}(q)\cdot\underline{S}$ is the corresponding flux evaluated at the surfaces $\underline{S}$. That q and $\underline{H}\cdot\underline{S}$ reside at different spatial positions is a central feature of the finite-volume concept. The three-dimensional undivided central-difference operator

$$\delta\psi_{ijk} = (\delta_I+\delta_J+\delta_K)\psi_{ijk} = (\psi_{i+1/2,j,k} - \psi_{i-1/2,j,k}) +$$
$$+ (\psi_{i,j+1/2,k} - \psi_{i,j-1/2,k}) + (\psi_{i,j,k+1/2} - \psi_{i,j,k-1/2})$$

expresses the net gain of flux into the cell and is fundamental to the conservation property and independent of any particular choice of spatial differencing.

4.2.4 Flux Evaluation

Since q_{ijk} is located in the centre of the cell, but $\underline{H}(q)$ must be expressed at its surfaces, some form of local interpolation of the neighboring discrete values q must be devised and a numerical quadrature of the surface integrals performed in order to carry out the discrete solution of Eq.(4.2). It is the particular type of interpolating function and quadrature that defines the specific spatial-difference scheme of the method. In our case the simplest, and perhaps most natural, function is

$$[\underline{H}\cdot\underline{S}]_{ijk}=[\underline{H}(\mu_I q_{ijk})\cdot\underline{S}_I+\underline{H}(\mu_J q_{ijk})\cdot\underline{S}_J+\underline{H}(\mu_K q_{ijk})\cdot\underline{S}_K] \,, \qquad (4.4a)$$

where μ is the averaging operator (here μ_I):

$$\mu_I \psi_{ijk} = 1/2(\psi_{i+1/2,j,k} + \psi_{i-1/2,j,k}) \,.$$

An alternative to this, since each face of the cell ijk lies between two dependent variables, is to compute the flux separately for each of the two neighboring dependent variables and then average the two results, i.e.

$$[\underline{H}\cdot\underline{S}]_{ijk}=[\mu_I\underline{H}(q_{ijk})]\cdot\underline{S}_I+[\mu_J\underline{H}(q_{ijk})]\cdot\underline{S}_J+[\mu_K\underline{H}(q_{ijk})]\cdot\underline{S}_K \,. \qquad (4.4b)$$

If the flux function $\underline{H}$ were linear, alternatives (4.4a) and (4.4b) would obviously be equivalent, but $\underline{H}$ is quadratic. We have tested both forms and found that in smooth regions of the flow the differences in the two results are imperceptible. They are larger at shock waves, however. We choose to work with (4.4b) because for the idealized case of one constant flow field ahead of and another behind a shock wave, and a cell face aligned to it, only scheme (4.4b) provides the correct jump in q across the shock. Equation (4.2) together with (4.4b) leads to a spatial-difference operator completely centered in all three coordinate directions, which is second-order-accurate in space if the variation in mesh size is reasonably smooth. These three-point differences lead to a simple program structure requiring no logic to decide whether to skew the differencing to one side or the other, which makes it amenable to a large degree of computer vectorization.

This finite-volume discretization bears some similarity to both the conventional finite-difference and finite-element discretizations. Its difference stencil is that of a finite-difference scheme, but it differs in that cell-averaged instead of point quantities are differenced, and as we shall see below this gives a significant distinction near a mesh singularity. Like the finite-element procedure, its formulation begins with the integral equation, and in fact we could present it in the context of a finite-element technique, but the resulting shape function is so peculiar that we think it warrants a presentation and name of its own.

4.2.5 Stability and Accuracy at Mesh Singularities

Notice that even if the underlying mesh transformation is singular, so that an edge of a cell contracts to a point, a surface collapses to a line or a point, or seven of its eight vertices become coplanar, this geometrical procedure still returns meaningful and accurate values for the areas and volume. The flux quanitites can therefore be defined, and, since (4.2) is balanced in the interior of the cell, where no coordinates are used, it remains finite even in the presence of these mesh singularities. And this is accomplished automatically without any special programming considerations. The same may not be true for the usual grid-point methods. Eriksson[4] analysed the further question of whether mesh singularities destroy the spatial accuracy of finite-volume and finite-difference schemes or their stability as they step forward in time. He found that without any modification the finite-volume technique remains stable in the presence of a singularity, but its accuracy decreases to somewhere between first and second order in space. Without alteration the finite-difference scheme is unstable even if the singularity is straddled. Stability can be restored, however, if a limiting form of the difference scheme is derived at the singular point and implemented in the computer code.

4.3 Relationship to Finite Differences

The finite-volume method is related to the finite difference method through the following identities, usually called the gradient, divergence, and curl theorems respectively in advanced calculus books

$$\int_{vol} \operatorname{grad} \Psi \, dv = \int_S \underline{n} \, \Psi \, ds \,,$$

$$\int_{vol} \operatorname{div} \underline{\Psi} \, dv = \int_S \underline{n} \cdot \underline{\Psi} \, ds \,,$$

$$\int_{vol} \operatorname{curl} \underline{\Psi} \, dv = \int_S \underline{n} x \underline{\Psi} \, ds \quad .$$

The first of these then show that in the finite volume context

$$\frac{\partial}{\partial x} \Psi = \{ \int_{S_I} \Psi \, \underline{n} \cdot \underline{e}_x ds + \int_{S_J} \Psi \, \underline{n} \cdot \underline{e}_x ds + \int_{S_K} \Psi \, \underline{n} \cdot \underline{e}_x ds \} / \Delta vol$$

with similar relations for $\partial/\partial y$ and $\partial/\partial z$.

From consideration of tensor calculus we have that the operator $\text{grad}=\underline{g}^k\partial/\partial x_k$ where the covariant coordinates x_k are identical to the edges of the finite volume X_I, X_J, X_K, and the contravariant base vectors $\underline{g}^k$ are parallel to $\underline{S}_I, \underline{S}_J, \underline{S}_K$. We then find that

$$\underline{g}^1 \frac{\partial \Psi}{\partial \xi} = \int_{S_I} \Psi \, \underline{n} ds/\Delta vol$$

and corresponding formulae for η and ζ. Lastly we have from the knowledge that the gradient of the surface ξ=constant is the normal to the surface

$$\underline{S}_I = \text{grad } \xi ,$$

and hence

$$\begin{aligned} \xi_x &= \underline{S}_I \cdot \underline{e}_x = SIX , \\ \xi_x &= \underline{S}_I \cdot \underline{e}_y = SIY , \\ \xi_z &= \underline{S}_I \cdot \underline{e}_z = SIZ . \end{aligned} \tag{4.5}$$

The volume element is determined by

$$\Delta vol = J \, d\xi \, d\eta \, d\zeta ,$$

where J is the Jacobian determinant of the coordinate transformation. As we saw before, it can also be calculated on strictly geometrical principles.

The condition that a cell is closed leads to the geometric identity

$$\int_S \underline{n} \, ds = 0 . \tag{4.6}$$

4.4 Numerical Conservation

The form of the discrete finite-volume method, as well as the conservative finite-difference form, guarantees that when the scheme is summed over all the grad points interior flux terms will cancel. This telescoping property defines conservative differencing, and indeed is part of the definition of a conservative scheme.

The standard finite-difference approach combines geometric and flow variables and subjects the resultant transformed variables to an algebraic treatment. This combined approach can give rise to numerical errors even when the flow is known exactly. Some of the sources of error are: 1) the geometric identity (4.6) is not satisfied precisely, 2) the calculation of derivatives requires approximations for q_ξ that may not be consistent with ξ_x and 3) the treatment of boundary points uses geometric quantities evaluated by one-sided differences that may be inconsistent with neighbouring interior values. These problems do not occur with the cell-centered finite-volume approach.

4.4.1 Uniform Free Stream

A good way to gauge these possible errors is to check if the scheme maintains the freestream at a constant state. If present, the errors mentioned above show their effect by altering the constant state. What is meant here is that the difference scheme should have the property that the discrete steady uniform flow should be an exact solution of the scheme. This can also be of practical significance in problems where a coarse grid is used in regions where the ultimately steady flow is a slight perturbation of an uniform parallel flow. When schemes do not have this property of preserving a constant state, the near-uniform flow in these problems may lead to relatively large errors due to the coarseness of the mesh. Conservative schemes usually maintain free-stream accuracy.

4.5 Cell Vertex Methods

The idea of formulating a finite volume scheme where the unknowns are located at the vertices of the cells instead of the cell centers originates probably with Ni[5]. Since then a number of investigators have extended and refined the concept. Even though the flow properties are located at the vertices, the equations are still balanced at the cell centers. This means that the residuals must be distributed in some manner to the vertices in order to obtain the updated quantities there. This distribution must be stable, and is an integral part of the scheme. It also opens the possibility for non-conservation.

Schemes have been devised that overcome these difficulties, see e.g. Refs.6 and 7, and it appears that cell vertex methods offer some advantage over cell centered schemes when the mesh is non-smooth.

4.6 Boundary Conditions for the Continuous Problem

A particular steady flow field is determined by the conditions imposed upon it at its boundaries, and usually the stability and accuracy of the discrete conditions are more difficult to analyse than the difference scheme itself. This means that in general the theory of boundary conditions for numerical computations is more empirical. In our case of a O-O mesh conforming to a wing/body combination, boundary conditions are enforced at the six outer surfaces of the computational space (Fig. 4.3). There are three distinct types: flow into or out of the far field, periodic conditions across coordinate cuts, and conditions on solid walls.

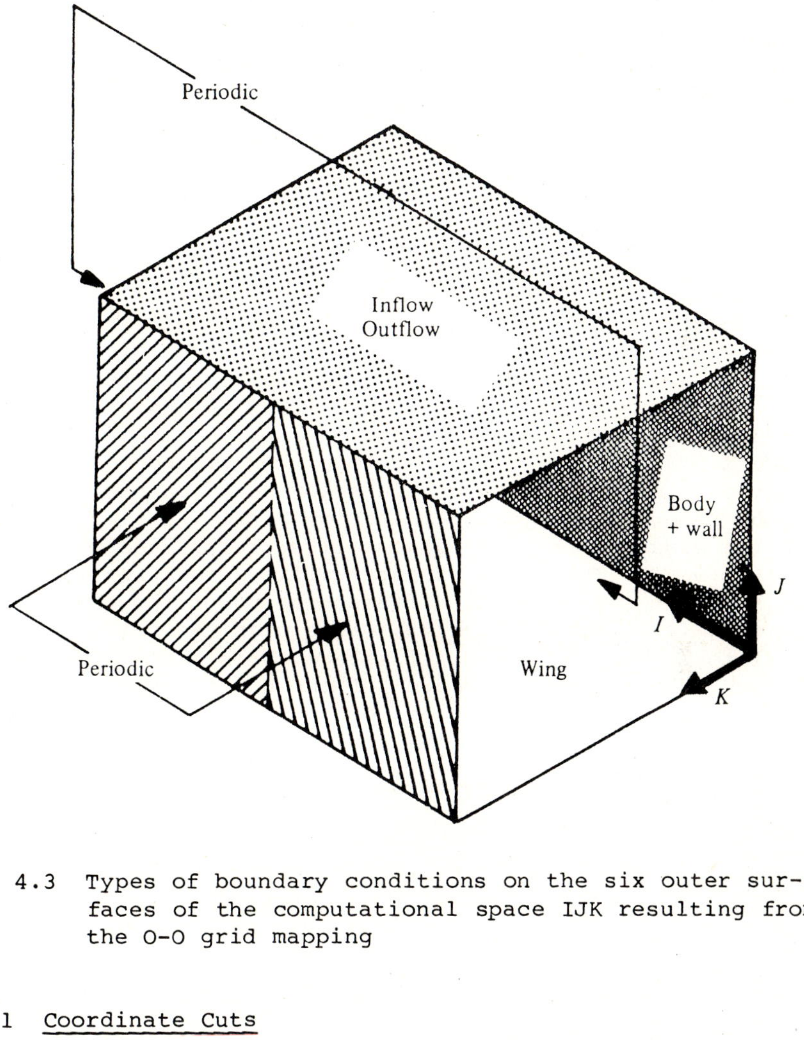

Fig. 4.3 Types of boundary conditions on the six outer surfaces of the computational space IJK resulting from the O-O grid mapping

4.6.1 Coordinate Cuts

Conditions on these boundaries are the least troublesome since at a cut the physical space folds on to itself and the condition on the flow at the computational boundary is periodicity. We remark that in the O-O topology these boundaries occur conveniently at the trailing edge and tip of the wing.

4.6.2 Solid Walls

Boundary conditions on solid walls in general can be imposed with regard only to the velocity components and the temperature or the heat flux[8]. Then, for inviscid flow the imposition of the boundary condition on the surface of the aerodynamic vehicles of interest in our work possesses two different but related aspects. The first is the one that no flow is allowed through a solid wall - the kinematic condition -, and the second is the so-called Kutta condition, which dictates that the flow separates from a sharp trailing edge which makes the flow problem unique in analogy to potential wing theory.

4.6.3 Zero-Flux Transport

The kinematic condition of zero transport $\underline{V}\cdot\underline{n}=0$ applies to two surfaces in the O-O wing mesh, the wing J=1 and the fuselage which merges together with the wall of symmetry K=1. Since the computational cells are aligned to both these surfaces, the physical conditon reduces the dependence of $H(q)_{body}$ to $H(p)_{body}$, and we are forced to determine a value for the pressure p on the vehicle surface by numerical means, usually by differencing some auxiliary equation in order to relate values in the field to those on the surface. Our procedure to obtain an estimate of p_{body} from the interior solution has been described before in the general case by Rizzi[9]. It is valid at both the wing and the body-plus-wall surfaces, and we summarize it briefly. The basis of our auxiliary relation for p_{body} begins with the streamline differentiation of the physical condition $(\partial/\partial t+\underline{V}\cdot\mathrm{grad})(\underline{V}\cdot\underline{n})\equiv 0$, where n is the unit vector normal to either the wing or fuselage. This expression, when combined with the inner product of the quasi-linear momentum equation and $\underline{n}$, and rearranged, becomes, for a stationary body,

$$\rho\underline{V}\cdot(\underline{V}\cdot\mathrm{grad})\underline{n} = \underline{n}\cdot\mathrm{grad}\ p \quad ,$$

where $$\mathrm{grad} \equiv \frac{\underline{S}_I}{|S_I|}\frac{\partial}{\partial x_I} + \frac{\underline{S}_J}{|S_J|}\frac{\partial}{\partial x_J} + \frac{\underline{S}_K}{|S_K|}\frac{\partial}{\partial x_K} \quad .$$

It relates ρ, $\underline{V}$ and the geometry of the surface to the normal derivative of p. When it is differenced to formally first-order accuracy the pressure on the surfaces is deduced from the interior values.

4.6.4 Inflow/Outflow Boundary

With a O-O mesh, flow in the farfield enters and leaves through the outermost J-surface. This is an artificial boundary in the sense that the actual flow in the physical domain is open, whereas the computational space must for practical reasons be closed. The numerical conditions, therefore, ideally should allow phenomena generated in the computational domain to pass through the boundary without undergoing significant distortion and without influencing the interior solution. In this way the maximum amount of transient energy escapes from the field, and the time-dependent solution can converge to the steady state. Engqvist and Majda[10] present a mathematical theory for the practical application of local absorbing boundary conditions at artificial boundaries.

One of the six faces of cell i,JL,k located in the mesh layer JL farthest from the body is coincident with the outer boundary; call the surface area of that face S_{JL} (see Fig. 4.4). The edges of the cell define the local curvilinear coordinate system (X_I, X_J, X_K) where the positive X_J direction points from the outer boundary surface into the domain, the other two being tangent to the surface. In this system the differential conservation equations equivalent to Eq.(4.1), but for isenthalpic flow are

$$\frac{\partial}{\partial t} q + A \frac{\partial q}{\partial X_I} + B \frac{\partial q}{\partial X_J} + C \frac{\partial q}{\partial X_K} = 0 \quad ,$$

where

$$A = \frac{\partial(\underline{H} \cdot \underline{S}_{JL})}{\partial q} \quad , \quad B = \frac{\partial(\underline{H} \cdot \underline{S}_{IL})}{\partial q} \quad , \quad C = \frac{\partial(\underline{H} \cdot \underline{S}_{KL})}{\partial q}$$

are the Jacobian matrices of the flux function. The matrix A works out to be

$$A = \begin{bmatrix} 0 & \alpha & \beta & \varepsilon \\ \alpha c^2 - uU + k\alpha V^2 & \alpha(1-k)u + U & \beta u - k\alpha v & \varepsilon u - k\alpha w \\ \beta c^2 - vU + k\beta V^2 & \alpha v - k\beta u & \beta(1-k)v + U & \varepsilon v - k\beta w \\ \varepsilon c^2 - wU + k\varepsilon V^2 & \alpha w - k\varepsilon u & \beta w - k\varepsilon v & \varepsilon(1-k)w + U \end{bmatrix} ,$$

where $\alpha = \underline{S}_{JL} \cdot \underline{e}_x$, $\beta = \underline{S}_{JL} \cdot \underline{e}_y$, $\varepsilon = \underline{S}_{JL} \cdot \underline{e}_z$,

and c is the local speed of sound.

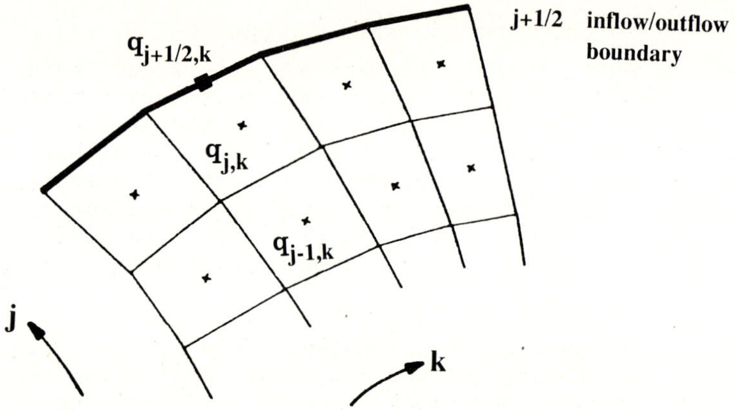

Fig. 4.4 Primitive variables $q_{j+1/2,k}$ are computed by specifying characteristic inflow variables and extrapolating characteristic outflow variables

The First Approximation in the hierarchical theory of Engqvist and Majda[10] amounts to specifying the characteristic variables of the corresponding one-dimensional problem which is well posed and maximally dissipative. Since we work only with their first approximation, it is easier to grasp the idea if we present our boundary condition in a heuristic development based directly on the characteristic variables instead of the formal theory. The presentation for our 4×4 system mirrors the one given by Gottlieb and Gustafsson[11] for the 3×3 system, but is more general because of our non-orthogonal coordinates.

We seek the characteristic variables of the corresponding one-dimensional problem local to a given cell

$$\frac{\partial}{\partial t} q + A \frac{\partial q}{\partial X_J} = 0 \quad ,$$

which means that we focus on a particular set of characteristic planes, those whose normals point along X_J and whose slopes in time are the eigenvalues λ of A. Solving $\det(A-\lambda I)=0$, we find

$$\lambda_1 = U \ , \quad \lambda_2 = U \ , \quad \lambda_3 = U-a_+ \ , \quad \lambda_4 = u-a_- \ ,$$

where $a_\pm = 1/2kU \pm 1/2[1/4k^2U^2+c^2(\alpha^2+\beta^2+\varepsilon^2)]$.

The left and right eigenvectors associated with these four eigenvalues make up the rows and columns of the transformation matrices T^{-1} and T respectively so that we obtain the diagonalized equation

$$\frac{\partial}{\partial t}\phi + \Lambda \frac{\partial \phi}{\partial X_J} = 0 \quad ,$$

where $\quad \phi = T^{-1}q \; , \qquad \Lambda = T^{-1}AT = \mathrm{diag}\{\lambda_1, \lambda_2, \lambda_3, \lambda_4\}$.

After the intermediate variables

$$\tilde{U} = \beta u+\alpha v \; , \qquad \tilde{V} = \varepsilon v+\beta w \; , \qquad \tilde{W} = \varepsilon u-\alpha w \quad ,$$

$$\xi=\alpha^2+\beta^2+\varepsilon^2 \; , \quad Q_\pm=kU-\theta_\pm \; , \quad R_\pm=\varepsilon Q_\pm-\zeta kw \; , \quad P_\pm=kw\alpha\pm\varepsilon c^2$$

have been defined for the sake of simplification, we find

$$T=\begin{bmatrix} k\tilde{U} & 0 & R_+ & R_- \\ kvU-\beta(kV^2+c^2) & \tilde{V} & uR_+ +\alpha P_+ & uR_- +\alpha P_- \\ -kuU+\alpha(kV^2+c^2) & \tilde{W} & vR_+ +\beta P_+ & vR_- +\beta P_- \\ 0 & \tilde{U} & \begin{matrix} wR_+ -k(\alpha u+\beta v)\alpha_+ \\ -(\alpha^2+\beta^2)c^2 \end{matrix} & \begin{matrix} wR_- k(\alpha u+\beta v)a_- \\ -(\alpha^2+\beta^2)c^2 \end{matrix} \end{bmatrix} ,$$

and

$$T^{-1}=\begin{bmatrix} \frac{\xi V^2-U^2}{d_1} & \frac{-\varepsilon\tilde{W}+\beta\tilde{U}}{d_1} & \frac{\varepsilon\tilde{V}-\alpha\tilde{U}}{d_1} & \frac{\alpha\tilde{W}-\beta\tilde{V}}{d_1} \\ \begin{matrix}[kw(U^2-\xi V^2) \\ +c^2(\varepsilon U-\xi w)]/d_2\end{matrix} & \begin{matrix}[+kw(-\beta\tilde{U}+\varepsilon\tilde{W}) \\ -\alpha\varepsilon c^2]/d_2\end{matrix} & \begin{matrix}[kw(\alpha\tilde{U}-\varepsilon\tilde{V}) \\ -\beta\varepsilon c^2]/d_2\end{matrix} & \begin{matrix}[-kw(\alpha\tilde{W}-\beta\tilde{V}) \\ +(\alpha^2+\beta^2)c^2]/d_2\end{matrix} \\ \frac{R_+}{d_3} & \frac{k\xi V^2-(U+\alpha_+)Q_+}{d_3} & \frac{-k\xi u+\alpha Q_+}{d_3} & \frac{-k\xi v+\beta Q_+}{d_3} \\ \frac{R_-}{d_4} & \frac{k\xi V^2-(U+\alpha_-)Q_-}{d_4} & \frac{-k\xi u+\alpha Q_-}{d_4} & \frac{-k\xi v+\beta Q_-}{d_4} \end{bmatrix} .$$

The factors d_1, d_2, d_3 and d_4 in the denominators are normalising coefficients so that $T^{-1}T$ equals the unit matrix.

For the one-dimensional case it is well known that the number of conditions to be imposed in a cell at the outer boundary should equal the number of characteristic directions that enter the computational domain. Four typical cases are depicted in the Fig. 4.5. With subsonic inflow our implementation is to set the three ingoing characteristic variables $\phi^{(1)}$, $\phi^{(2)}$ and $\phi^{(3)}$ to their free-stream values, linearly extrapolate the fourth $\phi^{(4)}$ from the computational field, and then solve for the original unknowns $q=T\phi$. At outflow it is $\phi^{(3)}$ that is given the values of undisturbed flow, and $\phi^{(1)}$, $\phi^{(2)}$ and $\phi^{(4)}$ are extrapolated from the computational field.

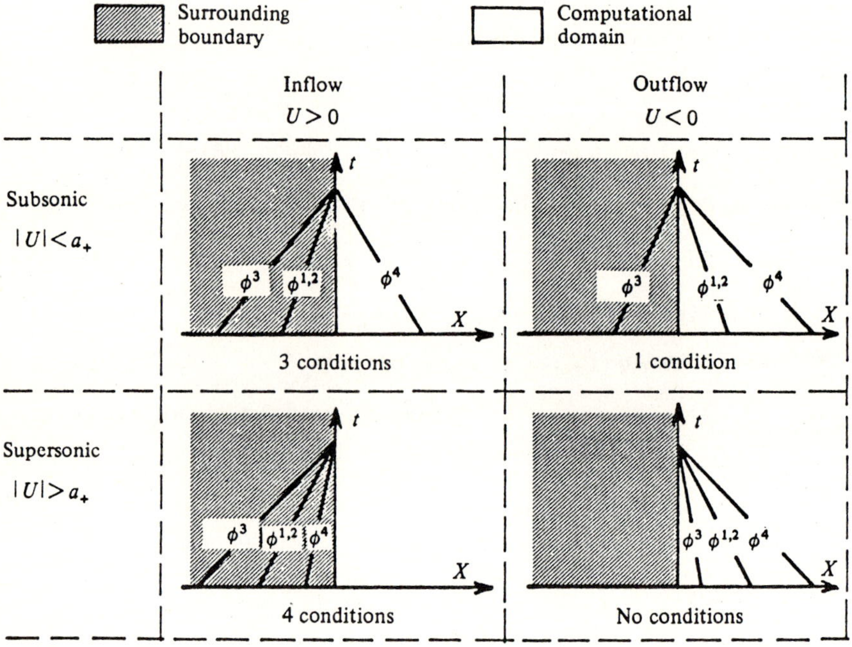

Fig. 4.5 The number of boundary conditions at inflow and outflow based on the ingoing characteristic variables ϕ^m, m=1,...,4

4.7 Discretization of the Flow Domain

We now come to projecting the independent variables to some finite dimensional space, i.e. we must discretize the flow domain. The finite volume concept has been presented for an arbitrary computational cell. The discretization of the domain therefore can be achieved by packing such cells throughout the intended region. The hexahedra cell in Fig.4.1 packs in a regular order, one on top of one another and one to either side. We can see that the edges of the cell make up a curvilinear coordinate system, and neighboring cells can be located by a Cartesian ordering system. Indeed highly structured grids like this are usually constructed from coordinate transformations from a rectilinear system.

Structured curvilinear grids most commonly are generated to conform to the solid body in the flowfield. With the help of a coordinate transformation a Cartesian mesh is mapped onto the desired physical space. While preserving the regular structure, this procedure also fits the mesh to the boundaries. Thus a body-conforming mesh spans the physical space and transforms it to a rectangular computational domain. The mesh generation is not trivial because one has to avoid cells being too greatly distorted in order to maintain accuracy and stability of the numerical scheme.

There are three basic types of grid topology termed O-type, C-type, or H-type. The first important step in grid generation is to choose the topology best suited for the given configuration. For a single airfoil an O-grid has been found to be useful, because of its good distribution of grid points near the wing surface, especially at the leading and trailing edges. A better resolution of the wake behind an airfoil is obtained with a C-grid. However, at the trailing edge a singularity can occur. Both types are suitable for the discretization in the spanwise direction of a wing. For very simple geometries like a ramp or for complex configurations a H-grid offers the most easily adaptible topology. By the use of interior branch cuts for the wings a single block H-grid can even be applied for a complete aircraft. But at rounded edges this mesh becomes severely singular. For 3-D calculations often combinations like H-O or C-O, i.e. distinct grid types in each direction, are applied.

Therefore, one can conclude that the advantages of a structured mesh are the flexibility of implementation of all classes of algorithms and the minimal usage of CPU and computer memory achieved by the efficient utilization of vector architectures. The principle disadvantage lies in the lack of flexibility for complex geometrical shapes when using a single block. The multi-block approach however, with the use of a structured mesh for each block, has been proven to be able to treat very complicated geometries. The one weakness of a block-structured grid is the awkwardness of automatically adapting the mesh.

The unstructured mesh gets its name from the irregular structure of the cells in the discretized flow field. Instead of using quadrilaterals or cubes as for a structured mesh, triangles in two dimensions and tetrahedra in three dimensions are used. This produces a maximum flexibility in treating complex geometries. Hence, unstructured meshes can be generated for arbitrary domains. Compared to structured meshes, it is much easier to add grid points.

The disadvantages of the unstructured meshes are the difficulties that arise during mesh generation due to their extreme generality. Further problems are the post-processing of the grid and the data, and the high computing costs. An unstructured mesh requires considerably more computer time and storage and, additionally, because of its lack of a regular pattern, it is less suited for vector computers. The unstructured grids are better suited for the finite element discretization due to their inherent irregularities, and hence the use of the most efficient numerical solution methods working with the finite volume technique is limited. It can be stated that almost all the advantages of the unstructured meshes are the disadvantages of the structured meshes and vice versa. However, for applications on complex configurations, hybrid meshes, i.e. combinations of both, are increasingly considered.

4.7.1 Resolution of Scales

In Sub-Chapter 3.10 above we brought up the matter of resolving fine-scale structure like shock waves, entropy layers, slipstreams etc. One principal aim in numerical fluid dynamics is to obtain short computation times together with a good resolution of the computational domain. Both are strongly related to the size of the mesh, in which the governing equations are discretized and solved. It is therefore desirable to have a grid with fine cells in regions where necessary, but coarse cells in the remaining flow field. Adaptive solution algorithms are an effective means to accomplish this because grid points move according to the developing numerical flow solution, and the point distribution is readjusted dynamically to concentrate points in the vicinity of large flow gradients. When using this technique the regions needing a further refinement are determined with the help of a sensor during the calculations. There are two distinct adaptivity criteria. The first is to estimate the truncation error of the discretized governing equations by means of a Richardson extrapolation. The second strategy is the choice of a physical value gradient as sensor to detect dominant flow features like entropy layers, shocks, wakes or vortex sheets. Currently, there are two major approaches to the adaptation technique. First the so-called local grid refinement or embedding technique and secondly the grid redistribution method.

i) Local grid refinement

The idea behind this technique is to add cells in the regions of strong gradients. Since the rest of the grid remains unchanged, the added cells cause an increase in computational effort. However, computations executed on a coarse mesh with some additional clustered cells certainly have a higher efficiency compared to calculations on a totally fine mesh. The common principle difficulty of these works is how to treat the fine/coarse mesh boundaries, i.e. how to retain flow conservation, accuracy and stability while at the same time keeping a clear structure of the flow solver algorithm.

Since unstructures meshes provide a natural environment for grid refinement, they are very suitable for this technique. In general, the local grid refinement can also be applied in an non-adaptive way.

ii) Grid redistribution

In this technique a fixed number of grid points is redistributed throughout the flow field. Contrary to the local grid refinement technique here the number of nodes are kept constant while the cells are rearranged by moving them around. Thereby changing their size and skewness. The major problems in this area are to avoid cells becoming too distorted and to ensure that they do not overlap.

Although the flexibility of unstructured meshes allows the application of alternative, more efficient, methods to get a better resolution of the flow field, investigations for the grid redistribution technique on unstructured meshes have also been carried out.

The adaption technique can be viewed as a natural development of the basic flowfield discretization. The idea is to obtain an optimal resolution of the scales in the flow with a minimal number of nodes. Mesh adaptivity, in general, is a very powerful tool to improve the resolution of fine scales. It comes closest in character to the traditional concept of tracking or fitting a front or discontinuity and is more easily implemented in an unstructured grid than in a structured one. Thompson[12] and Eiseman[13] have recently reviewed the current status of this technique.

4.7.2 Topology of Grid-Point Patterns

In addition to the accuracy of the results, the overall efficiency of the computational procedure also depends on the connectivity of the set of grid points. If the size of a given mesh cell varies substantially in comparison with its immediate neighbor, then a standard finite-difference approximation to a first derivative, formally accurate to second order $O(\Delta x^2)$ on a uniform grid with spacing Δx, falls to first-order accuracy. A finite-element approximation, however, can maintain its accuracy under these circumstances but may lose its superconvergence properties. Second-order accuracy is maintained for the standard finite-difference approximation provided the variation in size, i.e. the smoothness of the cell dilatation, is $\Delta x_2 = \Delta x_1 [1 + O(\Delta x_1)]$.

4.8 Finite-Volume Truncation Error[14]

At the core of the finite-volume method is the differencing of the flux on the left and right faces. The flux contains the metrics of the coordinate system and thus is equivalent to finite differences in the computational space. The formal accuracy of the difference approximation in this space typically is second order. But we really want to check the accuracy in the physical space since this is where we measure all the variables of the problem.

The usual analysis of the truncation error is based on Taylor series expansions, and is necessarily local. Thus if the mesh is graded, the solution may be formally second-order accurate throughout the domain, but the absolute value of the truncation error grows with the increase of the spacing due to the grading. A simple example illustrates the point.

Consider a one-dimensional point distribution function

$$x = x(u) \ .$$

The u-derivatives of a dependent variable f (x) then becomes

$$f_u = x_u f_x \ , \tag{4.7a}$$

$$f_{uu} = x_{uu} f_x + x_u (f_x)_u = x_{uu} f_x + x_u^2 f_{xx} \ , \tag{4.7b}$$

$$\begin{aligned} f_{uuu} &= x_{uuu} f_x + x_{uu}(f_x)_u + 2x_u x_{uu} f_{xx} + x_u^3 (f_{xx})_u = \\ &= x_{uuu} f_x + 3x_u x_{uu} f_{xx} + x_u^3 f_{xxx} \ . \end{aligned} \tag{4.7c}$$

A second-order difference approximation of f_x then becomes

$$D_{x_o} f = \frac{1}{x_u} \frac{f_{j+1} - f_{j-1}}{2\Delta u} f_x + \frac{1}{6} \frac{(\Delta u)^2}{x_u} f_{uuu} + \dots$$

This is not a useful estimate yet because the derivatives of f with respect to u are dependent upon the distribution of points. Change the distribution and the value of the derivative also changes. Therefore we transform via (4.7a-c) to

$$D_{x_o} = f_x + \frac{1}{6} \frac{(\Delta u)^2}{x_u} (x_{uuu} f_x + 3x_u x_{uu} \cdot f_{xx} + x_u{}^3 f_{xxx}) + \dots$$

If the distribution d of N points is

$$x(u) = d\left(\frac{u}{N}\right) \ , \qquad 0 \leqslant u \leqslant N \ ,$$

then $x_u = \frac{1}{N} d_u$, $\quad x_{uu} = \frac{1}{N^2} d_{uu}$, $\quad x_{uuu} = \frac{1}{N^3} d_{uuu}$.

So if the number of points in the grid is increased, but the distribution function d remains the same, the term f_{uuu} will be proportional to $1/N^3$. One concludes that the difference approximation is second-order regardless of the form of the point-distribution function because the truncation error goes to zero as $1/N^2$ with the increase in the number of points.

In two dimensions departure of the mesh from orthogonality also contributes to the truncation error. In this case consider the two-dimensional transformation x(u,v). The chain rule leads to the following equation system:

$$f_u = f_x x_u + f_y y_u , \qquad f_v = f_x x_v + f_y y_v .$$

We find f_x to be

$$f_x = \frac{f_u y_v - f_v y_u}{x_u y_v - x_v y_u} = \frac{f_u y_v - f_v y_u}{J} ,$$

where J is the Jacobian of the transformation.

The truncation errors also transform in the same way. The error T_x for the approximation to f_x is

$$T_x = \frac{y_v T_u - y_u T_v}{J} , \tag{4.8}$$

where T_u and T_v are the truncation errors for the difference expressions of f_u and f_v. Now all coordinate derivatives can be expressed using direction cosines of the angles of inclination, ϕ_u and ϕ_v, of the u and v coordinate lines. After substitution Eq(4.8) becomes

$$T_x = \frac{1}{\sin(\phi_u - \phi_v)} [\sin\phi_v \cos\phi_u \frac{T_u}{x_u} - \sin\phi_u \cos\phi_v \frac{T_v}{x_v}] ,$$

and varies inversely with the sine of the angle between the coordinate lines.

We now use two-point central-difference approximations for all derivatives and obtain an approximation of f_x

$$\tilde{f}_x = \frac{D_u f D_v y - D_v f D_u y}{D_u x D_v y - D_v x D_u y} ,$$

where $\quad D_u f := f_{u+\Delta u} - f_{u-\Delta u} , \qquad D_v f := f_{v+\Delta v} - f_{v-\Delta v} .$

The contribution from non-orthogonality can be isolated by considering the case of skewed parallel lines (see Fig.4.6) with

$$x_v = x_{vv} = x_{uv} = y_{uu} = 0 .$$

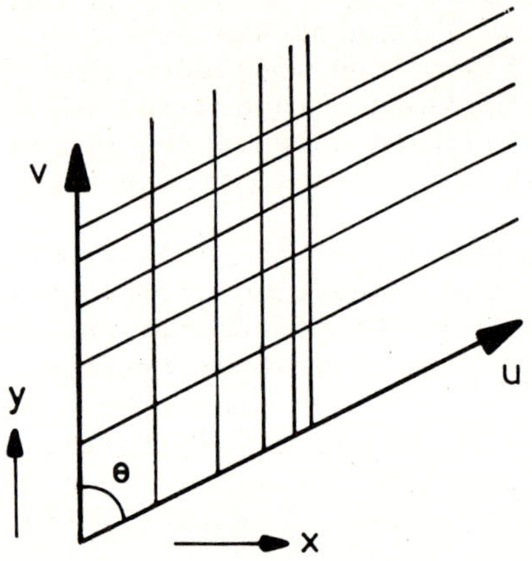

Fig. 4.6 Non-orthogonal mesh

The approximation of f_x then reduces to

$$\tilde{f}_x = f_x + \frac{1}{2}(-(\Delta u)^2 x_{uu}) f_{xx} + \frac{1}{2}(\Delta v^2 y_{vv} \frac{y_v}{x_u}) f_{yy} + \frac{1}{2}(-(\Delta u)^2 \frac{y_u}{x_u}) f_{xy} + \ldots =$$

$$= f_x - \frac{1}{2} \Delta u^2 x_{uu} f_{xx} + \frac{1}{2}(\Delta v^2 y_{vv} f_{yy} - \Delta u^2 u_{uu} f_{xy}) \cot\theta + \ldots , \quad (4.9)$$

since $y_u/x_u = \cot\theta$. The first term of the truncation error occurs even when the mesh is orthogonal, Eq.(4.7). The last two terms appear with departure from orthogonality. When $\theta < 45$ deg. they are of the same order as those from the non-uniform spacing. We conclude that moderate departure from orthogonality is acceptable when the rate of change of grid spacing is gradual. If the skewed mesh is uniform, the truncation error due to non-orthogonality vanishes. Because gradients are usually largest near body boundaries, it may be best to make the mesh orthogonal there.

Apart from formal accuracy is the matter of the diffraction of waves as they pass from a region of small mesh spacing to one of larger spacing. The analysis of Browning et al.[15] indicates that if their wavelength cannot be resolved on the coarse grid then distortion takes place. Recent results obtained with locally refined grids suggest that the question of diffraction caused by grid-size variation does not seem to be critical.

4.9 Multi-Block Meshes

A pattern of grid points ordered regularly, including those on the boundaries, simplifies the treatment of boundary conditions and reduces the overall computational work because it lays the ground-work for efficient communication between a cell and all its neighbors in the mesh. As we discuss later, this has important consequences for actual computations using computers having vector or parallel architectures. The smoothest and most regular patterns of grid point are those, according to Eiseman[13], that result from smooth coordinate mappings. The simplest and most regular patterns in this class arise when the points are connected only along coordinate directions, e.g. Cartesian grids. They minimize diffractions because of their smoothness and provide a consistent structured pattern that allows the neighbors of a given point to be identified in relation to each other. And this information leads to more efficient computations because knowing it eliminates having to compute the identity of the neighbors (indirect addressing). For vector computing it means reduced movement of the data in order to achieve proper alignment of the elements (gather/scatter commands).

Opposite to all of this are the so-called non-structured grids that are commonly used with finite-element methods. A non-structured grid can be simply a collection of points with no special order to them. The computational cell needs not to be a hexahedron and the variation in size from one to another needs not to be smooth. But because of the lack of inherent regularity, its neighbors must be identified through the additional computational work of indirect addressing. In its favour, non-structured grids are easier to construct, and local regions of refinement can be inserted more naturally.

Consider the example of a simple wing-fuselage combination in Fig. 4.7. Single mappings exist that relate a point P in the physical domain, specified by its rectilinear coordinates x, y, z, to its image in the space of the computational data, identified there by the coordinates ξ, η, ζ of its Cartesian grid. Through such mappings or coordinate transformations, one can produce regularly ordered and smoothly varying grids most simply and naturally by selecting a uniform spacing of points along each of the three Cartesian coordinates of the computational space. The connectivity in the resulting pattern of points then is preserved under the transformation to its image in the physical space and thus yields the grid of points forming a tessellation of hexahedral cells.

The differing types of grid patterns may be categorized according to characteristic local irregularities in their structure, where a coordinate direction joins or departs from a

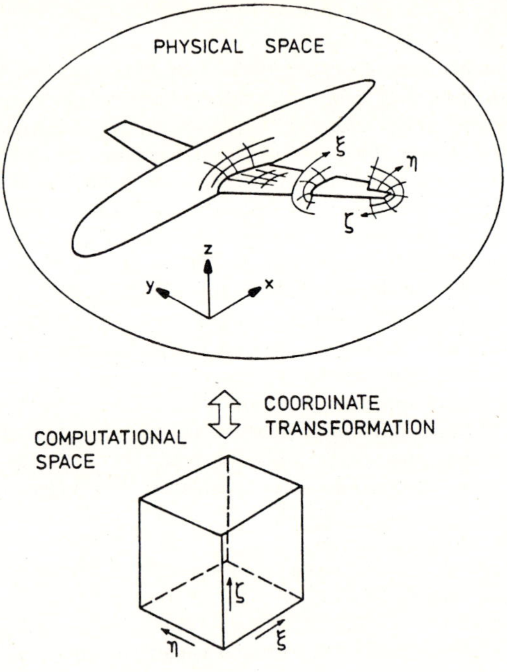

Fig. 4.7 Single global mapping of the physical space to a body-aligned computational coordinate space[16]

boundary or where two coordinate lines of the same family run together, i.e. a coordinate singularity. Figure 4.8 displays the tessellations most often used for the profile of an airplane wing. They can be seen as generalizations of the classical curvilinear or conformal coordinates. Providing the best resolution on thick rounded edges because they wrap around such features, the C and O types represent the entire profile by all or part of one computational coordinate line which may contain one or more singular points at sharp edges. The folding or wrapping of the transformation creates branch cuts which makes the mapped computational domain simply connected. The H type treats the image of the airfoil as a slit in a section of one computational coordinate. It yields a simply-connected grid but one with the poorest resolution for rounded leading or trailing edges. Related to it but with somewhat better resolution for blunt edges, the L type represents the airfoil as a combination of sections of computational surfaces belonging to different families. In so doing it forms a cavity in the grid, and the computational domain is multiply connected. Whereas the C and O tessellations are folded and contain singular points (lines in three dimensions), the H and L types create corners in the computational domain where none exist in its corresponding image in the physical space. These so-called fictitious corners may be seen as a form of coordinate singularity. Singularities, however, can be avoided entirely, but only by replacing the singular quadrilateral cells with some other type of polygon and thereby breaking the consistency of the pattern. In CFD the common preference is to retain consistency and accept the singularities.

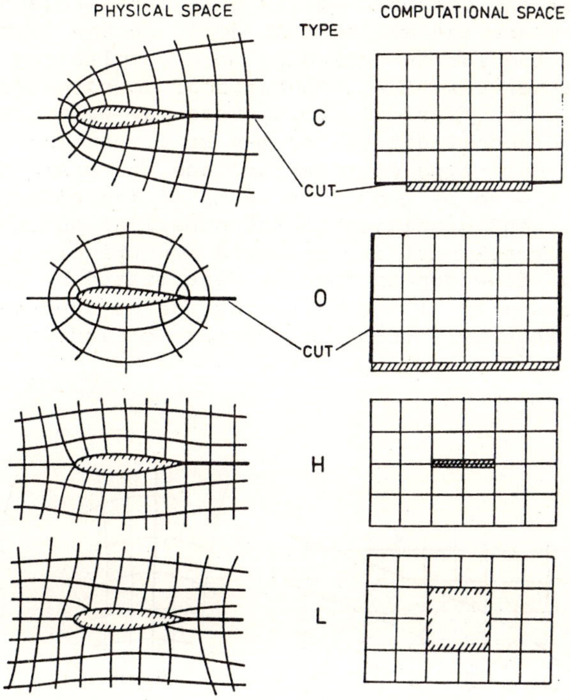

Fig. 4.8 Some of the common coordinate mapping types for grids around an airfoil[16]

In three dimensions Eriksson[4] describes the types of mappings suitable for fitting meshes to closed surfaces. If a consistent pattern of qudrilaterals is maintained for the surface mesh, one or more singularities in the mapping appear on the surface and persist out to the field mesh. Common types are polar and parabolic singular points. Eriksson[16] has analyzed the effect that these singular lines have in a numerical solution. He finds that a finite-difference scheme may lose stability at such a singularity, but the finite-volume scheme remains stable although it suffers a drop of about one-half order of accuracy.

The degree of geometrical complexity and detail that can be carried by the single global transformation in Fig. 4.8 is of course limited. To go beyond its limitations requires the flow domain be segmented into component sectors that together constitute the whole. The sectoring is usually done to produce a subdomain of the flow that a single coordinate transformation can represent adequately, thus achieving regularity and connectivity within the constituent or component grid for that subdomain. Each component grid shares all or part of a sector boundary with another component grid. The connectivity and smoothness in the pattern of points across the boundary depends on how the component grids are joined together. Two alternatives are the simple butt joint or the overlap joint (Fig. 4.9), the latter demanding more complicated interpolation. In either case the computational coordinates are discontinuous, but the physical coordinates may be smooth, may suffer metric discontinuities, or may be discontinuous themselves (Fig. 4.9). Each type of interface and its associated boundary conditions has to be judged in terms of its effect not only on the accuracy and stability of the numerical solution but also on the preservation of its conservation principle. Viewed as a whole, the flow domain then is discretized by an irregularly connected assembly of constituent grids or supercells, each of which is a regularly connected pattern of individual computational cells (Fig. 4.10).

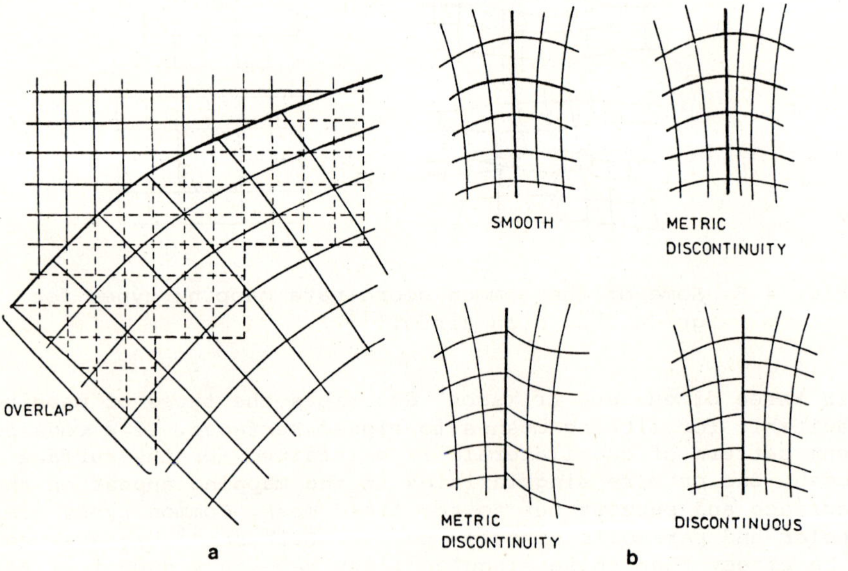

Fig. 4.9 Two types of supercell junctions: a) overlapping joint, and b) butt or patched joint

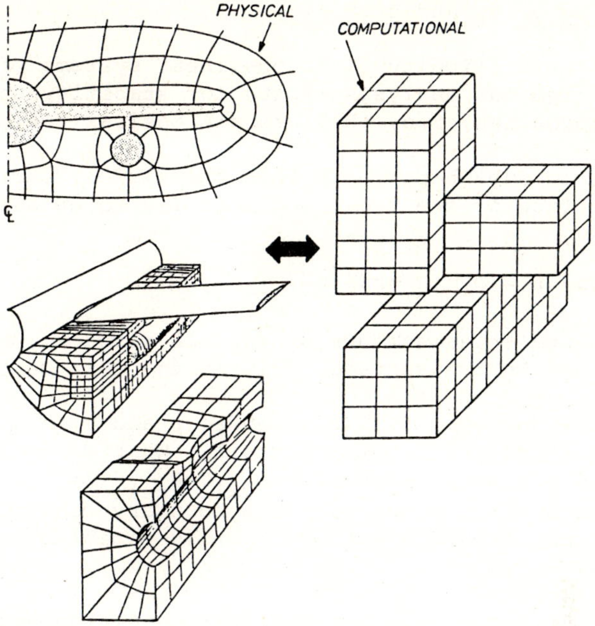

Fig. 4.10 Three constituent supercells comprising a boundary aligned mesh for wing-body-nacelle model

4.10 Boundary Conditions for the Discrete Problem

The O mesh mapping introduces three types of boundaries (see also the discussion in Sub-Chapter 4.6). We have proposed a concrete way in Sub-Chapter 4.6 to treat periodic, solid wall, and farfield boundary conditions (Fig. 4.11). Now we wish to look at these procedures more closely, particularly the accuracy and stability of the numerical or auxiliary boundary conditions.

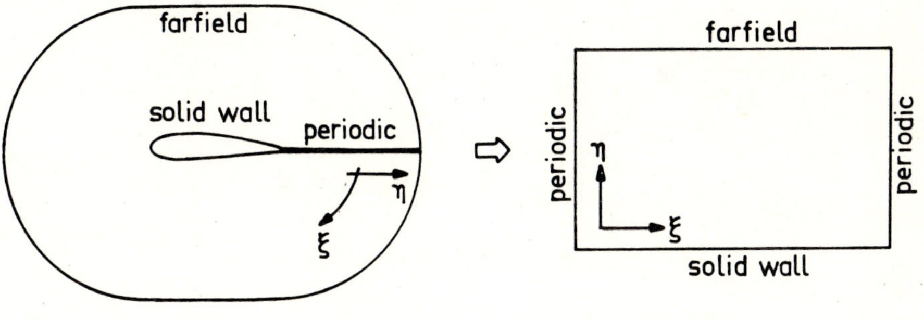

Fig. 4.11 Three types of boundaries introduced by the O mesh mapping

4.10.1 Accuracy and Stability

The central difficulty with non-linear systems of equations is that no rigorous analysis exists for stability and accuracy of these approaches, i.e. they are all basically empirical.

Recall from Sub-Chapter 4.6 that the airfoil surface is a streamsurface of the flow, and this led to an auxiliary condition

$$\underline{n}\cdot \text{grad } p = \rho \underline{V}\cdot(\underline{V}\cdot\text{grad})\underline{n}$$

valid on the streamsurface or airfoil surface. But we do not know values on the surface, only in the cell center.

So in Sub-Chapter 4.6.3 we expanded ρ and V in a Taylor series from the first cell to the airfoil. What we found then was that this auxiliary condition $P_{wall}=P_{cell}-\Delta\eta\ f(\rho,u,v_{cell})+O(\Delta\eta^2)$ is a first-order accurate boundary condition for a second-order accurate method. Is this satisfactory?

4.10.2 Empirical Rule for Boundary-Condition Accuracy

Boundary conditions can be one order of accuracy less than the discretization scheme used in the interior without degrading the overall second-order accuracy of the results!

This cannot be proved, but a heuristic argument, due to Kreiss, offers some motivation. Consider a wave disturbance propagating through an even mesh of spacing h (Fig. 4.12)

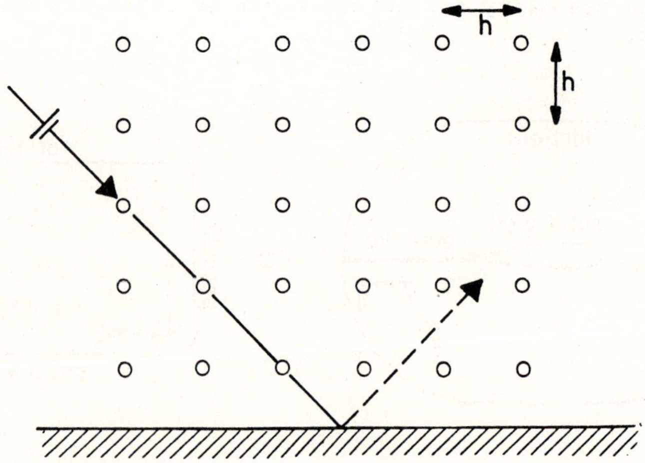

Fig. 4.12 Error in a wave reflecting from solid wall

At each grid point it undergoes an error of $O(h^3)$, but before it reaches the wall, it passes n grid points and the error accumulates to n $O(h^3)$. Now since $n \cdot h = O(1)$ by the time it arrives at the wall, the information content is only $O(h^2)$ and fully compatible with the boundary conditions $O(h^3)$.

Many years of computation also support the validity of this rule. Hence first-order boundary conditons are compatible with a second-order scheme.

4.10.3 Farfield Boundary Conditions

These are the most difficult because they are completely numerical i.e. the farfield is an artificial boundary. But on the other hand they are the most important for a hyperbolic wave propagation problem. A simple example can illustrate this point.

Example: Consider the linear one-dimensional propagation of a disturbance over a steady field i.e. the homogeneous equation

$$\frac{d}{dt} U + a \frac{dU}{dx} = 0 \text{ on } 0<x<1 \tag{4.10}$$

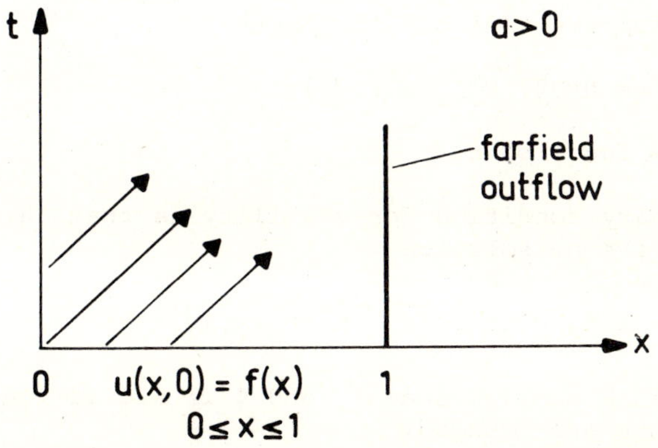

Fig. 4.13 Initial boundary value problem with outflow at x = 1

with inflow given at x=0, $u(o,t)=g(t)$, $t>o$. The function of the boundary condition at x=1 is to let the disturbance out. The essential question then is: does the energy norm (L2) of the disturbance decay? The answer is

$$\frac{d}{dt}\int_0^1 u^2 dx=\int_0^1 2u \frac{du}{dt} dx = -2a \int_0^1 u \frac{du}{dx} dx= -a\, u^2\Big|_0^1 = -a\, u^2(1)<0 \,, \tag{4.11}$$

i.e. the norm of the disturbance is diminishing. We see that its decay depends only on the boundary condition of the far-field.

This finding also motivates the use of Riemann invariants, or characteristic variables for the formulation of farfield conditions since they represent the uncoupled behaviour of the equation system.

But what if we chose simple extrapolation for the boundary condition? Is it stable? A rigorous theory for stability does exist for the scalar wave equation in 1-D, the so-called Gustafsson-Kreiss-Sundström (GKS) theory[17], which is an extension and improvement of the pioneering work by Godunov and Ryabenkii[18] on spectral theory for initial and boundary-value difference problems (see also Ref. 19).

The theory is very technical, but the above problem serves to illustrate it. Let U_j^n be a discrete approximation to Eq.(4.10). With an explicit scheme we have

$$U_j^{n+1} = G(U_{j-1}^n, U_j^n, U_{j+1}^n) , \qquad j=1,2 \;....\; J-1 \tag{4.12}$$

for some function G.

An additional boundary condition is required to determine U_J^{n+1}. Suppose this has the form

$$U_J^{n+1} = H(U_J^n,\ U_{J-1}^n,\ ....\ ) \tag{4.13}$$

for some function H.

A necessary condition for stability is that this problem should have no solution

$$U_j^n = Z^n \Phi_j \text{ with } |Z|>1, \text{ and } \sum_{j=1}^{\infty} |\Phi_j|^2 \Delta x < \infty \;.$$

Here Z is a complex scalar, and Φ is a grid function whose components are bounded.

To check for stability we seek a solution $U_j^n = \alpha\, Z^n \kappa^j$ where α, Z and κ are complex scalars. An essential condition for stability is that (4.12) should have no such solution with $|Z|>1$ and $|\kappa|=1$.

This is identical to the von Neumann condition for stability of the pure initial value problem. If this condition is satisfied then for $|Z|>1$, Eq.(4.12) has one solution $Z^n \kappa_1^j$ with $|\kappa_1|<1$ and a second solution $Z^n \kappa_2^j$ with $|\kappa_2|>1$. If $U_j^n = \alpha\, Z^n \kappa_1^j$ is substituted in Eq.(4.13) a necessary and sufficient condition for stability is that α be uniquely determined for all $|Z|>1$.

Some progress is being made in the extension of this theory to systems of linear equations in multiple dimensions. Otto and Thuné[20] have been applying it to the anlysis of boundary conditions used in the solution of the Euler equations by the centered Runge-Kutta scheme.

Example: Analysis of Boundary Conditions: Centered Finite Volume Scheme[20].

By decoupling the system of Euler equations, Otto and Thuné reduce the stability investigation to an analysis of the scalar problem (4.10). Applying the centered differencing scheme to it yields the one-stage form

$$u_j^{n+1} = (I + kaD_o + \frac{k^2}{2} a^2 D_o^2 + \frac{k^3}{4} a^3 D_o^3) u_j^n \; . \qquad (4.14)$$

Insertion of the normal mode ansatz $u_j^n = z^n w_j$ into the difference scheme (4.14) yields the resolvent equation

$$(z-1) w_j = (\lambda D_o + \frac{\lambda^2}{2} D_o^2 + \frac{\lambda^3}{4} D_o^3) w_j \; , \quad \lambda = \frac{ak}{2h} \; .$$

By assuming $w_j = \kappa^j$ in the resolvent equation they get the characteristic equation

$$(z-1)\kappa^j = (\lambda(\kappa-\kappa^{-1}) + \frac{\lambda^2}{2}(\kappa-\kappa^{-1})^2 + \frac{\lambda^3}{4}(\kappa-\kappa^{-1})^3) \; ,$$

where

$$z = 1 + \mu + \frac{\mu^2}{2} + \frac{\mu^3}{4} \; , \qquad \mu = \lambda(\kappa-\kappa^{-1}) \; , \qquad (4.15)$$

or

$$P(z,\kappa) \equiv (1-z)\kappa^3 + \lambda(\kappa^4-\kappa^2) + \frac{\lambda^2}{2}(\kappa^5-2\kappa^3+\kappa) + \frac{\lambda^3}{4}(\kappa^6-3k^4+3k^2-1) = 0 \; . \qquad (4.16)$$

For every fixed value of z equation (4.16) is a polynomial of degree 6, the roots of which determine the solution of the resolvent equation. The type of solution depends on whether or not the polynomial has multiple roots.

When the difference scheme is applied to the outflow boundary, complementary conditions are required at the grid points x_2, x_1, x_0. To design the numerical boundary conditions one extrapolates the characteristic variables. In this scalar case, they are simply u. Otto and Thuné[20] then insert the normal mode ansatz into these conditions, obtain the resolvent equation, and study the behavior of the roots k. They find that none of the roots can grow, and conclude that extrapolation is a stable condition.

4.11 References

1. Hirschel, E.H., Kordulla, W.: "Shear Flow in Surface-Oriented Coordinates". Vol.4 of Notes on Numerical Fluid Mechanics, Vieweg, Braunschweig/Wiesbaden, 1981.

2. Viviand, H.: "Conservative Forms of Gas Dynamic Equations". La Recherche Aerospatiale, No. 1974-1, 1974, pp. 65-68.

3. Rizzi, A.W.: "Computations of Rotational Transonic Flow". in Numerical Methods for the Computation of Inviscid Transonic Flow with Shocks, a GAMM Workshop, A.W. Rizzi and H. Viviand (eds.), Vol. 3 of Notes on Numerical Fluid Mechanics, Vieweg, Braunschweig/Wiesbaden, 1981.

4. Eriksson, L.-E.: "Generation of Boundary-Conforming Grids Around Wing-Body Configurations Using Transfinite Interpolation". AIAA J., Vol. 20, 1982, pp. 1313-1320.

5. Ni, R.H.: "A Multiple Grid Scheme for Solving the Euler Equations". Proc. AIAA 5th Computational Fluid Dynamics Conf., Palo Alto, 1981, Paper No. 81-0132, pp. 257-264.

6. Rossow, C.-C.: "Berechnung von Strömungsfeldern durch Lösung der Euler-Gleichungen mit einer erweiterten Finite-Volumen Diskretisierungsmethode". Doctoral thesis, Technical University Braunschweig, 1989, also DLR-FB89-38, 1989.

7. Kroll, N., Radespiel, R., Rossow, C.-C.: "Experience with Explicit Time-Stepping Schemes for Supersonic Flow Fields". NNFM Vol. 29, Vieweg, Braunschweig/Wiesbaden, 1990, pp. 252-261.

8. Hirschel, E.H., Groh, A.: "Wall Compatibility Conditions for the Solution of the Navier Stokes Equations". J. Comp. Phys, Vol. 53, No. 2, 1984, pp. 346-350.

9. Rizzi, A.W.: "Numerical Implementation of Solid-Body Boundary Conditions for the Euler Equations". ZAMM, Vol. 58, 1978, pp. T301-T304.

10. Engqvist, B., Majda, A.: "Absorbing Boundary Conditions for the Numerical Simulation of Waves", Math. Comp., Vol. 31, 1977, pp. 629-651.

11. Gottlieb, D., Gustafsson, B.: "On the Navier-Stokes Equations with Constant Total Temperature". Studies Appl. Math., Vol. 55, 1976, pp. 167-185.

12. Thompson, J.F.: "A Survey of Dynamically-Adaptive Grid in the Numerical Solution of Partial Differential Equations". Appl. Num. Math., Vol. 1, 1985, pp. 3-27.

13. Eiseman, P.R.: "Grid Generation for Fluid Mechanics Computations". Ann. Rev. Fluid Mech., Vol. 17, 1985, pp. 487- 522.

14. Thompson, J.F., Warsi, Z.U., Masten, C.W.: "Numerical Grid Generation - Foundations and Applications". North Holland, New York, 1985.

15. Browning, B., Kreiss, H.-O., Oliger, J.: "Mesh Refinement". Math Comp., Vol. 27, 1973, pp. 29-39.

16. Eriksson, L.-E.: "Transfinite Mesh Generation and Computer-Aided Analysis of Mesh Effects". Ph.D. Dissertation, Dept. Computer Science, Uppsala University, Sweden, 1984.

17. Gustafsson, B., Kreiss, H.-O., Sundström, A.: "Stability Theory of Difference Approximations for Mixed Initial Boundary Value Problem". II, Math. Comp., Vol. 2, 26, 1972, pp. 649-686.

18. Godunov, S.K., Ryabenkii, V.S.: "Theory of Difference Equations". Amsterdam, North Holland, 1964.

19. Sloan, D.M.: "On Boundary Conditions for the Numerical Solution of Hyperbolic Differential Equations". Intl. J. Num. Meth. Engr., Vol. 15, 1980, pp. 1113-1127.

20. Otto, K., Thuné, M.: "Stability of a Runge-Kutta Method for the Euler Equations on a Substructured Domain". Dept. of Scientific Computing, Report No. 109, Uppsala University, Uppsala, 1987.

V CENTERED DIFFERENCING

This chapter presents the construction of the approximation of the space derivatives by centered differences. The schemes must be able to treat discontinuities in the flowfield, i.e. shock waves or vortex sheets. Two ways to handle discontinuities are either to track the shock wave explicitly with an additional algorithm, or to capture it implicitly with the scheme.

Shock capturing is by far the most commonly applied technique today. The method is simple to program since a formula with the same structure is used all over the computational domain. No knowledge of the type and location of the discontinuities is in principal needed.

One important effort in the design of good capturing schemes goes into controlling the O(1) errors that occur around the discontinuities. If $q_t+F_x=0$ discretized by centered differences both in time and spaces, i.e. by the leap-frog method

$$q_j^{n+1} - q_j^{n-1} + \lambda\left(F(q_{j+1}^n)-F(q_{j-1}^n)\right) = 0 \ , \quad \lambda=\Delta t/\Delta x \ ,$$

the errors that are produced at a discontinuity will spread all over the computational domain. The propagation of these errors can be damped by adding artificial viscosity, as we shall see later.

The effects of artificial viscosity on such errors can be analyzed on linear models and have been studied in connection to errors in numerical boundary conditions. As a matter of fact linear problems are often worse than the non-linear ones. For non-linear problems dispersion is limited in spatial extent because characteristics run into the shock. Consequently, a balance is formed between the numerical dispersion and the dissipation at the shock.

Even if the errors are localized around the computed discontinuity the overall solution can be wrong, if the shock speed in the approximations is not correct. The speed in the differential equation case is determined by the Rankine-Hugoniot condition. This condition can be derived from integration by parts. It is important to design the schemes such that the corresponding discrete operation, summation by parts, can be performed. The scheme must be in conservation form. The derivative of the flux function, $(\partial/\partial x)F(q)$, must be discretized directly and not discretized in the form $F'(q)(\partial q/\partial x)$. For two level schemes the general conservation form in one space dimension is

$$q_j^{n+1}-q_j^n+\lambda\left(F(q_{j+r},\ldots,q_{j-r+1})-F(q_{j+r-1},\ldots,q_{j-r})\right)=0 \ .$$

If approximations in conservation form converge to piece-wise smooth functions, then these functions are weak solutions to the original differential equation. This is essentially the contents of an important theorem by Lax and Wendroff[1]. The conservation form induces a cancellation of errors and this is important for the speed of the discontinuity to be correct.

In Chapter III we mentioned that the weak formulation of the conservation law is somewhat too weak and allows for non-uniqueness in the form of unphysical expansion shocks. This may happen also for the discrete approximation. If e.g. in (3.5) $q_0(x)=Q_L, x<0$, $q_0(x)=Q_R, x>0$, $Q_L=-Q_R$, $Q_R>0$ the correct solution is

$$q = \begin{cases} Q_L & , \quad x<-t \ , \\ Q_R \, x/t & , \quad -t<x<t \ , \\ Q_R & , \quad x>t \end{cases}$$

with a rarefaction wave. The scheme (5.1) would produce

$$q = \begin{cases} Q_L & , \quad x<0 \ , \\ Q_R & , \quad x>0 \ , \end{cases}$$

which is a weak solution but not the correct one.

Discrete entropy conditions are needed in order to rule out unphysical solutions. One possibility is to impose a discrete analogue of the continuous entropy inequality[2]. Another is to let the artificial viscosity in the numerical scheme eliminate the possibilities of unphysical shocks.

The three global criteria discussed above, artificial viscosity, conservation form, and entropy conditions force the large errors in a capturing methods to be localized around a physically correct discontinuity at essentially the right location. The behaviour close to the discontinuity is different for different methods, and we shall look at this in more detail.

This chapter discusses flux-averaged methods. Chapter 6 presents upwind methods.

5.1 Flux-Averaged Methods

In the artificial viscosity method the capturing of shocks follows from a scheme that has a higher-order even derivative of a certain form as the leading term in its truncation error. This term then acts to damp the effects of the O(1) error at the shock as they spread into the smooth parts of the solution.

The schemes we wish to use within the artificial-viscosity approach are completely centered and do not explicitly use information about the direction of propagating signals. Gary[3] originally proposed a class of centered multistage schemes

$$q_j^{(0)} = q_j^n ,$$

$$q_j^{(1)} = q_j^n + \frac{1}{2} \lambda \left(F_{j+1}^n - F_{j-1}^n\right) + \frac{1}{2} \lambda \left(F_{j+1}^{(0)} - F_{j-1}^{(0)}\right) ,$$

$$q_j^{(k)} = q_j^n + \frac{1}{2} \lambda \left(F_{j+1}^n - F_{j-1}^n\right) + \frac{1}{2} \lambda \left(F_{j+1}^{(k-1)} - F_{j-1}^{(k-1)}\right) ,$$

$$q_j^{n+1} = q_j^{(k)} .$$

In the finite-volume notation, the flux $H_{j+1/2}$ at the face between cells j and j+1 is evaluated by averaging the values in the two cells

$$H_{j+1/2} = \frac{1}{2}(H_j + H_{j+1}) .$$

The centered flux difference for cell j then becomes

$$H_{j+1/2} - H_{j-1/2} = \frac{1}{2}(H_{j+1} - H_{j-1})$$

in complete analogy with the original suggestion of Gary[3].

During the early 1980s Jameson et al.[4], and Jameson[5] developed this concept into a broad class of centered multi-stage Runge-Kutta procedures suitable for problems in several space dimensions, which among others Rizzi and Eriksson[6] applied to the 3D problem of trans- sonic flow around a wing body combination. These schemes make no attempt to match the characteristic directions, but they benefit from the refinement of a new estimate of the non-linear function F on each stage, which at the same time increases the stability bound on the time step, if the optimum choice of weighting coefficients are used. Unlike the Lax-Wendroff schemes, the centered Runge-Kutta methods effectively uncouple the time integration from the space differencing, sometimes called the method of lines or the time-continuous approach. It has the desirable property that when a steady solution is reached, i.e. $q_j^{n+1} = q_j^n$, it is independent of the time step, and, as we shall see, the original steady difference operator is satisfied. But the scheme is non-dissipative, so numerical viscosity must be artificially added in order to capture shock waves.

The remaining sections deal with the stability of this method, and the construction of the artificial viscosity model.

5.2 Local Fourier Stability

We begin with the finite volume formulation in two dimensions for simplicity in a general quadrilateral mesh

$$\frac{\partial}{\partial t} \int_{\Omega_{jk}} q\, dvol + \iint_{\partial\Omega_{jk}} \underline{H}\cdot\underline{n}\, ds = 0 \quad . \tag{5.1}$$

In order to discretize (5.1) in space, we use the notation

$$\begin{aligned} &x_{j,k}=x(\xi_j,\eta_k),\; y_{j,k}=y(\xi_j,\eta_k),\; q_{j,k}=q(x_{j,k},y_{j,k},t),\\ &\underline{H}_{j,k}=\underline{H}(q_{j,k}) \; , \end{aligned} \tag{5.2}$$

and define the operators $D_\xi, D_\eta, \mu_\xi, \mu_\eta$ by

$$\begin{aligned} &D_\xi q_{j,k}=q_{j+1/2,k}-q_{j-1/2,k} \; , \qquad D_\eta q_{j,k}=q_{j,k+1/2}-q_{j,k-1/2},\\ &\mu_\xi q_{j,k}=1/2(q_{j+1/2,k}+q_{j-1/2,k}),\; \mu_\eta q_{j,k}=1/2(q_{j,k+1/2}+q_{j,k-1/2}). \end{aligned} \tag{5.3}$$

Now we approximate (5.1) by the Mean-Value theorem

$$\begin{aligned} &\frac{\partial}{\partial t} \int_{\Omega_{jk}} q\; dvol = \frac{d}{dt}\{q_{j,k}\cdot VOL_{jk}\} \; ,\\ &\iint \underline{H}\cdot\underline{n}\; ds = D_\xi \iint_{S_\xi} \underline{H}\cdot\underline{n}_\xi ds + D_\eta \iint_{S_\eta} \underline{H}\cdot\underline{n}_\eta ds = \\ &= D_\xi[(\mu_\xi\, \underline{H}_{j,k}\cdot)\underline{n}_\xi]_{jk} + D_\eta[(\mu_\eta\, \underline{H}_{j,k}\cdot)\underline{n}_\eta]_{jk} \; , \end{aligned} \tag{5.4}$$

which gives a semi-discrete centered scheme involving q at the mesh points (j,k),(j+1,k),(j-1,k),(j,k+1),(j,k-1) and x,y at the mesh points (j+1/2,k+1/2),(j+1/2,k-1/2),(j-1/2,k+1/2), (j-1/2,k-1/2). This means that the desired flow variables are given in the centers of the cells formed by the required x,y-coordinates.

It can be shown that the semi-discrete scheme defined by (5.1) and (5.4) is 2nd-order accurate if the mapping $x(\xi,\eta),y(\xi,\eta)$ is sufficiently smooth. Furthermore, the scheme is conservative and consistent, i.e. it is exact for $q_{j,k}$=constant. In Chapter 4 we mentioned that this scheme has many "good" properties in terms of stability and error at mesh singularities.

We study the stability of scheme (5.1) and (5.4) by a local Fourier analysis. It begins by assuming that the mapping is smooth locally, and that the metrics $\underline{S}_\xi\; \underline{S}_\eta$ are frozen at the values at the mesh point j, k, i.e.

$$\begin{aligned}
SIX_{j,k} &= \quad y_{j,k+1/2} - y_{j,k-1/2} \quad , \\
SIY_{j,k} &= -(x_{j,k+1/2} - x_{j,k-1/2}) \quad , \\
SJX_{j,k} &= -(y_{j+1/2,k} - y_{j-1/2,k}) \quad , \\
SJY_{j,k} &= \quad (x_{j+1/2,k} - x_{j-1/2,k}) \quad , \\
VOL_{j,k} &= \quad \text{area of cell } (j,k) \ .
\end{aligned}$$

The situation is depicted in Fig. 5.1.

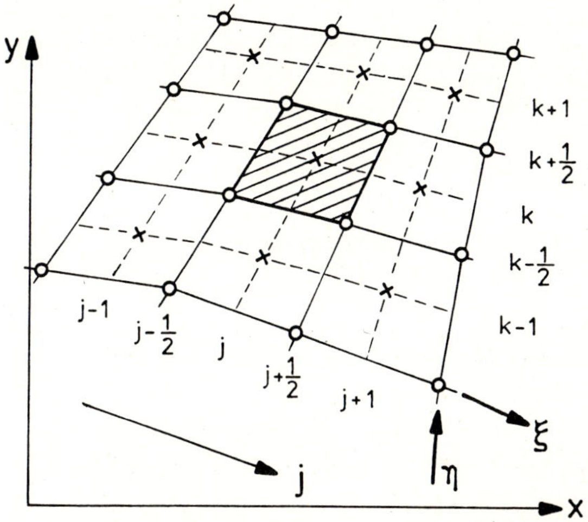

Fig. 5.1 Finite Volume Semi-Discretization

Next we discretize ξ and η uniformly according to $\xi_j=\xi_o+j\Delta\xi$, $\eta_k=\eta_o+k\Delta\eta$ and integrate (5.1) over the region $D_{j,k}=\{(\xi,\eta)\,|\,\xi_{j-1/2} \le \xi \le \xi_{j+1/2},\ \eta_{k-1/2} \le \eta \le \eta_{k+1/2}\}$.

The centered semi-discrete scheme becomes

$$\begin{aligned}
\frac{d}{dt}\, q_{j,k} + \frac{1}{VOL_{i,j}} \Big[& SIX_{j+1/2,k} F(\tfrac{1}{2}q_{j,k} + \tfrac{1}{2}q_{j+1,k}) + \\
& + SIY_{j+1/2,k} G(\tfrac{1}{2}q_{j,k} + \tfrac{1}{2}q_{j+1,k}) - \\
& - SIX_{j-1/2,k} F(\tfrac{1}{2}q_{j,k} + \tfrac{1}{2}q_{j-1,k}) - \\
& - SIY_{j+1/2,k} G(\tfrac{1}{2}q_{j,k} + \tfrac{1}{2}q_{j-1,k}) + \\
& + SJX_{j,k+1/2} F(\tfrac{1}{2}q_{j,k} + \tfrac{1}{2}q_{j,k+1}) + \\
& + SJY_{j,k+1/2} G(\tfrac{1}{2}q_{j,k} + \tfrac{1}{2}q_{j,k+1}) - \\
& - SJX_{j,k-1/2} F(\tfrac{1}{2}q_{j,k} + \tfrac{1}{2}q_{j,k-1}) - \\
& - SJY_{j,k-1/2} G(\tfrac{1}{2}q_{j,k} + \tfrac{1}{2}q_{j,k-1}) \Big] = 0 \ .
\end{aligned} \tag{5.5}$$

We then assume that the matrices are held constant over the mesh i.e. SIX, SIY, SJX, SJY, VOL are independent of j,k. If we also assume that $q_{j,k}$ varies smoothly on the given mesh, we may linearize the flux functions F(q), G(q) around $q_{j,k}$ according to

$$F(q)=f(q_{j,k}) + A(q-q_{j,k}) \ , \quad G(q) \cong g(q_{j,k}) + B(q-q_{j,k}) \ ,$$
$$A=F'(q_{j,k}) \ , \quad B=G'(q_{j,k}) \ , \tag{5.6}$$

where A,B are the Jacobian matrices evaluated at $q_{j,k}$. From (5.5) and (5.6) we then obtain the linear constant-coefficient local approximation

$$\begin{aligned} \frac{d}{dt}q_{j,k} + \frac{1}{2VOL}\,(SIX\cdot A + SIY\cdot B)(q_{j+1,k}-q_{j-1,k}) \,+ \\ + \frac{1}{2VOL}\,(SJX\cdot A + SJY\cdot B)(q_{j,k+1}-q_{j,k-1}) = 0 \ , \end{aligned} \tag{5.7}$$

which we can treat by Fourier analysis if we assume periodic spatial behaviour. We thus search for solutions to (5.7) of the type

$$q_{j,k}(t)=qe^{st}\kappa_1^{\,j}\kappa_2^{\,k}, \quad \kappa_1=e^{i\theta_1}, \quad \kappa_2=e^{i\theta_2}, \quad -\pi\leq\theta_1\leq\pi, \quad -\pi\leq\theta_2\leq\pi, \tag{5.8}$$

which gives us the eigenvalue problem

$$\left[sI- \frac{i\sin\theta_1}{VOL}(SIX\cdot A+SIY\cdot B) - \frac{i\sin\theta_2}{VOL}(SJX\cdot A+SJY\cdot B)\right]q = 0. \tag{5.9}$$

A non-trivial solution to (5.9) can only be found if the determinant of the matrix inside the outer brackets in (5.9) is zero. This condition can then be written

$$\det\left[sI-i\left(\sin\theta_1 \frac{SIX}{VOL}+\sin\theta_2 \frac{SJX}{VOL}\right)A-i\left(\sin\theta_1 \frac{SIY}{VOL}+\sin\theta_2 \frac{SJY}{VOL}\right)B\right]=0. \tag{5.10}$$

Writing $\alpha=\sin\theta_1 \frac{SIX}{VOL}+\sin\theta_2 \frac{SJX}{VOL}$, $\beta=\sin\theta_1 \frac{SIY}{VOL}+\sin\theta_2 \frac{SJY}{VOL}$ (5.11)

simplifies (5.10) to

$$\det\left[sI-i(\alpha A + \beta B)\right] = 0 \ . \tag{5.12}$$

Since the original conservation law (5.1) is hyperbolic, the matrices A and B by definition satisfy the condition

$$\det(\alpha A + \beta B - \lambda) = 0 \rightarrow \{\lambda_p(\alpha,\beta)\}_{p=1}^{n} \text{ all real} \ , \quad \alpha,\beta \text{ real} \tag{5.13}$$

which means that the eigenvalues $s_p(\theta_1, \theta_2)$ of (5.12) are all purely imaginary and given by

$$s_p(\theta_1, \theta_2) = -i\lambda_p(\alpha, \beta) \ , \quad p=1,\dots,n \tag{5.14}$$

for all components n of the system. Thus continuous-time transients will oscillate indefinitely, indicating that the centered difference is non-dissipative.

For the Euler equations the Jacobian matrices A and B are known analytically as well as the eigenvalues $\lambda_p(\alpha, \beta)$. This means that the local spectral radius at mesh point (j,k)

$$\rho_{j,k} = \max_{p,\theta_1,\theta_2} \left| s_p(\theta_1, \theta_2) \right| \tag{5.15}$$

is easily estimated by analytic means.

The eigenvalues $\lambda(C)$ of the linear combination $C=\alpha A+\beta B$ are

$$\lambda_1 = \gamma \ , \quad \lambda_2 = \gamma + \sqrt{\gamma^2+\delta c} \ , \quad \lambda_3 = \gamma - \sqrt{\gamma^2+\delta c} \ , \tag{5.16}$$

where $\gamma=\alpha u+\beta v$, $\delta=\alpha^2+\beta^2$.

Thus we have

$$s_1 = i\gamma \ , \quad s_2 = i(\gamma+\sqrt{\gamma^2+\delta c}) \ , \quad s_3 = i(\gamma-\sqrt{\gamma^2+\delta c}) \ ,$$

where

$$\gamma=(\sin\theta_1 \frac{SIX}{VOL} + \sin\theta_2 \frac{SJX}{VOL})u + (\sin\theta_1 \frac{SIY}{VOL} + \sin\theta_2 \frac{SJY}{VOL})v \ ,$$

$$\delta=(\sin\theta_1 \frac{SIX}{VOL} + \sin\theta_2 \frac{SJX}{VOL})^2 + (\sin\theta_1 \frac{SIY}{VOL} + \sin\theta_2 \frac{SJY}{VOL})^2 \ .$$

The Courant, Friedrichs, Levy-condition on stability is

$$\Delta t \cdot |s|_{max} \leq CFL \ , \tag{5.18}$$

where the numerical value of CFL is given by the particular time integration scheme, e.g. CFL=2.8 for the fourth-order Runge-Kutta method. Condition (5.18) scales the time step so that its product with the maximum eigenvalue is within the stability bound of the scheme i.e.

$$\Delta t \leq \frac{CFL}{|s|_{max}} \ . \tag{5.19}$$

This calls for an estimate on $|s|_{max}$. One such reasonable estimate is

$$|s|_{max} \leq \tilde{\gamma} + \sqrt{\tilde{\gamma}^2 + \tilde{\delta} c} \, ,$$

where
$$\tilde{\gamma} = \left(\left|\frac{SIX}{VOL}\right| + \left|\frac{SJX}{VOL}\right|\right)|u| + \left(\left|\frac{SIY}{VOL}\right| + \left|\frac{SJY}{VOL}\right|\right)|v| \, ,$$
$$\tilde{\delta} = \left(\left|\frac{SIX}{VOL}\right| + \left|\frac{SJX}{VOL}\right|\right)^2 + \left(\left|\frac{SIY}{VOL}\right| + \left|\frac{SJY}{VOL}\right|\right)^2 \, . \qquad (5.20)$$

Equations (5.19) and (5.20) define the stability condition for the finite volume scheme.

Under the same assumptions, but now in three dimensions, a similar Fourier analysis of scheme (5.7) indicates, and it is confirmed by actual numerical tests, that the stability limit on the step size is $\Delta t \leq CFL \min_{ijk}(\Delta t_1)$, where

$$\Delta t_1 = \frac{\Omega_{ijk}}{1/2\kappa(Q_I+Q_J+Q_K)+[1/4\kappa^2(Q_I^2+Q_J^2+Q_K^2)+c^2(S_I^2+S_J^2+S_K^2)]^{1/2}} \qquad (5.21)$$

with $Q_I=|\underline{V}\cdot\underline{S}_I|$, $Q_J=|\underline{V}\cdot\underline{S}_J|$, $Q_K=|\underline{V}\cdot\underline{S}_K|$, $c^2=1/2\kappa(2h_o-u^2-v^2-w^2)$.

5.3 Local Time-Step Scaling

On a mesh with widely varying mesh spacing in the physical domain it is well known that the numerical time-integration of semi-discrete schemes like (5.4) by explicit methods is difficult due to the corresponding variation of the local spectral radius. The maximum time step is usually determined by the smallest mesh spacing which means that the computational work needed to integrate the equations to a certain fixed time level can be excessive. If a steady solution is sought, the situation may be even worse. As we shall see from the study of the convergence to steady state, large variations in the mesh spacing can give rise to local eigensolutions which are virtually undamped.

For the computation of steady solutions it has been found that scaling the equations so that the local spectral radius becomes roughly uniform throughout the mesh is highly beneficial in terms of convergence to steady state. This concept is generally known as the "local time step" scaling. The "local time step" scaling of the original semi-discrete scheme (5.1) and (5.4) is now defined by

$$VOL_{jk}\, \rho_{jk} \frac{d}{dt} q_{jk} + D_\xi[(\mu_\xi \underline{H}_{jk}\cdot)\underline{n}_\xi] + D_\eta[(\mu_\eta \underline{H}_{jk}\cdot)\underline{n}_\eta] = 0 \, .$$

which evidently scales the problem so that the local spectral radius is unity everywhere. Clearly this type of scaling does not affect the steady soluton of the original scheme, but it does destroy the time accuracy of the calculation because the time integration now is not consistent with the original partial differential equation.

5.4 Artificial-Viscosity Model

The purely convective difference operator in (5.7) written now as $F_c(q_{jk})=-(AD_\xi+BD_\eta)q_{jk}$ suffers a number of drawbacks. It is well known, even for linear problems, when the boundary conditions are unable to prevent it, that centered-differences admit as a solution so-called sawtooth or plus-minus waves, i.e. waves with the shortest wavelength $L \sim 2\Delta X$ that the mesh can support. This just reflects the fact that the truncation error of these schemes is entirely dispersive and not dissipative. When the problem is non-linear there arises an aliasing phenomenon whereby short waves interact with each other, vanish, and reappear again as distorted long waves. But these defects in general could be dealt with satisfactorily by discrete filtering techniques if it were not for further deficiencies in the differential Euler equations themselves.

When shocks are to be captured in a non-linear flow field, the conservation equations admit non-unique weak solutions (see Chap. III), and an entropy condition has to be supplied in order to obtain the physically correct weak soluton[1]. A standard way to invoke an entropy condition is to model the true physical process inside a shock by the addition of a small viscosity term to the convective differences[7]. But even if the flow is smooth, the fundamental question of the existence and uniqueness of a steady-state solution to these equations is not fully answered. In non-linear transport there is a mechanism by which energy migrates from long-wavelength motion to progressively shorter and shorter scales until in reality it is removed from the flow by molecular viscosity.

The differential Euler equations possess no such viscosity, so that this energy will just pile up in the small scales. In the discrete representation this energy would migrate to the smallest scale resolvable on the mesh and then return transformed to large-scale motion via aliasing, which is clearly non-physical and would appear to make a steady state unattainable[8].

Within the context of the inviscid-flow equations, our best recourse against all of these deficiencies, albeit crude, is to attenuate waves more and more severely as their wavelength decreases, so that none migrate out and alias back, but in such a way as not to alter completely the inviscid character of the solution. This is the so-called artifical-viscosity model[9]. The idea of course is to mimic the short-wave dissipation by the real physical viscosity, and its justification is simply that in inviscid flow short-wave motion is of such low amplitude that whether removed or not it has no important effect on the overall flow character. In actual flow simulations this model is judged with a view to the crispness of shock profiles and the thinness of vortex sheets in weak solutions, and the amount of entropy produced or equivalently the variation in total pressure through regions of smooth flow.

We prefer to introduce artificial viscosity into our system at the same time level as the transport process by adding to it damping terms whose magnitude lies in or below the range of the truncation error of the discrete approximation[10]. Our total difference operator $F(q)$ therefore consists of: (i) the convective part $F_C(q)$ that results from discretizing the Euler equations in space by the centred finite-volume scheme, and (ii) the viscosity part $F_D(q)$. The semidiscrete approximation (5.4) can then be written

$$\frac{d}{dt} q_{jk} = F_C(q_{jk}) + F_D(q_{jk}) = F(q_{jk}) \; . \tag{5.22}$$

The total discrete artificial viscosity operator $F_D(q_{jk})$ includes its own artificial boundary conditions, described below, and comprises both linear and non-linear terms according to $F_D(q_{jk}) = g(q_{jk}) + Dq_{jk}$, where D is a constant matrix. The non-linear expression $g(q_{jk})$ is designed to provide dissipation at discontinuities, whereas the linear one is formulated to suppress spurious solutions (sawtooth waves) and to control the migration of energy from large to subgrid scales.

5.4.1 Non-Linear Artificial Viscosity

For all cells in the interior of the domain the non-linear artificial viscosity is expressed in three dimensions by

$$q_{ijk} = \chi\{\delta_I[s_I(q_{ijk})\delta_I] + \delta_J[s_J(q_{ijk})\delta_J] + \delta_K[s_K(q_{ijk})\delta_K]\}q_{ijk}, \tag{5.23}$$

where χ is a constant in the range 0 to 0.1, and s_I, s_J and s_K are coefficients that depend on the solution field through the pressure p according to $s_I \propto |\delta_I^2 p_{ijk}|$, $s_J \propto |\delta_J^2 p_{ijk}|$ and $s_K \propto |\delta_K^2 p_{ijk}|$. These coefficients are normalized by their maximum value so that their magnitudes lie between 0 and 1. Their purpose is to sense non-smooth flow and increase the filtering of large gradients so that in effect an entropy condition is enacted.

5.4.2 Linear Artificial Viscosity

Our model for linear artificial viscosity uses the fourth-difference operator

$$Dq_{ijk} = -\gamma(\delta_I^4 + \delta_J^4 + \delta_K^4)_{ijk} \qquad (5.24)$$

at all interior cells where γ is a constant in the range 0 to 0.02. The first non-linear term g_{ijk}, Eq.(5.23), acts as an entropy condition at shocks, whereas the second linear term Dq_{ijk} acts as a global filter to suppress spurious solutions like "saw-tooth" waves.

5.4.3 Boundary Conditions

At the boundaries of the computational domain, the semi-discrete scheme (5.22) must be supplemented by suitable boundary conditions. There are essentially three different types of boundary conditions for the Euler equations – periodic conditons, solid-wall conditions, and far-field conditions. Periodic conditions are the simplest and apply at coordinate cuts where two different parts of the boundary of the computational box are connected. The boundary conditions are then defined by the "connection" rules for the variables at these boundaries. Since the finite-volume scheme is "staggered" in the sense that the variables $q_{j,k}$ are defined in the centers of the mesh cells in the computational domain, there are no coinciding variables at coordinate cuts, only coinciding mesh cell walls. This means that the periodic conditions are easy to implement both for the convective terms and the artificial viscosity terms.

At solid walls outside of which q_{ijk} cannot be defined naturally, (5.22) must be modified by what we call artificial boundary conditions, and here enters a degree of arbitrariness that causes the results of one method to differ markedly from another. For the artificial viscosity terms in (5.22), we use special non-centered differences at solid walls that ensure a positive dissipation even in the cells next to these boundaries.

Eriksson[11] has found that the quadratic form q^THq, where $Hq=g(q)$ for fixed s_I, s_J and s_K, provides a useful guideline for the appropriate conditions at such boundaries (see also Ref.12).

The purpose of the total artificial viscosity operator is to drain off energy as time increases. How this is accomplished can be shown best by considering this locally linearized term Hq separate from the convective term. In the absence of F_C, system (5.22) behaves as $dq/dt=Hq$. Let us define the "energy" quantity $q^2=q^Tq$ of the discrete dependent variables. Since the time derivative of this energy should be negative, i.e. $dq^2/dt=2q^T(dq/dt)<0$, we determine the condition $q^THq<0$ for the quadratic form. Now if at boundaries with an artificial boundary condition, we simply set the corresponding sensors s_I, s_J, s_K in (5.23) to zero, we find that the quadratic form

$$
\begin{aligned}
q^THq &= \sum_{i=1}^{NI} \sum_{j=1}^{NJ} \sum_{k=1}^{NK} q_{ijk} H q_{ijk} = \\
&= -\chi \sum_{i=1}^{NI-1} \sum_{j=1}^{NJ} \sum_{k=1}^{NK} s_{I_{i+1/2,jk}} (q_{i+1,j,k} - q_{i,j,k})^2 - \\
&\quad -\chi \sum_{i=1}^{NI} \sum_{j=1}^{NJ-1} \sum_{k=1}^{NK} s_{J_{i,j+1/2,k}} (q_{i,j+1,k} - q_{i,j,k})^2 - \\
&\quad -\chi \sum_{i=1}^{NI} \sum_{j=1}^{NJ} \sum_{k=1}^{NK-1} s_{K_{i,j,k+1/2}} (q_{i,j,k+1} - q_{i,j,k})^2
\end{aligned}
$$

is always negative, and hence energy dissipates even in the boundary cells. We therefore believe this to be a good choice of artificial boundary condition for the non-linear artificial viscosity.

Consider next the linear fourth-difference artificial viscosity (Eq. 5.24). At boundaries we also must alter this expression by some suitable boundary procedure, and we seek to do so in a way that guarantees positive dissipation in all cells. Guided again by its quadratic form, we use no data outside the computational domain, but instead incorporate non-centered differences for the boundary cells together with scheme (5.24) at the interior ones in order to obtain the total discrete linear dissipative operator D with the property

$$
\begin{aligned}
q^TDq &= \sum_{i=1}^{NI} \sum_{j=1}^{NJ} \sum_{k=1}^{NK} q_{ijk} D q_{ijk} = \\
&= -T \sum_{i=2}^{NI-1} \sum_{j=1}^{NJ} \sum_{k=1}^{NK} (q_{i+1,j,k} - 2q_{i,j,k} + q_{i-1,j,k})^2 - \\
&\quad -T \sum_{i=1}^{NI} \sum_{j=2}^{NJ-1} \sum_{k=1}^{NK} (q_{i,j+1,k} - 2q_{i,j,k} + q_{i,j-1,k})^2 - \\
&\quad -T \sum_{i=1}^{NI} \sum_{j=1}^{NJ} \sum_{k=2}^{NK-1} (q_{i,j,k+1} - 2q_{i,j,k} + q_{i,j,k-1})^2 .
\end{aligned}
$$

Example

As an example we may consider the two-dimensional case[13] of the dissipative terms in (5.22) at j=1 and j=2:

$$[(\delta_\xi s_{Jj,k}\delta_\xi+\delta_\eta s_{Kj,k}\delta_\eta)q_{j,k}]_{j=1} =$$

$$= s_{J\ 3/2,k}(q_{2,k}-q_{1,k}) - s_{J\ 1/2,k}(q_{1,k}-q_{0,k}) + \qquad (5.25)$$

$$+ s_{J\ 1,k+1/2}(q_{1,k+1}-q_{1,k}) - s_{K,k-1/2}(q_{1,k}-q_{1,k-1}) ,$$

$$[-T(\delta_\xi^4+\delta_\eta^4)q_{j,k}]_{j=1} =$$

$$= T(-q_{-1,k}+4q_{0,k}-6q_{1,k}+4q_{2,k}-q_{3,k}) + \qquad (5.26)$$

$$+ T(-q_{1,k-2}+4q_{1,k-1}-6q_{1,k}+4q_{1,k+1}-q_{1,k+2}) ,$$

$$[-T(\delta_\xi^4+\delta_\eta^4)q_{j,k}]_{j=2} =$$

$$= T(-q_{0,k}+4q_{1,k}-6q_{2,k}+4q_{3,k}-q_{4,k}) + \qquad (5.27)$$

$$+ T(-q_{2,k-2}+4q_{2,k-1}-6q_{2,k}+4q_{2,k+1}-q_{2,k+2}) .$$

If the boundary j=1/2 is a solid wall, the variables $q_{0,k}$ and $q_{-1,k}$ cannot be defined in a natural way. We then replace (5.25) by

$$S_{J3/2,k}(q_{2,k}-q_{1,k})+S_{K1,k+1/2}(q_{1,k+1}-q_{1,k})-S_{K1,k-1/2}(q_{1,k}-q_{1,k-1}).$$

Eq.(5.26) is replaced by

$$T(-q_{1,k}+2q_{2,k}-q_{3,k})+ (-q_{1,k-2}+4q_{1,k-1}-6q_{1,k}+4q_{1,k+1}-q_{1,k+2}) ,$$

and finally (5.27) is replaced by

$$T(2q_{1,k}-5q_{2,k}+4q_{3,k}-q_{4,k}) +$$

$$+ T(-q_{2,k-2}+4q_{2,k-1}-6q_{2,k}+4q_{2,k+1}-q_{2,k+2}) .$$

To show that this local modification of the artificial viscosity terms ensures a positive dissipation, we evaluate the quadratic form

$$S = \sum_{j=1}^{\infty} \sum_{k=-\infty}^{+\infty} q_{j,k}^T D\, q_{j,k} , \qquad (5.28)$$

where D denotes the complete artificial viscosity operator. From (5.24), (5.25), (5.26), (5.27) and (5.28) we obtain after some manipulations

$$
\begin{aligned}
S= &-\sum_{j=1}^{\infty}\sum_{k=-\infty}^{+\infty} \alpha_{j+1/2,k}(q_{j+1,k}-q_{j,k})^T(q_{j+1,k}-q_{j,k})-\\
&-\sum_{j=1}^{\infty}\sum_{k=-\infty}^{\infty} \beta_{j,k+1/2}(q_{j,k+1}-q_{j,k})^T(q_{j,k+1}-q_{j,k})-\\
&-T\sum_{j=2}^{\infty}\sum_{k=-\infty}^{+\infty} (q_{j-1,k}-2q_{j,k}+q_{j+1,k})^T(q_{j-1,k}-2q_{j,k}+q_{j+1,k})-\\
&-T\sum_{j=2}^{\infty}\sum_{k=-\infty}^{+\infty} (q_{j,k-1}-2q_{j,k}+q_{j,k+1})^T(q_{j,k-1}-2q_{j,k}+q_{j,k+1})\leq 0 \ ,
\end{aligned}
\qquad (5.29)
$$

if $S_{Jj,k}\geq 0$, $S_{Kj,k}\geq 0$, $T\geq 0$. In other words, the modification of the viscosity terms at the boundary ensures that the overall viscosity operator is negative semi-definite. This type of boundary modification is obviously possible to implement at any boundary in the computational domain. The attractive feature of this artificial viscosity model, easily seen in (5.29), is that, if q_{ijk} is bilinear in i,j and k, the total operator D acting on q_{ijk} always returns zero, even at the boundaries. Compare this to the second-difference operator H that returns zero only if q_{ijk} is a constant.

At far-field boundaries the boundary conditions for the convective terms of the Euler scheme (5.22) are of non-reflecting type, derived by a local linearization and eigenvector decomposition procedure. The basic idea is that the ingoing characteristic variables (normal to the boundary) are set to their corresponding far-field values whereas the outgoing characteristic variables are obtained by extrapolation from the variables inside the boundary. With this procedure extra variables outside the boundary can be defined so that the centered finite-volume scheme can be applied at the interior points. The boundary conditions for the viscosity terms in (5.22) are here of the same type as for a solid wall.

5.5 Time Integration and Convergence to Steady State

With the above boundary conditions and artificial-viscosity model now included in $F=F_C+F_D$, our complete difference operator, the problem (5.22) we want to solve becomes, with spatial indices ijk suppressed, a unique system of non-linear ordinary differential equations

$$\frac{dq}{dt} = F(q) \ , \qquad q(0)= q^0 \ . \qquad (5.30)$$

For a given mesh size one can look upon this as a large system of ordinary differential equations, the so-called semidiscrete representation. Ultimately of course the problem must be solved in discrete time, but it is instructive to look at this form (5.30) first. Our goal is to integrate (5.30) forward in time until a steady state is reached, but without concern for time accuracy. The central issue to be addressed is the stability of the integration, not in the strict formal sense but in the more limited one of convergence to steady state, that is to say we do not attempt to establish a uniform stability bound on the time integration in the limit of finer and finer meshes. Instead, what we ask of any candidate used to march (5.30) to a steady state q^* for a given grid are two criteria: 1) that after many time steps the transients dq^*/dt become vanishingly small, and 2) that q^* satisfy the steady discrete operator $F(q^*)=0$ independently of the time step size.

One class of explicit integration technique is the three-stage two-step scheme in parameter $\theta=1/2$ or 1:

$$\begin{aligned} q' &= q^n + \Delta t F(q^n), \\ q'' &= q^n + \Delta t\left[(1-\theta)F(q^n) + \theta F(q')\right], \\ q^{n+1} &= q^n + \Delta t\left[(1-\theta)F(q^n) + \theta F(q'')\right] \end{aligned} \qquad (5.31)$$

over the discrete time step from q^n to q^{n+1}.

Equations (5.30) and (5.31) are non-linear, and it is possible to study their stability only after the complete operator F has been locally linearized. Think of q as some transient perturbation superimposed upon the steady state q^*. The linearization of (5.30) then leads to a homogeneous equation for the transients

$$\frac{dq}{dt} = Aq, \quad \text{where } A = \frac{\partial F}{\partial q} . \qquad (5.32)$$

Now q decays if all of the eigenvalues λ of A lie to the left of the imaginary axis. This is required for stability of the semi-discretization (5.30). The linearized fully discrete system for the perturbations q becomes

$$q^{n+1} = \{I+\Delta t\ A + \theta(\Delta t\ A)^2 + \theta^2(\Delta t\ A)^3\}q^n = Cq^n . \qquad (5.33)$$

Scheme (5.33) is stable if $\|C\|<1$, and a useful estimate for this is that the spectral radius $\rho(C)$ of C satisfies

$$\rho(C)<1 .$$

Thus the domain of absolute stability of this linear scheme is given by

$$D = \{\text{complex } z: \left|1+z+\theta z^2+\theta z^3\right|\}, \tag{5.34}$$

where $z = \Delta t\lambda$ and λ is complex.

It is easily verified that for $\theta=1/2$ D contains the imaginary axis between -2i and 2i and a certain region to the left of the imaginary axis. This means that the scheme is conditionally stable for our centered finite-volume approximation with a CFL-number of at most 2. The fact that D also contains a rather large region to the left of the imaginary axis means that there is some room for the damping caused by boundary conditions and artificial viscosity.

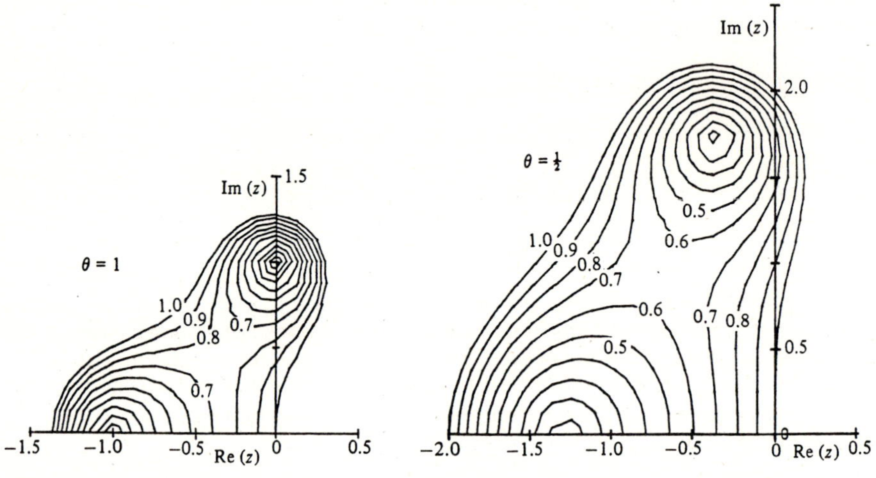

Fig. 5.2. Contours of constant amplitude of the amplification factor $|\sigma|$ of scheme (5.31)

Contour levels of D in intervals of 0.1 are plotted versus z in the complex plane of Fig. 5.2.

Since, if we neglect artificial viscosity, all λ are imaginary, we see that the CFL condition is encountered by the mode ($L=4\Delta X$) associated with the eigenvalue of largest modulus, and is satisfied, if $\Delta t|\lambda|_{max}$ is less than the CFL number, 2 for $\theta=1/2$ and 1.2 for $\theta=1$. But, except for very long waves or extremely short ones ($L\sim 2\Delta X$), all modes are temporally damped by the time integration, and to a greater degree as θ goes from 1/2 to 1. Compare this to the corresponding plot (Fig.

5.3) for the fourth-order time-accurate Runge-Kutta scheme, which offers significantly less temporal damping but a larger stability bound (CFL=2.8).

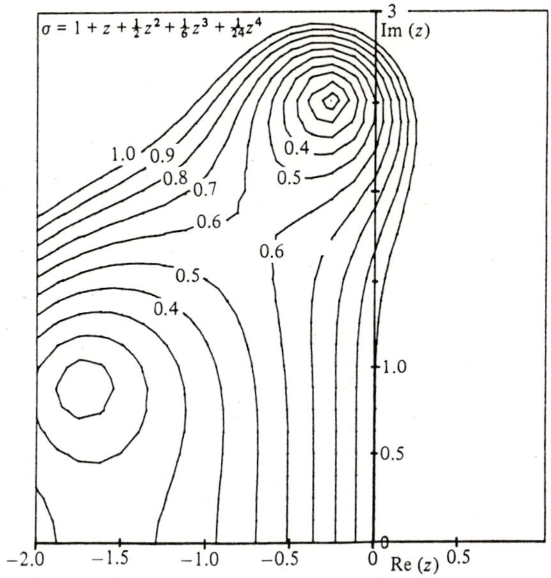

Fig. 5.3. Contours of constant amplitude of the amplification factor $|\sigma|$ of the fourth-order Runge-Kutta scheme

5.5.1 Steady-State Operator

After the flow has been marched forward by scheme (5.32) to a state where all time perturbations q cease, the question that remains is: does this state $q=q^{n+1}=q^n$, given by the integration scheme (5.32), satisfy the (linearized) steady operator $Aq=0$ identically? If all time variations are absent, the relation

$$q=\{I+\Delta t\ A+\ \theta(\Delta t\ A)^2 + \theta^2(\Delta t\ A)^3\}q$$

follows from (5.33), and hence $\Delta t[I+\theta\Delta t\ A+\theta^2(\Delta t\ A)^2]Aq=0$ holds. The answer then is yes, because the characteristic polynomial $1+\theta z+\theta^2 z^2$ of the bracketed term cannot be zero for any $z=\Delta t\ \lambda$ within the scheme's stability region, since substituting its roots $z=(-1\pm i\sqrt{3})/2\theta$ in (5.34) yields $\sigma=1$, which is just outside the stability bound. And we see that the solution to the discrete steady operator does not depend upon the time step used to reach it. This property of scheme (5.31) relies heavily on the fact that the same discrete operator is used in each of its stages. In schemes where this is not the case, for example the MacCormack scheme, the commutativity of the skewed forward and backward differences has to be considered, and the question becomes much more difficult to answer.

5.5.2 Eigenspectrum of Centered Schemes

The preceeding Fourier analysis tells us a good deal about the character of A under the simplifying assumptions of no boundaries and zero artificial viscosity, but we would like to know how it changes as these assumptions are gradually removed. Consider boundaries first, since the release of energy through them is known to play a crucial role in the convergence to steady state. Lomax, Pulliam and Jespersen[14] have taken this simplified linear one-dimensional example, added boundary conditions at each end, and obtained the discrete operator A. Since in one dimension the order of A is low, they were able to calculate the entire eigenvalue spectrum of A, and found that the effect of introducing boundary conditions is to shift the eigenvalues a small distance horizontally to the left of the imaginary axis. The boundary conditions therefore provide a very important damping of transients.

For the third step of our study we consider a much more realistic problem, the two-dimensional counterpart of (5.31), where the complete discrete operator F includes the boundary conditions and artificial viscosity described above and is formed on a non-uniform O-type mesh for transonic flow around an airfoil. The matrix A, the local linearization of F around the state obtained by scheme (5.32) after 15 steps from the free stream, is too large now to determine the full eigenvalue spectrum exactly, but by a Krylov subspace method we have been able to compute a signature of the spectrum[15].

Fig. 5.4a presents the approximate spectrum that we computed for A and confirms the expected shift to the left by the boundary conditions and artificial viscosity. But what is disturbing is the eigenvalue very close to the origin, not only because it is damped very slightly, but because it indicates that A is poorly conditioned. This situation is a direct result of the non-uniformity of the mesh, and can be alleviated by the well-known technique of advancing the solution with the local Δt_ℓ instead of the minimum time step. To demonstrate its effect we scale the matrix A by multiplying its ℓth row by Δt_ℓ and present the resulting spectrum in Fig. 5.4b. Notice that the smallest-modulus eigenvalue now is also shifted to the left, the overall condition is improved, and the discrete-time solution of (5.32) can be expected to decay to a steady state. Eriksson and Rizzi[15] further show that without artificial viscosity some eigenvalues move to the right of the imaginary axis and the computation fails to converge. Artificial viscosity is an essential feature of the numerical model for inviscid transonic flow.

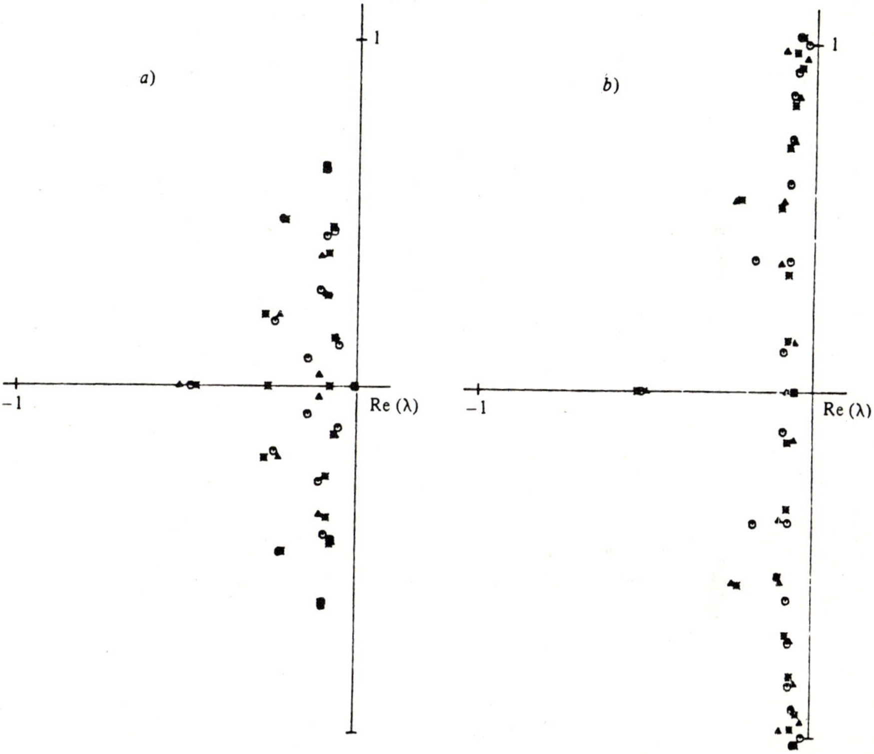

Fig. 5.4. Effect of local time-step scaling on spectrum of linearized system. 32×7 grid around NACA 0012 airfoil, $M_\infty = 0.8$, $\alpha = 0^\circ$, non-linear dissipation added. System linearized after 15 time steps with free-stream initial conditions, a) Unscaled system. b) Scaled system

5.6 References

1. Lax, P.D., Wendroff, B.: "Systems of Conservation Laws". Comm. Pure. Math., Vol.23, 1960, pp. 217-237.

2. Engqvist, B., Osher, S.: "One-Sided Difference Approximations for Nonlinear Conservation Laws". Math. Comp., Vol.36, 1981, pp. 321-351.

3. Gary, J.: "On Certain Finite Difference Schemes for Hyperbolic Systems". Math. Comp., Vol.18, 1964, pp. 1-18.

4. Jameson, A., Schmidt, W., Turkel, E.: "Numerical Solutions of the Euler Equations by Finite Volume Methods Using Runge-Kutta Time-Stepping Schemes". AIAA Paper 81-1259, 1981.

5. Jameson, A.: "The Evolution of Computational Methods in Aerodynamics". J. Appl. Mech., Vol.50, 1983, pp. 1052-1070.

6. Rizzi, A., Eriksson, L.-E.: "Transfinite Mesh Generation and Damped Euler Equations". AIAA Paper 81-0999, 1981.

7. MacCormack, R.W., Paullay, A.J.: "The Influence of the Compuational Mesh on Accuracy for Initial Value Problems with Discontinuous or Nonunique Solutions". Computers & Fluids, Vol. 2, 1974, pp. 339-361.

8. Lomax, H.: "Some Prospects for the Future of Computational Fluid Dynamics". AIAA J., Vol.20, 1982, pp. 1033-1043.

9. Pulliam, T.H.: "Artificial Dissipation Models for the Euler Equations". AIAA-Paper 85-0438, 1985.

10. Rizzi, A., Eriksson, L.-E.: "Computation of Flow Around Wings Based on the Euler Equations". J. Fluid Mech., Vol.148, 1984, pp. 45-71.

11. Eriksson, L.-E.: "Transfinite Mesh Generation and Computer-Aided Analysis of Mesh Effects". Ph.D. Dissertation, Dept. Computer Science, Uppsala Univ., Sweden, 1984.

12. Olsson, P.: "Flow Calculations Using Explicit Methods on a Data Parallel Computer". Report No. 117/1989, Uppsala Univ., 1989.

13. Eriksson. L.-E.: "Boundary Conditions for Artificial Dissipation Operators". FFA TN 1984-53, Stockholm 1984.

14. Lomax, H., Pulliam, T.H., Jespersen, D.C.: "Eigensystem Analysis Techniques for Finite-Difference Equations". AIAA-Paper No. 81-1027, 1981.

15. Eriksson, L.-E., Rizzi, A.: "Computer-Aided Analysis of the Convergence to Steady State of a Discrete Approximation to the Euler Equations". J. Comp. Phys., Vol.57, 1985, pp. 50-128.

VI PRINCIPLES OF UPWINDING

Since the Euler equations do only contain first derivatives in space and time they obviously have no terms expressing the presence of damping in time and space. So if we construct a numerical scheme for their solution we have to keep in mind the fact that a numerical error creeping somehow into the iterative solution process might grow over all bounds leading to the blow up of the scheme since it is not damped. The art of the program designer is to find a mean to incorporate numerical damping into the discrete approximation of the Euler equations which is small enough to reproduce the original equations as faithful as possible, but large enough to keep the course of iterations in a well ordered time evolution towards the steady state. One tool for this purpose is the addition of a higher derivative of the flow variables, multiplied by a suited coefficient, to each line of the Euler equations. This is called the artificial viscosity approach, and is the topic of the preceding Chapter V.

Another way for introducing the desired numerical damping is to leave the conventional approximation of a partial space derivative by a centered difference

$$f_{\xi i} = f_{i+1} - f_{i-1}$$

in favour of asymmetric differences

$$f_{\xi i} = f_{i+1} - f_i ,$$

or $f_{\xi i} = f_i - f_{i-1}$.

The first is a forward difference, the second a backward difference. If the coefficient multiplying the numerical difference analogue is used for the decision which of both possibilities has to be chosen, this technique is called upwinding. This latter feature can be relaxed by a weighted sum of both, forward and backward differences, the weights of which are determined by an analysis of the coefficients multiplying the original partial derivatives. This approach is called biased upwinding. In this chapter an introduction to both techniques is given.

6.1 Initial Considerations

Since it is very hard, if not impossible, to analyze the Euler equations as they stand mathematically, we seek a simpler equation which simulates at least the basic features of the Euler equations. For this purpose we first drop two of the space dimensions and simply write down the Euler equations in symbolic form

$$\dot{D}U + E_\xi = 0 \; . \tag{6.1}$$

This is called the model equation. As in the original equations E is assumed being a flux depending on the flow variable(s) U:

$$E = E(U) \; . \tag{6.2}$$

Whether U is assumed being a vector of more than one quantity or not is not important at present. In the latter case the model equation is called a scalar equation. In one dimension the Jacobian mapping determinant D boils down to the space increment Δx of the physical abscissa. As mentioned earlier, $\partial/\partial\xi$ is an undivided difference operator, whereby it is assumed that the spacing in the computational domain is equidistant. Better access to mathematical analysis is obtained by applying the chain rule of differentiation to the flux difference E_ξ:

$$\dot{D}U + E_U U_\xi = 0 \; . \tag{6.3}$$

Now the model equation is in quasi-linear form, with E_U being the Jacobian element or, as we will see later, the eigenvalue of the equation. Let us focus now our attention on the numerical solution of the model equation. For this purpose we try to solve the possibly simplest three-point Dirichlet boundary value problem in Fig. 6.1.

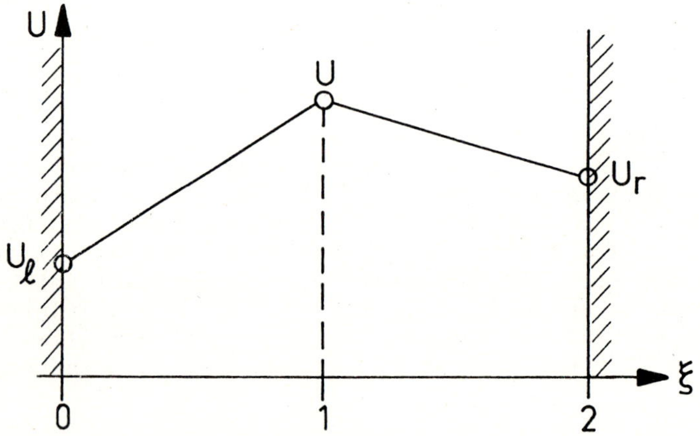

Fig. 6.1 Dirichlet boundary value problem for the model equation. U_ℓ and U_r are fixed boundary values, U is floating

The computational grid consists of three equidistantly distributed points. The points labelled "left" (ℓ) and "right" (r) carry boundary values U_ℓ and U_r, which are constant with respect to time. The unprimed function value U assigned to the point in the middle may float with respect to time. Our goal is to construct a numerical scheme with the following constraints

a) fulfill the model equation at its best,

b) stable course of iterations,

c) correct steady state solution for $t\to\infty$.

For this purpose we write down the manifold of all possible schemes in semi-discrete form. Since we have the choice using either the left difference $U-U_\ell$, or the right difference U_r-U, or an average of both, we represent the numerical difference approximating the space derivative by a weighted sum of both:

$$D\dot{U} + E_U[(U_r-U)a + (U-U_\ell)(1-a)] = 0 , \tag{6.4}$$

where "a" is a parameter controlling the preference of either the left or the right difference. Rewriting the equation gives rise to an ordinary differential equation for the function value in the middle of the computational interval

$$D\frac{dU}{dt} + E_U[U(1-2a) + U_r a + U_\ell(a-1)] = 0 , \tag{6.5}$$

where the Jacobian element $E_{\bar{U}}$ is calculated from the function value at the center point. The separation of variables gives

$$\frac{dU}{U(1-2a) + U_r a + U_\ell(a-1)} = -\frac{E_U dt}{D} .$$

The solution is

$$\ell n[U(1-2a) + U_r a + U_\ell(a-1)] = -\frac{1-2a}{D}\int E_U dt . \tag{6.6}$$

If the time interval of successive iterations is sufficiently small, the Jacobian element does not vary considerably and can be taken as a constant. For this case the solution is

$$U = \frac{1}{1-2a}[U_\ell(1-a) - aU_r + Ae^{-(1-2a)E_U t/D}] , \tag{6.7}$$

where A is a constant of integration. The first observation is that "a" may not be 1/2. From the exponent we read that the expression 1-2a should have the same sign as the Jacobian element E_U in order to prevent the solution from unbounded growth with respect to time (the denominator D is always positive).

The latter requirement can be expressed by

$$1-2a = b\,\frac{|E_U|}{E_U}\,, \qquad b>0\,, \tag{6.8}$$

where "b" is a new positive parameter. Rearrangement yields

$$a = \frac{1}{2}\left(1-b\,\frac{|E_U|}{E_U}\right)\,. \tag{6.9}$$

The solution now reads:

$$U = \frac{E_U}{2b|E_U|}\left[U_\ell\left(1+b\,\frac{|E_U|}{E_U}\right) - U_r\left(1-b\,\frac{|E_U|}{E_U}\right) + Ae^{-(|E_U|/D)t}\right]\,.$$

Having iterated the solution for a sufficiently long time, we finally can put $t=\infty$ in order to obtain the steady state solution

$$U(t=\infty) = \frac{1}{2}\left[U_\ell\left(\frac{E_U}{|E_U|b}+1\right) - U_r\left(\frac{E_U}{|E_U|b}-1\right)\right]\,. \tag{6.10}$$

Going back to the model equation we see immediately that the steady state is obtained when the partial time derivative approaches zero

$$\dot{U}(t=\infty) = 0\,, \tag{6.11}$$

which in turn has the consequence that the space derivative U_ξ should vanish. So the function U should be at least piecewise constant in space for the steady state. This in turn suggests that the parameter "b" should be unity:

$$b = 1\,,$$

such that the centered function value U equals either the left value U_ℓ, with U_r being simply ignored, or the right value U_r, with U_ℓ being ignored. The choice depends on the sign of the Jacobian element. With

$$a = \frac{1}{2}\left(1-\frac{|E_U|}{E_U}\right)$$

the upwind version of the model equation can be recast into

$$D\dot{U} + \frac{1}{2}\left[(U_r-U)(E_U-|E_U|) + (U-U_\ell)(E_U+|E_U|)\right] = 0\,, \tag{6.12}$$

which is known as the Courant/Isaacson/Rees (CIR) scheme[1].

6.2 Foundation of Upwinding

We are going now to give the idea of upwinding a more rigorous foundation. Starting again with the original model equation the time dependent update is given by

$$U(t+\Delta t) = U(t) + \int_t^{t+\Delta t} \dot{U}\, dt \,. \tag{6.13}$$

The integrand can be read directly from the model equation as the space derivative of the flux E devided by the Jacobian mapping determinant

$$U(t+\Delta t) = U(t) - \frac{1}{D} \int_t^{t+\Delta t} E_\xi dt \,. \tag{6.14}$$

Since we are interested only in steady state solutions on non-moving grids, the mapping determinant does not depend on time and has therefore been put outside the integral. We see immediately that we are stuck with a problem: Usually the space derivative of the flux is only known at the initial time level t, which is exactly at the lower bound of the interval of integration. The fundamental theorem of integration, however, says that the integrand must be known inside the limits of inte-

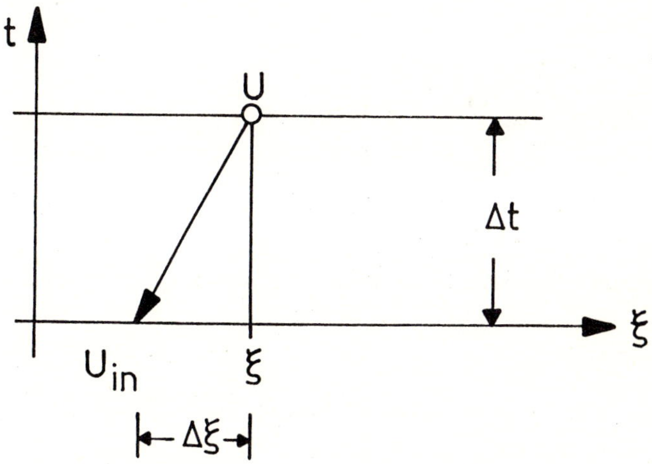

Fig. 6.2 Introduction of the concept of characteristics. The inclined line through the update point indicates the direction of the U-differentiation

gration for a stable numerical update. For example, the trapezoidal rule would require the integrand to be known at the time level $t+\Delta t/2$. So we need a device to propagate the integrand in time in order to perform a stable update without violating the fundamental theorem of integration. For this purpose we perform a linear backward Taylor expansion in the neighbourhood of the update point with coordinates $\xi, t+\Delta t$ to obtain the initial value labelled "in" at a point on the abscissa with coordinates $\xi-\Delta\xi, t$, see Fig. 6.2.

The backward Taylor series expansion reads

$$U_{in} = U - \dot{U}\Delta t - U_\xi \Delta\xi \ . \tag{6.15}$$

Upon the definition of a wave speed

$$\Delta\xi = \dot{\xi}\Delta t \ ,$$

we obtain after rearrangement

$$U = U_{in} + (\dot{U} + U_\xi \dot{\xi})\Delta t \ .$$

At this stage no term is known since neither the partial derivatives in time and space are known at the new time level nor the unprimed function value. Also the place where the initial value has to be taken from is not yet known, as well as the time step which will be specified later. However, from the model equation the time derivative can be read directly as

$$\dot{U} = -\frac{E_\xi}{D} = -\frac{E_U}{D} U_\xi \ , \tag{6.16}$$

such that we get

$$U = U_{in} + U_\xi(\dot{\xi} - \frac{E_U}{D})\Delta t \ . \tag{6.17}$$

Now the coefficient of the unknown space derivative can be made vanish by the definition of the wave speed being an eigenvalue

$$\dot{\xi} = \frac{E_U}{D} \ .$$

The result for the updated value is simply

$$U = U_{in} \ ,$$

where the initial value U_{in} has to be taken from the point with coordinates $\xi-\dot{\xi}\Delta t, t$. The line with the slope $\dot{\xi}^{-1}$ through the update point is called a characteristic.

For the numerical evaluation linear interpolation of the characteristic base point value labeled "in" comes immediately to our mind. The initial data are supposed to be given at equidistantly distributed points in the computational domain, Fig. 6.3.

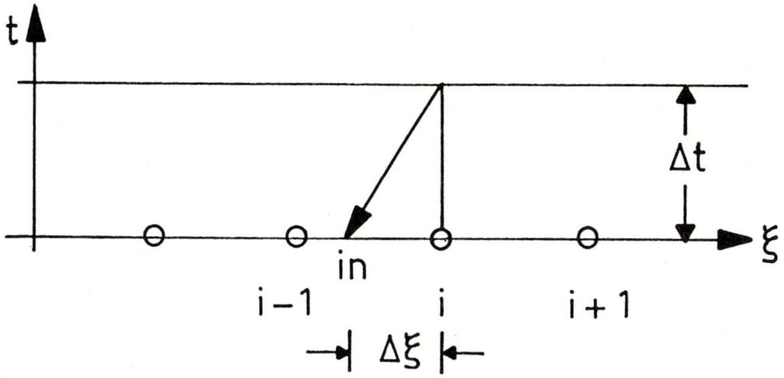

Fig. 6.3 Base point interpolation of initial values (in)

The spacing can be taken to be unity without loss of generality.

From Fig. 6.3 we see immediately that the time step has to be restricted such that

$$|\Delta\xi| \equiv |\dot{\xi}|\,\Delta t < 1$$

in order to make the interpolation point "in" ly inside the left (right) interval adjacent to the update point for $\dot{\xi}$ being a positive (negative) eigenvalue. This restriction is called the Courant-Friedrich-Lewy criterion. If the model equation is assumed being a vector equation then the absolute largest eigenvalue of the system of equations has to be taken for the time step specification.

$$\Delta t = \frac{CFL}{|\dot{\xi}|_{max}} \quad , \qquad CFL<1 \quad ,$$

where CFL is a dimensionless number. Since we use the time increment only as a vehicle to converge the solution to the steady state as fast as possible, the time step may vary from point to point.

The linear interpolation reads

$$U_{in} = U_i - \frac{\dot{\xi}_i \, CFL}{|\dot{\xi}_i|_{max}} \cdot \begin{cases} (U_i - U_{i-1}) , & \dot{\xi} > 0 \\ (U_{i+1} - U_i) , & \dot{\xi} < 0 \end{cases} , \tag{6.18}$$

which can be rewritten as the CIR scheme

$$U_{in} = U_i - \frac{CFL}{|\dot{\xi}_i|_{max}} \left[(\dot{\xi} + |\dot{\xi}|)_i (U_i - U_{i-1}) + (\dot{\xi} - |\dot{\xi}|)_i (U_{i+1} - U_i) \right], \tag{6.19}$$

with CFL<1/2 this time.

We understand that all right-hand side terms are taken at the present time level, while the left-hand side function value is the value at point i at the new time level. Furthermore from the dimensional analysis of the last equation we see that the geometric mapping determinant D occuring in the calculation of the eigenvalues cancels out. So D needs not to be calculated in a variable time step scheme. The technique with distributed time steps is called pseudo-unsteady method.

6.3 A Local Solution to the Model Equation

The preceeding investigations on the numerical solution of the model equation

$$\dot{U} + \dot{\xi} \, U_{\xi} = 0 \tag{6.20}$$

can be augmented by a characteristic coordinate transformation converting the partial differential equation to an ordinary differential equation. For this purpose the function U is assumed to depend on a new coordinate g

$$U = f(g) + C ,$$

where C is a constant. A local solution can be found by a linear two-dimensional ansatz for the coordinate g:

$$g = \xi - \dot{\xi} t .$$

For the constant coefficient case any function of g is a global solution of the model equation. If the eigenvalue varies both in space and time, any function of g is a local solution valid in a sufficiently small domain which includes the update point.

If the function f(g) is linear, we again obtain the CIR scheme mentioned previously. If the function f(g) is, say, a higher-order polynomial formal accuracy is increased both in space and time simultaneously. In the latter case the characteristics g=const. still are straight lines because the eigenvalue is kept constant.

6.4 Conservative Upwinding

Up to now only quasi-linear upwinding was considered. It turns out that the quasi-linear difference analogue of the governing model equation results always uniquely in the CIR scheme regardless which concept is used for its derivation. The only improvement conceivable would be the replacement of the simple first-order differences by higher-order differences. This issue will be discussed later.

Unfortunately the CIR scheme does not reflect the numerical conservation form, which should look like the following operator

$$\dot{U}_i + E_{i+1/2} - E_{i-1/2} = 0 \ .$$

In this case the summation of the algebraic difference equations over any arbitrary interval imin<i<imax makes all interior fluxes cancel, such that the total time variation of the function inside the interval depends only on the boundary fluxes, which is the condition of numerical conservation. This reflects numerically correct Stokes' integral. The CIR scheme

$$\dot{U} + \frac{1}{2}\left[(\dot{\xi}+|\dot{\xi}|)_i(U_i-U_{i-1}) + (\dot{\xi}-|\dot{\xi}|)_i(U_{i+1}-U_i)\right] = 0 \qquad (6.21)$$

however, is not in numerical conservation form, since it cannot be rearranged to a flux difference. With a slight modification the desired form can be obtained. For this purpose the subscript of the eigenvalue is dropped and the CIR flux difference regrouped:

$$\frac{1}{2}\left[(\dot{\xi}+|\dot{\xi}|)U_i+(\dot{\xi}-|\dot{\xi}|)U_{i+1}-(\dot{\xi}+|\dot{\xi}|)u_{i-1}-(\dot{\xi}-|\dot{\xi}|)U_i\right]=E_{i+1/2}-E_{i-1/2}. \qquad (6.22)$$

Now only the right flux

$$E_{i+1/2}= \frac{1}{2}\left[(\dot{\xi}+|\dot{\xi}|)U_i+ (\dot{\xi}-|\dot{\xi}|)U_{i+1}\right] \qquad (6.23)$$

needs to be considered since the left flux follows immediately by replacing the subscript i by i-1. The art of constructing a conservative scheme is reduced to indicating by suitable subscripts at which local position the eigenvalues should be calculated. This choice is, in contrast to the CIR scheme, not at all unique. Many schemes are possible among which there are some well known methods:

- $$E_{i+1/2} = [(\dot{\xi}+|\dot{\xi}|)_i U_i + (\dot{\xi}-|\dot{\xi}|)_{i+1} U_{i+1}] , \qquad (6.24)$$

which is a Steger-Warming type flux, Ref. 2.

- $$E_{i+1/2} = \frac{1}{2} [\dot{\xi}_i U_i + \dot{\xi}_{i+1} U_{i+1} + |\dot{\xi}|_{i+1/2}(U_i - U_{i+1})] , \qquad (6.25)$$

here the eigenvalue at i+1/2 is calculated from a symmetric average, for example the arithmetic mean of the function values at i and i+1. This kind of flux is used frequently by Harten[3], Yee (see e.g. Ref. 4) and Chakravarthy[5].

- $$E_{i+1/2} = \frac{1}{2} [(\dot{\xi}+|\dot{\xi}|)_{i+1/2} U_i + (\dot{\xi}-|\dot{\xi}|)_{i+1/2} U_{i+1}] , \qquad (6.26)$$

again the eigenvalues are calculated from some average of the initial data given at points i and i+1. This form is a linearized version of the Godunov scheme[6].

The previous schemes were derived from linear averages based on the identity

$$\dot{\xi} U \equiv \frac{E_U}{D} U .$$

There is another group of schemes which makes directly use of the functional relation of the flux $E = E(U)$:

- $$E_{i+1/2} = E\left(\frac{U_i + U_{i+1}}{2}\right) + \frac{|\dot{\xi}|_{i+1/2}}{2}(U_i - U_{i+1}) , \qquad (6.27)$$

which is a semi-homogeneous flux with an explicitly added artificial viscosity required for stability.

- $$E_{i+1/2} = E\left(\frac{U_i + U_{i+1}}{2}\right) + \frac{1}{2} (|\dot{\xi}|_i U_i - |\dot{\xi}|_{i+1} U_{i+1}) , \qquad (6.28)$$

which again is a semi-homogeneous flux with an artificial viscosity term taken from the Steger/Warming type flux.

- $$E_{i+1/2} = E\{\frac{1}{2} [U_i + U_{i+1} + \left(\frac{\dot{\xi}}{|\dot{\xi}|}\right)_{i+1/2} (U_i - U_{i+1})]\} , \qquad (6.29)$$

which is finally a fully homogeneous flux. This is the principal form of Godunov type schemes.

Two of the presented schemes are of outstanding importance: The Steger-Warming type flux difference generates a desirable positive contribution to the matrix diagonal at places where the eigenvalue changes from positive to negative, which always happens at shock waves. Thus sufficient diffusion is available for calculations with extremely strong captured shock waves. The other schemes do not share this property. Godunov type schemes do allow the most economic conversion to implicit formulations since in contrast to the other schemes the Jacobian matrix has to be calculated only at one place i+1/2 for each flux. Furthermore they do imply the exploitation of simplifications based on the homogeneous property of the Euler equations for cheap implicit solvers.

6.5 Accuracy of Three-Point Schemes

Up to now we have considered the class of lowest-order numerical schemes for the solution of the model equation. The flux difference was constructed by invoking the initial data of three points, the update point and two neighbours, one to the left and one to the right of the update point. It has been accepted in silence by the aerodynamic community to check the accuracy by the truncation error estimated by a Taylor series expansion of the quasi-linear discretized scheme in the computational space rather than in the physical space simply ignoring the usually non-equidistant distribution of discretization points in the latter. Assuming the eigenvalue being greater than zero the CIR scheme produces the scheme

$$U_i^{n+1} - U_i^n + \Delta t \dot{\xi}(U_i^n - U_{i-1}^n) = 0 \ ,$$

where n is the index of the time level:

n → initial or old data ,
n+1 → updated or new data .

Since we are interested only in the steady state result, ignoring the time accuracy of the evolution of the solution, the Taylor series expansion needs only be applied to the space difference, while the time difference needs not to be considered. The backward Taylor expansion about the point i reads

$$U_{i-1} = U_i - U_{\xi i} + \frac{U_{\xi\xi i}}{2} - \dots \ , \tag{6.30}$$

thereby assuming that the point spacing is unity. The numerical space derivative is

$$U_{\xi\ num} = U_i - U_{i-1} = U_{\xi i} - \frac{U_{\xi\xi i}}{2} . \qquad (6.31)$$

It can be seen that the scheme is only first-order accurate because the coefficient of the second derivative is different from zero. We see that a three-point scheme for the solution of the present model equation can be at best only first-order accurate. Later we will construct higher-order schemes by extending the support of the flux difference to more than three points.

6.6 Stability Considerations for Three-Point Schemes

The pseudo-unsteady CIR schemes reads for positive eigenvalues

$$U_i^{n+1} = U_i^n - \frac{CFL\ \dot{\xi}}{|\dot{\xi}|_{max}} (U_i^n - U_{i-1}^n) . \qquad (6.32)$$

Since the coefficient of the bracket is a dimensionless positive number less than unity the updated value is an asymetric mean of the initial values at the points i and i-1

$$U_i^{n+1} = U_i^n(1-a) + aU_{i-1}^n , \quad 0<a<1 .$$

Therefore the new value is bounded by the inequality

$$\max(U_i^n, U_{i-1}^n) > U_i^n > \min(U_i^n, U_{i-1}^n) .$$

No new extremum can be generated during the course of the pseudo unsteady iterations. Therefore convergence to the steady state without unbounded values is guaranteed. It is exactly this property which makes the classic first-order scheme so attractive for the derivation of higher-order schemes, which in general are constructed by adding high-order corrections in regions of monotonic distributed initial values. These correction terms are then switched off near extrema of the point function U in order to prevent an unbounded growth of the solution.

6.7 The Finite-Volume Cell-Face Concept

Here we try to translate the results found in the previous paragraphs to the full set of the Euler equations written in the discretized form for a finite volume. As we have seen this reduces to finding an asymetric average of the flow variables at the cell face separating two neighbouring finite volumes in order to form subsequently the fluxes there. For this purpose we consider a cell face which is, say a ξ=const. surface, separating at the position i+1/2,j,k the two finite volume cells located at i,j,k and i+1,j,k as shown in Fig. 6.4.

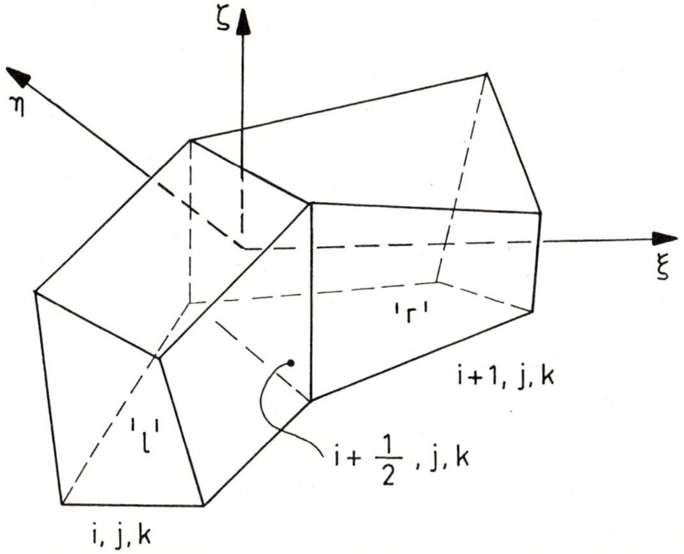

Fig. 6.4 Two adjacent finite volumes separated by a cell face at which the fluxes are to be calculated. Each cell carries a constant set of flow variables

Our goal is to find the flow variables at the cell face from an average of the left (right) flow variables carried by the cell i,j,k (i+1,j,k). In a higher-order than first-order scheme the flow varaiables named "left" ("right") need not necessarily be the flow variables labeled i,j,k (i+1,j,k). Later we will see how the "left ("right") state should be specified in order to obtain a high-order scheme. We assume that the "left" ("right") state is constant in either of both cells. This means that the flow variables form a step function in the cell face with respect to the abscissa ξ but are otherwise constant. Therefore the characteristic base-point interpolation simply reduces to a truth function which decides upon which of both function values "left" or "right" should enter into the averaging procedure. This can easily be visualized with the help of the following Fig. 6.5.

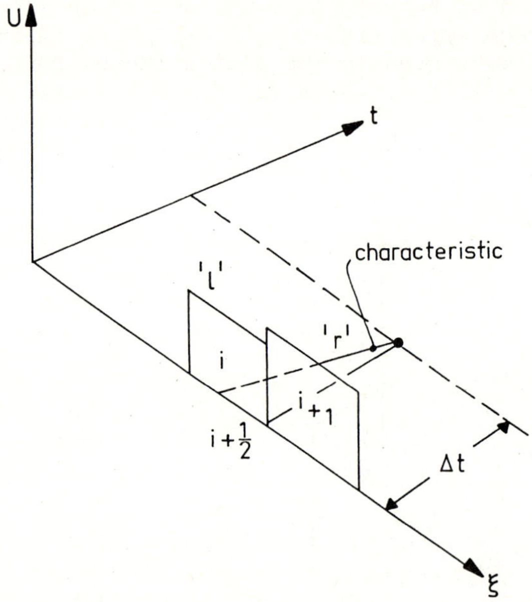

Fig. 6.5 Distribution of flow variables in a finite volume scheme. The characteristic points to the cell from which initial data are to be taken for the flux calculation

If the characteristic calculated from the eigenvalue using some average of the flow variables given at cell centers i and i+1 points to the "left" ("right") cell, then the "left" ("right") flow variable state vector is used for calculating the flux at the cell face considered. This is mathematically expressed by the truth function

$$U_{i+1/2}=(\rho,u,v,w,p)^T=\frac{1}{2}[U_\ell+U_r+\left(\frac{\dot{\xi}}{|\dot{\xi}|}\right)_{i+1/2}(U_\ell-U_r)]\ , \quad (6.33)$$

or, in order to avoid difficulties with the denominator, which may be zero

$$U_{i+1/2}=\frac{1}{2}\,[(1+\text{sign}\ \dot{\xi}_{i+1/2})U_\ell+(1-\text{sign}\ \dot{\xi}_{i+1/2})U_r]\ . \quad (6.34)$$

Finally a second observation should be considered: Since the flow variables distribution is constant in each of both cells, any uniquely calculated average in the cell face is all the same regardless of the position in that cell face. Therefore the average of the flow variables is not a function of the surface aligned tangential coordinates η and ζ. So the tangential operators

$$\frac{\partial}{\partial\eta}=\frac{\partial}{\partial\zeta}=0$$

do not produce any contribution to the numerical analogue of the Euler equations in the quasi-linear form.

The derivatives in ξ-direction, however, do not vanish. The operator

$$\frac{\partial}{\partial\xi} \neq 0$$

in the quasi-normal direction is a finite partial derivative the value of which is not easy to calculate in the case of a step function like the present. This will lead us in a natural way to the definition of a so called RIEMANN-problem of gas dynamics. The Euler equations look for the present problem, with the tangential derivatives dropped as mentioned, above like:

$$\begin{aligned}
&D\dot{\rho} + \rho_\xi \xi_o + \rho(u_\xi \xi_x + v_\xi \xi_y + w_\xi \xi_z) = 0 \; , \\
&\rho(D\dot{u} + u_\xi \dot{\xi}_o) + p_\xi \xi_x = 0 \; , \\
&\rho(D\dot{v} + v_\xi \dot{\xi}_o) + p_\xi \xi_y = 0 \; , \qquad\qquad (6.35) \\
&\rho(D\dot{w} + w_\xi \dot{\xi}_o) + p_\xi \xi_z = 0 \; , \\
&D\dot{p} + p_\xi \dot{\xi}_o + \gamma p(u_\xi \xi_x + v_\xi \xi_y + w_\xi \xi_z) = 0 \; .
\end{aligned}$$

For clearness it is repeated that D is the mapping determinant. Later we will see that it needs not to be calculated explicitly. $\dot{\xi}_o$ is the normal velocity multiplied by the cell face area. $\xi_{x,y,z}$ are the three Cartesian components of the cell face area vector calculated as usual with a finite-volume approach. The coefficients of the flow data multiplying the space derivatives are assumed as being taken from some average of the flow variables given at cell i and i+1, say the arithmetic mean.

The choice of the non-conservative equations for the analysis to follow is because they do allow a much simpler approach than the conservative equations. This is based on the consideration that the conservative equations are nothing but the result of a linear superposition of the non-conservative equations. So any linear analysis of the non-conservative equations can be transmitted with ease to the conservative equations.

6.8 The Riemann Problem at a Finite-Volume Cell Face

By inspecting the last set of equations we recognize some severe mathematical problems. Since in the finite-volume concept the cell face separates two constant sets of flow variables by a step function, the partial space derivative in the quasi-normal direction cannot be calculated since it is undetermined. The same also holds for the time derivative. It cannot be integrated to a differential representation for an updated new flow variable value since the initial values inside the step are undetermined. So the goal of solving a RIEMANN-problem like the present is to eliminate out of the Euler equations as well the partial time derivatives as the partial space derivatives and try to find a set of equations being free of derivatives which allow the evaluation of the flow variable vector exactly inside the step. The work for this purpose can be considerabely simplified by transforming the non-conservative Euler equations to the new variables speed of sound s and an entropy variable S.

The Euler equations for the latter set of variables have been already provided in Chapter II. They read with the tangential derivatives being dropped and using the notation of the last set of equations:

$$\begin{aligned}
&\frac{D}{A}\dot{s} + \lambda_o s_\xi + \frac{\gamma-1}{2} s(xu_\xi + yv_\xi + zw_\xi) = 0 , \\
&\frac{D}{A}\dot{u} + \lambda_o u_\xi + \frac{s^2 x}{\gamma-1}\left(2\frac{s_\xi}{s} - \frac{1}{\gamma} S_\xi\right) = 0 , \\
&\frac{D}{A}\dot{v} + \lambda_o v_\xi + \frac{s^2 y}{\gamma-1}\left(2\frac{s_\xi}{s} - \frac{1}{\gamma} S_\xi\right) = 0 , \\
&\frac{D}{A}\dot{w} + \lambda_o w_\xi + \frac{s^2 z}{\gamma-1}\left(2\frac{s_\xi}{s} - \frac{1}{\gamma} S_\xi\right) = 0 , \\
&\frac{D}{A}\dot{S} + \lambda_o S_\xi = 0 .
\end{aligned} \tag{6.36}$$

6.9 The Characteristic Derivative

The characteristic derivative is a device to eliminate the partial time derivatives out of the last set of the Euler equations. For this purpose we use the backward Taylor expansion of first order of the flow variables along a characteristic line with the components in space time being $(d\xi, dt)^T$. The initial values indicated by the subscript "in" are obtained from the yet unknown flow variables at the cell face by the following operation

$$U_{in} = U - \dot{U}\Delta t - U_\xi \Delta\xi , \tag{6.37}$$

where the partial time derivative and the partial space derivative are thought of being taken at the new level $t+\Delta t$. The function value U is also thought to be the new flow vector at the cell face, while the flow variables vector labeled "in" is the known initial flow variables vector which is taken either from the left state if the increment $\Delta\xi$ is positive, or from the right state if the increment $\Delta\xi$ is negative. Note that the interpolation problem is reduced to a simple left-right decision since the left and the right state is constant and therefore is not a function of the quasi-normal abscissa ξ. Fig. 6.6 clarifies the geometric situation.

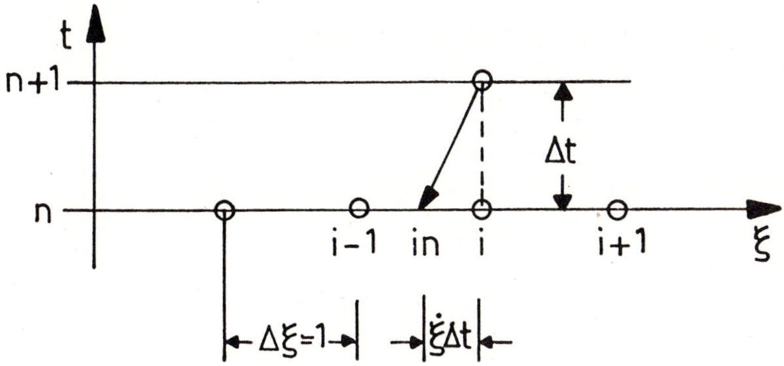

Fig. 6.6 Cut through two adjacent finite volumes in the ξ-t plane

The partial time derivative is easily obtained form the last equation:

$$\dot{U} = \frac{1}{\Delta t} (U - U_{in} - U_{\xi} \Delta\xi) . \tag{6.38}$$

The increment $\Delta\xi$ can be expressed by a speed multiplied by the time increment:

$$\Delta\xi = \dot{\xi} \; \Delta t .$$

From a dimensional analysis we find easily that $\dot{\xi}$ is not really a physical speed since its dimension is only an inverse time. In order to enter a physical wave speed with the proper dimension, we introduce by definition the wave speed λ via

$$\dot{\xi} = \lambda \frac{A}{D} .$$

The time derivatives of the flow variables are now

$$\begin{aligned}
\dot{s} &= \frac{s-s_{in}}{\Delta t} - \lambda \frac{A}{D} s_\xi , \\
\dot{u} &= \frac{u-u_{in}}{\Delta t} - \lambda \frac{A}{D} u_\xi , \\
\dot{v} &= \frac{v-v_{in}}{\Delta t} - \lambda \frac{A}{D} v_\xi , \\
\dot{w} &= \frac{w-w_{in}}{\Delta t} - \lambda \frac{A}{D} w_\xi , \\
\dot{S} &= \frac{S-S_{in}}{\Delta t} - \lambda \frac{A}{D} S_\xi .
\end{aligned} \tag{6.39}$$

Now the Euler equations look the following way

$$\begin{aligned}
&\frac{D}{A\Delta t}(s-s_{in}) + (\lambda_o-\lambda)s_\xi + \frac{\gamma-1}{2}s(xu_\xi+yv_\xi+zw_\xi) = 0 , \\
&\frac{D}{A\Delta t}(u-u_{in}) + (\lambda_o-\lambda)u_\xi + \frac{s^2x}{\gamma-1}(2\frac{s_\xi}{s} - \frac{1}{\gamma} S_\xi) = 0 , \\
&\frac{D}{A\Delta t}(v-v_{in}) + (\lambda_o-\lambda)v_\xi + \frac{s^2y}{\gamma-1}(2\frac{s_\xi}{s} - \frac{1}{\gamma} S_\xi) = 0 , \\
&\frac{D}{A\Delta t}(w-w_{in}) + (\lambda_o-\lambda)w_\xi + \frac{s^2z}{\gamma-1}(2\frac{s_\xi}{s} - \frac{1}{\gamma} S_\xi) = 0 , \\
&\frac{D}{A\Delta t}(S-S_{in}) + (\lambda_o-\lambda)S_\xi = 0 .
\end{aligned} \tag{6.40}$$

We see that all the time derivatives are replaced by ordinary differences. There remains to eliminate the partial space derivatives, which are undetermined at the cell face because of the step function in the flow variable vector.

6.10 The Scalar Invariant

Here we follow the method of RIEMANN which is mentioned in a book by Courant and Hilbert, Ref. 7, on page 313. Riemann provides a tool for eliminating unknowns out of a system of linear equations. The method also can be applied to our system of five equations, the unknowns of which are taken to be the five partial space derivatives

$$s_\xi,\ u_\xi,\ v_\xi,\ w_\xi,\ S_\xi .$$

The consideration applied is the following. Each line of our system of equations is zero. Therefore each line can be multiplied by an arbitrary coefficient without changing the system of equations. Since each line is still zero, also the sum of the equations multiplied by the set of coefficients a,b,c,d,e must be zero. Let us do now this intermediate operation:

$$\frac{D}{A\Delta t}\left[a(s-s_{in})+b(u-u_{in})+c(v-v_{in})+d(w-w_{in})+e(S-S_{in})+\right.$$
$$+\left[a(\lambda_o-\lambda)+\frac{2s}{\gamma-1}(bx+cy+dz)\right]s_\xi +$$
$$+\left[asx\frac{\gamma-1}{2}+b(\lambda_o-\lambda)\right]u_\xi +$$
$$+\left[asy\frac{\gamma-1}{2}+c(\lambda_o-\lambda)\right]v_\xi +$$
$$+\left[asz\frac{\gamma-1}{2}+d(\lambda_o-\lambda)\right]w_\xi +$$
$$+\left[e(\lambda_o-\lambda)+\frac{s^2}{(\gamma-1)}(bx+cy+dz)\right]S_\xi = 0 \,. \qquad (6.41)$$

If we can manage to make the coefficients of the partial space derivatives zero, the Euler equations will then be recast in ordinary difference form, containing no more undetermined partial derivatives:

$$a(s-s_{in})+b(u-u_{in})+c(v-v_{in})+d(w-w_{in})+e(S-S_{in}) = 0 \,. \qquad (6.42)$$

6.11 Characteristic Condition

The space derivatives contained in the scalar invariant can be eliminated by putting their coefficients zero:

$$a(\lambda_o-\lambda)+\frac{2s}{\gamma-1}(bx+cy+dz) = 0 \,,$$
$$asx\frac{\gamma-1}{2}+b(\lambda_o-\lambda) = 0 \,,$$
$$asy\frac{\gamma-1}{2}+c(\lambda_o-\lambda) = 0 \,, \qquad (6.43)$$
$$asz\frac{\gamma-1}{2}+d(\lambda_o-\lambda) = 0 \,,$$
$$e(\lambda_o-\lambda)-\frac{s^2}{(\gamma-1)}(bx+cy+dz) = 0 \,.$$

6.12 Eigenvalues and Invariants

Here we solve the last five algebraic equations for the coefficients and the still unknown wave speed λ. Putting tentatively

$$\lambda = \lambda_o ,$$

the coefficient b must be

$$b = -\frac{cy+dz}{x} .$$

This fulfills the first and the last line of Eq. 6.43. All the other equations are fulfilled if the coefficient a is put zero:

$$a = 0 .$$

The invariant for the zeroth eigenvalue is then

$$\begin{aligned} & c[x(v-v_o)-y(u-u_o)] + \\ & + d[x(w-w_o)-z(u-u_o)] + \\ & + ex(S-S_o) = 0 . \end{aligned} \tag{6.44}$$

where the subscript "in" has been replaced by the subscript "o", in order to indicate that this result is assigned to the zeroth eigenvalue. Since this equation is always true regardless of the choice of the coefficients c,d,e the entities they multiply must be zero individually. Therefore, we obtain three difference equations for the threefold zeroth eigenvalue:

$$\begin{aligned} & \lambda = \lambda_o : \\ & x(v-v_o)-y(u-u_o) = 0 , \\ & x(w-w_o)-z(u-u_o) = 0 , \\ & S-S_o = 0 . \end{aligned} \tag{6.45}$$

Another solution to the set of the five algebraic equations for the coefficients can be found the following way. The second, third and fourth equation is fulfilled by

$$b = a\,\frac{sx(\gamma-1)}{2(\lambda-\lambda_o)}\,,$$

$$c = a\,\frac{sy(\gamma-1)}{2(\lambda-\lambda_o)}\,, \qquad (6.46)$$

$$d = a\,\frac{sz(\gamma-1)}{2(\lambda-\lambda_o)}\,.$$

Inserting these values into the first line of Eq. 6.43 gives

$$a(\lambda_o-\lambda) + \frac{s^2a}{\lambda-\lambda_o}\,(x^2+y^2+z^2) = 0. \qquad (6.47)$$

Since $x^2+y^2+z^2=1$, because $(x,y,z)^T$ is the unit normal, and $a \neq 0$, a simple quadratic equation for the wave speed λ is obtained

$$(\lambda_o-\lambda)^2 = s^2\,.$$

Its solution is

$$\lambda_1 = \lambda_o + s\,,$$

and

$$\lambda_2 = \lambda_o - s\,.$$

For the first eigenvalue λ_1 the coefficients are

$$b = ax\,\frac{\gamma-1}{2}\,,$$

$$c = ay\,\frac{\gamma-1}{2}\,,$$

$$d = az\,\frac{\gamma-1}{2}\,, \qquad (6.48)$$

$$e = -\,\frac{as}{2\gamma}\,.$$

The invariant for the first eigenvalue is therefore

$$\lambda = \lambda_1 = \lambda_o+s: \qquad (6.49)$$

$$s-s_1 + \frac{\gamma-1}{2}[x(u-u_1)+y(v-v_1)+z(w-w_1)] - \frac{s}{2\gamma}(S_o-S_1) = 0\,.$$

Here use has been made of the solution for the zeroth eigenvalue $S=S_o$.

The second eigenvalue generates the coefficients

$$b = - ax \frac{\gamma-1}{2} ,$$

$$c = - ay \frac{\gamma-1}{2} ,$$

$$d = - az \frac{\gamma-1}{2} , \qquad (6.50)$$

$$e = - \frac{sa}{2\gamma} .$$

The invariant for the second eigenvalue is therefore

$$\lambda = \lambda_2 = \lambda_0 - s: \qquad (6.51)$$

$$s-s_2 - \frac{\gamma-1}{2}[x(u-u_2)+y(v-v_2)+z(w-w_2)] - \frac{s}{2\gamma}(S_0-S_2) = 0 .$$

We have now found five characteristic difference equations for the five flow variables at the cell face which are needed for the determination of the fluxes there. Our goal is now to solve these equations for the cell face flow variables.

6.13 A Simple Linear Riemann Solver

If we add the invariants associated with the first and the second eigenvalue, we obtain immediately the cell face speed of sound:

$$s = \frac{1}{2} \{s_1+s_2+ \frac{\gamma-1}{2}[x(u_1-u_2)+y(v_1-v_2)+z(w_1-w_2)] + $$
$$+ \frac{s}{2\gamma}(2S_0-S_1-S_2)\} . \qquad (6.52)$$

The entropy at the cell face is simply

$$S = S_0 .$$

There remains the evaluation of the cell face velocity components. The work for this purpose can be minimized by the introduction of two new quantities, which are defined by

$$r_1 = s_1 + \frac{\gamma-1}{2} [x(u_1-u_0)+y(v_1-v_0)+z(w_1-w_0)] +$$
$$+ \frac{s}{2\gamma}(S_1-S_0) , \qquad (6.53)$$

$$r_2 = s_2 - \frac{\gamma-1}{2} [x(u_2-u_0)+y(v_2-v_0)+z(w_2-w_0)] -$$
$$- \frac{s}{2\gamma}(S_2-S_0) . \qquad (6.54)$$

The speed of sound is then

$$s = \frac{1}{2}(r_1+r_2) .$$

The invariants associated with the first and the second eigenvalue become with the definition of r_1 and r_2:

$$\begin{aligned} s-r_1 + \frac{\gamma-1}{2}\left[x(u-u_o)+y(v-v_o)+z(w-w_o)\right] &= 0 , \\ s-r_2 - \frac{\gamma-1}{2}\left[x(u-u_o)+y(v-v_o)+z(w-w_o)\right] &= 0 . \end{aligned} \qquad (6.55)$$

After subtraction we arrive at

$$r_2-r_1 + (\gamma-1)\left[x(u-u_o)+y(v-v_o)+z(w-w_o)\right] = 0 .$$

With a little phantasy we see that the first two invariants associated with the zeroth eigenvalue are solved trivially by the ansatz

$$\begin{aligned} u &= u_o + xa , \\ v &= v_o + ya , \\ w &= w_o + za . \end{aligned} \qquad (6.56)$$

From the previous equation we find with the latter three definitions that the coefficient a must be

$$a = \frac{r_1-r_2}{\gamma-1} ,$$

where use has been made of the fact that the absolute length of the unit vector $(x,y,z)^T$ is one:

$$x^2 + y^2 + z^2 = 1 .$$

For forming the fluxes, the density and the pressure are needed. For this purpose the definition of the speed of sound and the entropy is repeated here for clearness:

$$p = e^S \rho^\gamma = \frac{\rho s^2}{\gamma} , \qquad (6.57)$$

from which the density follows to be

$$\rho = \left(\frac{s^2 e^{-S}}{\gamma}\right)^{\frac{1}{\gamma-1}} .$$

The entropy is evidently

$$S = \ln \frac{p}{\rho^{\gamma}} .$$

There is obviously a piece of arbitrariness contained in this simple Riemann solver:

The only coefficient occuring in the formulas is the speed of sound multiplying the differences of entropies. This coefficient can be calculated from any suitable mean of the left and right state, for example the arithmetic mean. Therefore this Riemann solver is called approximate rather than exact. The next paragraphs will deal with exact iteration-free Riemann solvers. The present result has been worked out to its full extend since it forms the basis of all the other Riemann solvers to follow.

Finally a remark is made on the quantities r_1 and r_2, which will also occur in the next paragraphs. These two quantities play an important role for the characteristic boundary condition formulation. If, say, we want to fulfill the solid body boundary condition at a cell face, then there the normal velocity must vanish:

$$\lambda_o = ux + vy + wz = 0 . \tag{6.58}$$

With the definition adopted above this means that

$$xu_o + yu_o + zu_o + a = 0 ,$$

or with the result for a being inserted

$$r_2 - r_1 = (\gamma-1)(xu_o + yv_o + zw_o) .$$

If the solid body is on the right side of the cell face then no flow variables are available for the quantity r_2 associated with the second eigenvalue, which points inside the body.

Therefore the last equation is solved for r_2:

$$r_2 = r_1 + (xu_o + yv_o + zw_o) .$$

In the other case, solid body on the left side of the cell face, the quation must be solved for r_1. After this manipulation r_1 and r_2 are used for the calculation of the speed of sound and the velocity components. The entropy formula remains untouched since it does not contain r_1 or r_2.

6.14 A Near Exact Riemann Solver

As mentioned above there is a piece of arbitrariness contained in the linear Riemann solver concerning the coefficient S. The particular choice of flow variables s, u, v, w, S makes it easy to remove this drawback. By inspection of the first equation of the previous paragraph it can be seen immediately that it can be solved in an exact way for the speed of sound. A simple rearrangement gives

$$s = \frac{s_1+s_2+\frac{\gamma-1}{2}\left[x(u_1-u_2)+y(v_1-v_2)+z(w_1-w_2)\right]}{2+\frac{1}{2\gamma}(S_1+S_2-2S_o)} \quad . \qquad (6.59)$$

The remainder of the Riemann solver remains just the same as the simple linear Riemann solver except that the above formula is used whenever the speed of sound is used as a coefficient. The present Riemann solver can be said to be exact in the sense that all characteristic equations are solved properly. It is one of the few requiring no iterations for this purpose.

There is a remark necessary concerning the uniqueness of the present Riemann solver. Although the solution is exact with the present definition of entropy any other definition of the latter indroduces a slight non-uniqueness. Instead of the quantity S defined above any convex function of a new argument, say B, can be used for the representation of the entropy:

$$S = f(B) \; .$$

The only condition to be imposed on the function f is convexity or in other words, if B increases then also the entropy must increase. The possible non-uniqueness consists in the twofold way an entropy increment can be calculated:

either

$$dS = df(B) \; ,$$

or

$$dS = \frac{df}{dB} \cdot dB = f'dB \; . \qquad (6.60)$$

While the first formula is an identity, the second leaves us with some latitude which comes from the definition of the function f.

An example enlightens the situation. Let us use the logarithm function for f:

$$S = \ln B \; .$$

As an example we want to form the entropy difference

$$S_1 - S_o = \ln B_1 - \ln B_o = \ln \frac{B_1}{B_o} .$$

This is an identity. But the other choice is also admissible

$$S_1 - S_o = dS = \frac{dB}{B} = \frac{B_1 - B_o}{B_o} . \tag{6.61}$$

The fact that the denomiator is taken from the zeroth eigenvalue state, comes from

$$S - S_o = \frac{B - B_o}{B} = 0 ,$$

from which follows that $B=B_o$ at the cell face. Both ways of forming the entropy differences are admissible. But the results are of course different. To remove this arbitrariness uniqueness properties can be imposed, say of the form that the Riemann solver should return identical results for different sets of flow variables. Working out such techniques is however beyond the scope of this book.

We suspect that all so called exact Riemann solvers do lack a clear uniqueness. The only exact and unique solver is for sure the isentropic Riemann solver described in the next paragraph.

6.15 The Isentropic Riemann Solver

If the flow is free of shocks and of any sources of vorticity, then the assumption of isentropy is justified. In this case the entropy is constant throughout the flow field (homentropic flow):

$$S = \text{const}, \quad dS = 0 .$$

The quantities r_1 and r_2 defined in the previous paragraphs assume then the simple form

$$r_1 = s_1 + \frac{\gamma-1}{2}[x(u_1-u_o)+y(v_1-v_o)+z(w_1-w_o)] ,$$

$$r_2 = s_2 - \frac{\gamma-1}{2}[x(u_2-u_o)+y(v_2-v_o)+z(w_2-w_o)] .$$

The speed of sound at the cell face is

$$s = \frac{1}{2}(r_1+r_2) ,$$

and the velocity components at the cell face are

$$u = u_o + xa ,$$
$$v = v_o + ya , \tag{6.63}$$
$$w = w_o + za ,$$

where a is

$$a = \frac{r_1 - r_2}{\gamma - 1} .$$

This Riemann solver is both exact and unique.

6.16 An Osher-Type Riemann Solver

Osher and Solomon[8] suggest a near exact Riemann solver, which resembles much the near exact Riemann solver mentioned in Sub-Chapter 6.14. Analyzing the derivation of their Riemann solver, it turns out that their choice of the integration path is nothing but an introduction of a new entropy function of the form

$$S = \ln B^n ,$$

where n is given a value such as to simplify the final result. The entropy increment is

$$dS = n \frac{dB}{B} .$$

Because of

$$S - S_o = \frac{n}{B} (B - B_o) = 0 ,$$

it turns out that trivially the condition holds

$$B = B_o .$$

The cell-face speed of sound is therefore

$$s = \frac{s_1 + s_2 + \frac{\gamma-1}{2} \left[x(u_1-u_2)+y(v_1-v_2)+z(w_1-w_2) \right]}{2 + \frac{n}{2\gamma} \left(\frac{B_1+B_2}{B_o} - 2 \right)} , \tag{6.64}$$

see also Sub-Chapter 6.14. With the particular choice of

$$n = 2\gamma ,$$

the result is simplified to

$$s = \frac{s_1 + s_2 + \frac{\gamma-1}{2}\,[x(u_1-u_2)+y(v_1-v_2)+z(w_1-w_2)]}{\frac{B_1+B_2}{B_o}} \,. \qquad (6.65)$$

From the definition

$$S = \ln \frac{p}{\rho^\gamma} = \ln B^{2\gamma} \,,$$

we see that B is

$$B = \left(\frac{p^{1/\gamma}}{\rho}\right)^{\frac{1}{2}} \,.$$

For completeness the entities r_1 and r_2 are rewritten for the new argument B:

$$r_1 = s_1 + \frac{\gamma-1}{2}[x(u_1-u_o)+y(v_1-v_o)+z(w_1-w_o)] - s\left(\frac{B_1}{B_o}-1\right) \,,$$

$$r_2 = s_2 + \frac{\gamma-1}{2}[x(u_2-u_o)+y(v_2-v_o)+z(w_2-w_o)] - s\left(\frac{B_2}{B_o}-1\right) \,. \qquad (6.66)$$

The cell face velocity components are as before

$$u = u_o + xa \,,$$

$$v = v_o + ya \,, \qquad (6.67)$$

$$w = w_o + za \,,$$

with

$$a = \frac{r_1 - r_2}{\gamma - 1} \,.$$

Engqvist and Osher[9] provide a tool to divide a flux difference over a cell face into three pieces each aligned with the three eigenvalues, the sign of which serves as an indicator to which side of the cell face each of the three individual fractions of the total flux difference should be assigned to. Let the flow be from left to right, then the flux difference across the cell face can be split into three parts

$$E_r - E_\ell = E_2 - E_1 = E_2 - E'' + E'' - E' + E' - E_1 \,.$$

The intermediate state indicated by a dash (') is assumed to be placed somewhere on the first characteristic, and the flux difference $E'-E_1$ is assigned to the first eigenvalue λ_1. The intermediate state indicated by a double dash (") is assumed to be placed somewhere on the second characteristic, and the flux difference E_2-E'' is assigned to the second eigenvalue. Finally the flux difference $E''-E'$ is associated with the zeroth eigenvalue. We see that the fluxes associated with the extreme eigenvalues are simply those calculated from either the left or the right state depending on the sign of the eigenvalues λ_1 and λ_2. So they can be assumed to be known from the flow variable vectors given by the left and right state. Our goal is now to find the flow quantities of the two intermediate states. They are defined by the following rules:

The intermediate state on the first characteristic is connected with the state labeled 1 by isentropy. The intermediate state on the second characteristic is connected with the state labeled 2 also by isentropy. So the entropies at the intermediate states are given by the initial values

$$B' = B_1 ,$$

$$B'' = B_2 .$$

The speeds of sound at the intermediate states must then be

$$s' = s \frac{B_1}{B_o} ,$$

$$s'' = s \frac{B_2}{B_o} ,$$

where the speed of sound at the cell face itself has been provided in the previous formula. From the condition that the normal speed is constant in between the two intermediate states, as postulated by Osher, we obtain the velocity vectors at the intermediate states directly as

$$u' = u'' = u ,$$

$$v' = v'' = v ,$$

$$w' = w'' = w ,$$

where u, v, w are given by the above formula. The invariance of the normal speed in between the two intermediate states includes the condition that the pressure must be constant in between the two intermediate states. This can be read from the Riemann invariants using the formula for the pressure increment derived previously:

$$dp = \frac{\rho s^2}{\gamma-1}(2\,\frac{ds}{s} - \frac{1}{\gamma}\,dS) = \frac{2\rho s^2}{\gamma-1}(\frac{ds}{s} - \frac{dB}{B}) = 0 \,, \qquad (6.68)$$

from which follows:

$$\frac{s''-s'}{s} - \frac{B''-B'}{B_o} = 0 \,.$$

Using the formulae provided above we see immediately the identity

$$B_2 - B_1 - (B_2-B_1) \equiv 0$$

being fulfilled trivially.

6.17 A Linear Riemann Solver Using Primitive Variables

Here we try to convert the simple linear Riemann solver to the primitive variables ρ, u, v, w, p. These variables might be more desirable since they are direct entries to the calculation of the conventional Euler fluxes. We start by developing the cell face density about the state of the zeroth eigenvalue

$$\rho = \rho_o+\rho-\rho_o = \rho_o+d\rho \,.$$

As was stated earlier, the differential of the density is

$$d\rho = \frac{\rho}{\gamma-1}\,(2\,\frac{ds}{s} - dS) \,, \qquad (6.69)$$

such that the cell face density can be written as

$$\rho = \rho_o + \frac{\rho}{\gamma-1}\,[2\,\frac{s-s_o}{s} - (S-S_o)] \,. \qquad (6.70)$$

From the simple linear Riemann solver we know that

$$S = S_o \,,$$

and

$$s = \frac{1}{2}\,(r_1+r_2) \,,$$

from which follows

$$\rho = \rho_o + \frac{\rho}{(\gamma-1)s}\,(r_1+r_2-2s_o) \,.$$

Written out in full length the cell face density reads

$$\rho = \rho_o + \frac{\rho}{(\gamma-1)s}\{s_1-s_o-\frac{s}{2\gamma}(S_1-S_o) + \\ + \frac{\gamma-1}{2}[x(u_1-u_o)+y(v_1-v_o)+z(w_1-w_o)]+ \\ + s_2-s_o-\frac{s}{2\gamma}(S_2-S_o)+ \\ + \frac{\gamma-1}{2}[x(u_2-u_o)+y(v_2-v_o)+z(w_2-w_o)]\} . \qquad (6.71)$$

We see that in the brackets differences are contained the diffentials of which are

$$ds - \frac{s}{2\gamma} dS .$$

In order to evaluate them we remember the definition of the entropy:

$$S = \ln \frac{p}{\rho^\gamma} .$$

Therefore we obtain

$$dS = \frac{dp}{p} - \gamma \frac{d\rho}{\rho} .$$

The definition of the speed of sound is

$$s^2 = \gamma \frac{p}{\rho} .$$

Upon differencing both sides we have

$$2sds = \frac{\gamma}{\rho}(dp - \frac{p}{\rho} d\rho) , \qquad (6.72)$$

or

$$ds = \frac{\gamma}{2\rho s}(dp - \frac{p}{\rho} d\rho) .$$

With these findings the following expression can be formed

$$ds - \frac{s}{2\gamma}dS = \frac{\gamma}{2\rho s}(dp - \frac{p}{\rho} d\rho) - \frac{s}{2\gamma}(\frac{dp}{p} - \gamma\frac{d\rho}{\rho}) = \frac{\gamma-1}{2\rho s} dp . \qquad (6.73)$$

The cell face density is therefore

$$\rho = \rho_o + \frac{1}{2s^2}\{p_1-p_o+\rho s[x(u_1-u_o)+y(v_1-v_o)+z(w_1-w_o)] + \\ + p_2-p_o+\rho s[x(u_2-u_o)+y(v_2-v_o)+z(w_2-w_o)]\} . \qquad (6.74)$$

This result suggests the definition of new quantities R_1 and R_2 which can be written as

$$R_1 = \frac{1}{2s^2} \{p_1 - p_o + \rho s[x(u_1-u_o)+y(v_1-v_o)+z(w_1-w_o)]\} ,$$

and

$$R_2 = \frac{1}{2s^2} \{p_2 - p_o + \rho s[x(u_2-u_o)+y(v_2-v_o)+z(w_2-w_o)]\} .$$

The density is then

$$\rho = \rho_o + R_1 + R_2 .$$

The velocity components are as before

$$u = u_o + xa ,$$

$$v = v_o + ya ,$$

$$w = w_o + za ,$$

but the coefficient a assumes now the the form

$$a = \frac{s}{\rho} (R_1 - R_2) .$$

This can easily be verified by comparison of the new definition of R with the old one of r defined for the simple linear Riemann solver. There remains to evaluate the cell face pressure. From

$$S - S_o = dS = \frac{dp}{p} - \gamma \frac{ds}{\rho} = \frac{p-p_o}{p} - \gamma \frac{\rho-\rho_o}{\rho} = 0 , \qquad (6.75)$$

we find

$$p = p_o + s^2(\rho-\rho_o) = p_o + s^2(R_1+R_2) .$$

At this stage a remark is in order. The previous Riemann solver again is not exact and contains a piece of arbitrariness. We see that each line requires first the coefficients s^2 and ρs upon which subsequently the flow values are calculated. These coefficients are not directly available at the cell face and are therefore calculated from an average of the flow values given at the two adjacent cells. Usually the arithmetic mean is taken for this purpose. But this is not an unique choice. If the flow values in the adjacent cells are not to far apart from each other, or in other words, if the jumps in the step functions are sufficiently small, then any reasonable bounded mean value will do a good job. But if the jumps become large such as at shock waves, different Riemann solvers may generate different results at such places.

6.18 The Exact Non-Conservative Riemann Solver

In order to remove the non-uniqueness of the linear Riemann solver a simple remedy is to cycle more than once through the Riemann solution using the formulas of the previous paragraph. Starting with the arithmetic mean for the coefficients we obtain a first set of cell face variables from which we can calculate again the eigenvalues for the truth function evaluation and a new set of coefficients. This iteration is then repeated until the change of the cell face flow variables falls below some prescribed error bound. Studies with random generated left and right states revealed that usually fast convergence can be expected. If, however, the states are too disparate from each other, it may happen that the eigenvalues change sign from one iteration to another such that the iteration ends up in a limit cycle or even does not converge. In practical applications this often happens at hypersonic flow conditions in low pressure regions in the vicinity of shocks.

6.19 An Alternative Osher-Type Approximate Riemann Solver

This Riemann solver is very similar to that of Engqvist and Osher[9], except that the paths of integration along the first and the second characteristic are interchanged, see also Ref. 10. Figure 6.7 clears up the difference to the Engqvist-Osher approximate Riemann solver.

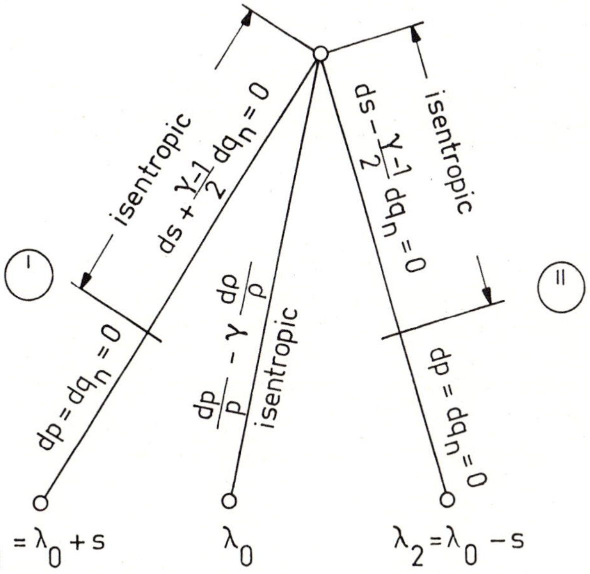

Fig. 6.7 Intergration paths for an alternate Osher-type Riemann solver

With all the knowledge of the previous paragraphs the integration can be written directly as (for the primes see Fig. 6.7):

$$p = p_o \left(\frac{\rho}{\rho_o}\right)^{\gamma} ,$$

$$\frac{2(s-s')}{\gamma-1} + x(u-u') + y(v-v') + z(w-w') = 0 ,$$

$$p' - p_1 = 0 ,$$

$$x(u'-u_1) + y(v'-v_1) + z(w'-w_1) = 0 , \qquad (6.76)$$

$$\frac{2(s-s'')}{\gamma-1} - x(u-u'') - y(v-v'') - z(w-w'') = 0 ,$$

$$p'' - p_2 = 0 ,$$

$$x(u''-u_2) + y(v''-v_2) + z(w''-w_2) = 0 .$$

Since $\frac{p'}{p} = \left(\frac{\rho'}{\rho}\right)^{\gamma} = \frac{p_1}{p_o}\left(\frac{\rho_o}{\rho}\right)^{\gamma}$,

the density of the left intermediate state is

$$\rho' = \rho_o\left(\frac{p_1}{p_o}\right)^{1/\gamma} , \quad \text{similarly} \quad \rho'' = \rho_o\left(\frac{p_2}{p_o}\right)^{1/\gamma} .$$

Therefore

$$s' = \left(\gamma \frac{p'}{\rho'}\right)^{0.5} = \left(\gamma \frac{p_1}{\rho'}\right)^{0.5} , \text{ and } \quad s'' = \left(\gamma \frac{p''}{\rho''}\right)^{0.5} = \left(\gamma \frac{p_2}{\rho''}\right)^{0.5} .$$

The cell face speed of sound is easily obtained as

$$s = \frac{1}{2}\{s'+s'' - \frac{\gamma-1}{2}[(u_2-u_1)x + (v_2-v_1)y + (w_2-w_1)z]\} . \qquad (6.77)$$

The density is

$$\rho = \rho_o\left(\frac{s}{s_o}\right)^{2/(\gamma-1)} .$$

The cell face velocity components are calculated similarly as in the previous paragraph:

$$
\left.\begin{array}{l} u = u_o + xa \\ v = v_o + ya \\ w = w_o + za \end{array}\right\} \quad a = \frac{2(s'-s)}{\gamma-1} + x(u_1-u_o) + y(v_1-v_o) + z(w_1-w_o) \ ,
$$

$$
\begin{aligned}
u' &= u + \frac{2x}{\gamma-1}(s-s') \ , & u'' &= u + \frac{2x}{\gamma-1}(s-s'') \ , \\
v' &= v + \frac{2y}{\gamma-1}(s-s') \ , & v'' &= v + \frac{2y}{\gamma-1}(s-s'') \ , \\
w' &= w + \frac{2z}{\gamma-1}(s-s') \ , & w'' &= w + \frac{2z}{\gamma-1}(s-s'') \ .
\end{aligned} \tag{6.78}
$$

6.20 Asymmetric Osher-Type Approximate Riemann Solvers

In the preceeding paragraphs we have found two symmetric arrangements of paths of integration. With a little phantasy we can now generate another two Osher-type Riemann solvers using asymmetric paths of integration. They are visualized in the following Fig. 6.8.

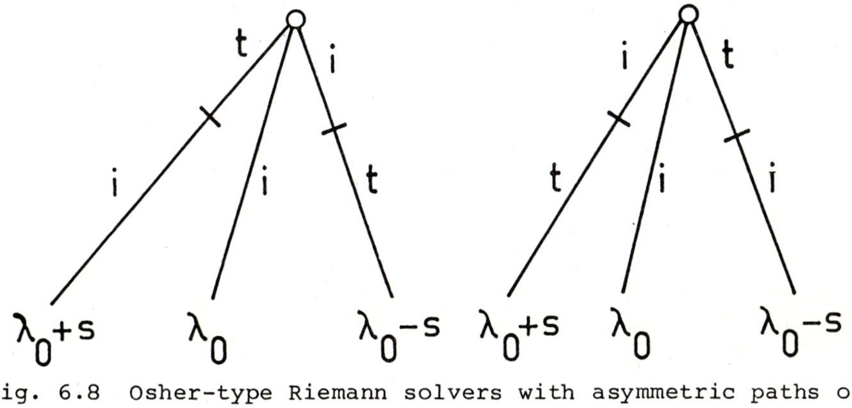

Fig. 6.8 Osher-type Riemann solvers with asymmetric paths of integration

The meaning of the letters i and t is "isentropic" and "trivial solution". The isentropic solution is characterized by

$$
p = \overset{*}{p}\left(\frac{\rho}{\overset{*}{\rho}}\right)^{\gamma} \ ,
$$

and the trivial solution by

$$
p = \overset{*}{p} \ ,
$$

and $\quad q_n = \overset{*}{q}_n \ ,$

where the star indicates a reference state and q_n is the normal speed

$$q_n = ux + vy + wz .$$

In order to make the two arrangements of paths of integration independent of the flow direction, a particular logic has to be incorporated into both present Riemann solvers which relates the arrangement of the leading and the trailing characteristic to the sign of the zeroth eigenvalue. Otherwise the solutions for the flow direction from left to right and that from right to left will be different. The evaluation of both Riemann solvers follows exactly the lines of the previous paragraphs.

6.21 A Linear Newton-Type Riemann Solver

Adding the difference equations for the Riemann invariants along the first and the second characteristic the following expression is obtained

$$p + \frac{\rho s}{2}[(u_2-u_1)x + (v_2-v_1)y + (w_2-w_1)z] = \frac{p_1+p_2}{2} . \tag{6.79}$$

The cell-face pressure can be obtained from the isentropic integration along the zeroth characteristic

$$p = p_o\left(\frac{\rho}{\rho_o}\right)^{\gamma} .$$

The coefficient ρs taken at the cell face is therefore

$$\rho s = \rho\left(\gamma\,\frac{p}{\rho}\right)^{1/2} = (\gamma\, p\rho)^{1/2} = \left(\gamma\, p_o\,\frac{\rho^{\gamma+1}}{\rho_o^{\gamma}}\right)^{1/2} . \tag{6.80}$$

Inserting these two expressions into the previous equation gives a non-linear equation for the cell-face density

$$\rho^{\gamma} + \frac{1}{2}\left(\gamma\,\frac{\rho_o^{\gamma}}{p_o}\right)^{1/2}[(u_2-u_1)x + (v_2-v_1)y + (w_2-w_1)z]\rho^{\gamma+1/2} -$$

$$- \frac{p_1+p_2}{2p_o}\,\rho_o^{\gamma} = 0 . \tag{6.81}$$

This equation can be rewritten as

$$f = \rho^{\gamma} + A\rho^{\gamma+1/2} - B = 0 ,$$

where A and B are local constants. A linear Taylor expansion of the function f around a value of the density in the neighbourhood of the expected exact solution gives an improved value of the density

$$\rho = \overset{*}{\rho} - \frac{\overset{*}{f}}{\overset{*}{f}_\rho} , \tag{6.82}$$

where $\overset{*}{f}$ is

$$\overset{*}{f} = \overset{*}{\rho}^{\gamma} + A\overset{*}{\rho}^{\gamma+1/2} - B ,$$

and $\overset{*}{f}_\rho$ is

$$\overset{*}{f}_\rho = \gamma\overset{*}{\rho}^{\gamma-1} + \frac{\gamma+1}{2} A\overset{*}{\rho}^{\gamma-1/2} .$$

The initial guess for the density is somewhat arbitrary and can be taken as the arithmetic mean of the left and right density

$$\overset{*}{\rho} = \frac{\rho_\ell + \rho_r}{2} .$$

Now, after the cell-face density is known, all other thermodynamic quantities are known. The velocity components can be obtained the same way as we did with the non-conservative linear Riemann solver.

6.22 A Quadratic Newton-Type Riemann Solver

The extension of the previous Riemann solver to a somewhat better accuracy is obtained by the quadratic Taylor expansion of the function f of the previous sub-chapter:

$$\overset{*}{f} + \overset{*}{f}_\rho \Delta\rho + \frac{\overset{*}{f}_{\rho\rho}}{2} \Delta\rho^2 = 0 , \tag{6.83}$$

with

$$\Delta\rho = \rho - \overset{*}{\rho} ,$$

and

$$\overset{*}{f}_{\rho\rho} = \gamma(\gamma-1)\overset{*}{\rho}^{\gamma-2} + \frac{\gamma^2-1}{2} A\overset{*}{\rho}^{\gamma-3/2} . \tag{6.84}$$

The quadratic equation has the solution

$$\rho = \overset{*}{\rho} \pm \left(\left(\frac{\overset{*}{f}_{\rho}}{\overset{*}{f}_{\rho\rho}} \right)^2 - \frac{2\overset{*}{f}}{\overset{*}{f}_{\rho\rho}} - \frac{\overset{*}{f}_{\rho}}{\overset{*}{f}_{\rho\rho}} \right)^{1/2} . \tag{6.84}$$

Since the density never is negative, the minus sign is discarded.

6.23 A Linear Conservative Riemann Solver

Since conservative methods usually require working with the conservative variables, the Riemann problem at the cell face should be solved also with the conservative rather than with the primitive variables (see also Sub-Chapter 2.7). Furthermore, if the homogeneous property of the Euler equations is exploited for particular methods it is a must that the entries to the Riemann solver are the conservative variables. So it is our goal here to convert the linear non-conservative Riemann solver to the conservative variables. The difference representation connecting the state of the primitive variables at the cell face with the state assigned to the zeroth eigenvalue is repeated here for clearness:

$$\rho-\rho_o = R_1+R_2, \quad u-u_o = \frac{xs}{\rho}(R_1-R_2), \quad v-v_o = \frac{ys}{\rho}(R_1-R_2),$$

$$w-w_o = \frac{zs}{\rho}(R_1-R_2), \qquad p-p_o = s^2(R_1+R_2) . \tag{6.85}$$

The analoguous differences for the conservative variables read

$$\ell = \ell_o+\ell-\ell_o =$$

$$= (\rho u)_o+\rho u-(\rho u)_o = \ell_o+d(\rho u) = \ell_o+u(\rho-\rho_o)+\rho(u-u_o) =$$

$$= \ell_o+u(R_1+R_2)+xs(R_1-R_2), \qquad \ell = \ell_o+(u+sx)R_1+(u-sx)R_2 .$$

Similarly:

$$m = m_o+(v+sy)R_1+(v-sy)R_2, \qquad n = n_o+(w+sz)R_1+(w-sz)R_2, \tag{6.86}$$

$$e = e_o+ e-e_o =$$

$$= e_o+ de = e_o+d\left(\frac{p}{\gamma-1} +\rho \frac{q^2}{2}\right) = e_o+ \frac{dp}{\gamma-1} + \rho(udu+vdv+wdw)+ \frac{q^2}{2} d\rho =$$

$$= e_o+ \frac{p-p_o}{\gamma-1} + \rho\left[u(u-u_o)+v(v-v_o)+w(w-w_o)\right] + \frac{q^2}{2}(\rho-\rho_o) =$$

$$= e_o+\left(\frac{s^2}{\gamma-1} + \frac{q^2}{2}\right)(R_1+R_2) + s(ux+vy+wz)(R_1-R_2) . \tag{6.87}$$

The first coefficient of Eq. (6.87) is easily identified as the total enthalpy H. The second coefficient is nothing but the normal velocity or the zeroth normalized eigenvalues λ_o multiplied by the speed of sound:

$$e = e_o + (H+s\lambda_o)R_1 + (H-s\lambda_o)R_2 \ .$$

Now all conservative variables are available. The quantities R_1 and R_2, however, also do need the transformation to conservative variables. This is easily achieved, since the content of the quantities $R_{1,2}$ are plain differences of the non-conservative variables, the conversion of which to the difference representation of the conservative variables was mentioned previously, see Chapter II. The result is

$$R_1 = \frac{1}{2s^2}\{(\rho_1-\rho_o)(\frac{\gamma-1}{2}q^2-s\lambda_o) + (\ell_1-\ell_o)[sx-(\gamma-1)u] +$$

$$+ (m_1-m_o)[sy-(\gamma-1)v] + (n_1-n_o)[sz-(\gamma-1)w] + (e_1-e_o)(\gamma-1)\} \ ,$$

$$R_2 = \frac{1}{2s^2}\{(\rho_2-\rho_o)(\frac{\gamma-1}{2}q^2+s\lambda_o) - (\ell_2-\ell_o)[sx+(\gamma-1)u] -$$

$$- (m_2-m_o)[sy+(\gamma-1)v] - (n_2-n_o)[sz+(\gamma-1)w] + (e_2-e_o)(\gamma-1)\} \ . \tag{6.88}$$

6.24 <u>The Exact Conservative Riemann Solver</u>

As in the case of the non-conservative variables there is a piece of non-uniqueness in the linear conservative Riemann solver. Again the choice of the averaging procedure for obtaining the coefficients (speed of sound and velocity components) will cause at least a slight dependency of the characteristic averaged variables on the averaging for the coefficients. Usually the arithmetic mean is taken for this purpose. The coefficients can be improved by iteration using the newest available cell face flow variables and the newest available eigenvalues at this place for the evaluation of the truth function, which decides upon which of both states – left or right – is entry to the Riemann solver. This way an exact Riemann solver can be constructed. For improving the initial guess for starting the iterations the result of the simple isentropic Riemann solver can be taken. Again the change of sign of one of the three eigenvalues during the course of iterations may make them run into a limit cycle in extreme cases.

6.25 <u>Roe's Average</u>

In order to avoid iterating the conservative Riemann solver for improving about the coefficients and to remove the arbitrariness of the choice of averaging the coefficients of the latter there is a desire to impose some sort of uniqueness on these coefficients. For this purpose Roe, Ref. 11, postulates the identity

$$E_U U_\xi \equiv E_\xi \ ,$$

or in a slightly different formulation with the Jacobian element being interpolated somewhere in between the left and the right state

$$[E_{U\ell}(1-\xi)+\xi E_{Ur}](U_r-U_\ell) \equiv E_r-E_\ell \ , \tag{6.89}$$

where the coordinate ξ in the computational space is chosen such that the identity holds. Let us consider a representative flux of the momentum equations:

$$(\rho uv)_r-(\rho uv)_\ell = \left(\frac{\ell m}{\rho}\right)_r-\left(\frac{\ell m}{\rho}\right)_\ell = \tag{6.90}$$

$$= \frac{m}{\rho}(\ell_r-\ell_\ell)+\frac{\ell}{\rho}(m_r-m_\ell)-\frac{\ell m}{\rho^2}(\rho_r-\rho_\ell) = u(\ell_r-\ell_\ell)+v(m_r-m_\ell)-uv(\rho_r-\rho_\ell).$$

We apply the ansatz:

$$u = u_\ell(1-\xi)+\xi u_r \ , \qquad v = v_\ell(1-\xi)+\xi v_r \ ,$$

which is nothing but a linear interpolation between the left and the right state with a yet unknown coordinate ξ. After inserting the ansatz into the above difference representation we obtain after rearrangement:

$$(\rho uv)_r-(\rho uv)_\ell =$$

$$= u_r v_r[\rho_r\xi(2-\xi)+\xi^2\rho_\ell] - u_\ell v_\ell[\rho_r(1-\xi)^2+\rho_r(1-\xi^2)] + \tag{6.91}$$

$$+ [\rho_r(1-\xi)^2 - \rho_\ell\xi^2](u_\ell v_r+u_r v_\ell) \ .$$

The coefficient of the sum of the cross products of the velocity components u and v can be made vanish by solving the quadratic equation

$$\rho_r(1-\xi)^2 = \rho_\ell\xi^2 \ ,$$

with the result

$$1-\xi = \pm\left(\frac{\rho_\ell}{\rho_r}\right)\xi^{0.5} \ .$$

From $\xi = \dfrac{1}{1\pm\left(\frac{\rho_\ell}{\rho_r}\right)^{0.5}}$

we see that the solution with the denominator never being zero is

$$\xi = \frac{\sqrt{\rho_r}}{\sqrt{\rho_\ell} + \sqrt{\rho_r}} .$$

The identity postulated initially can be verified by inserting this value into the last difference equation. It is clear that an unique averaging for the velocity components should be

$$u = \frac{u_\ell\sqrt{\rho_\ell} + u_r\sqrt{\rho_r}}{\sqrt{\rho_\ell} + \sqrt{\rho_r}} , \quad v= \frac{v_\ell\sqrt{\rho_\ell} + v_r\sqrt{\rho_r}}{\sqrt{\rho_\ell} + \sqrt{\rho_r}} , \quad w= \frac{w_\ell\sqrt{\rho_\ell} + w_r\sqrt{\rho_r}}{\sqrt{\rho_\ell} + \sqrt{\rho_r}} . \tag{6.92}$$

So far we have obtained a unique average of the velocity components. By inspection of the conservative linear Riemann solver we see that we also need an unique average for the speed of sound occuring in the coefficients there. Unfortunately the speed of sound is not a conservative variable. The key point lies in convective differences of the energy equation, which reads, written in the conservative variables ρu and the total enthalpy,

$$\Delta(\rho uH) = \rho u(H_r-H_\ell) + H[(\rho u)_r- (\rho u)_\ell] .$$

Now, since H depends on the velocity square as

$$H = \frac{s^2}{\gamma-1} + \frac{q^2}{2} ,$$

it is clear that the total enthalpy H should be interpolated at the same abscissa we used for the velocity components too. Therfore, Roe's average of the total enthalpy is

$$H = \frac{\sqrt{\rho_\ell}\, H_\ell + \sqrt{\rho_r}\, H_r}{\sqrt{\rho_\ell} + \sqrt{\rho_r}} . \tag{6.93}$$

The speed of sound should then be calculated from

$$s^2 = (\gamma-1)(H - \frac{q^2}{2}) .$$

In principle all coefficients needed for the Riemann solver are averaged uniquely this way. For completeness, however, also the unique average of the density is evaluated here. For this purpose we write down a representative conservative difference of the continuity equation:

$$\Delta\rho u = (\rho u)_r - (\rho u)_\ell =$$

$$= u(\rho_r - \rho_\ell) + \rho(u_r - u_\ell) = \tag{6.94}$$

$$= \frac{u_\ell\sqrt{\rho_\ell} + u_r\sqrt{\rho_r}}{\sqrt{\rho_\ell} + \sqrt{\rho_r}}(\rho_r - \rho_\ell) + \rho(u_r - u_\ell) =$$

$$= (u_\ell\sqrt{\rho_\ell} + u_r\sqrt{\rho_r})(\sqrt{\rho_r} - \sqrt{\rho_\ell}) + \rho(u_r - u_\ell) =$$

$$= (\rho u)_r - (\rho u)_\ell + (u_r - u_\ell)(\rho - \sqrt{\rho_\ell\rho_r}) .$$

The identity of the conservative difference is guaranteed if the unique average for the density is

$$\rho = \sqrt{\rho_\ell\rho_r} .$$

6.26 Riemann Solvers Based on Fluxes: The Steger-Warming Fluxes[2]

If the flow variables with subscripts 0,1,2 are redefined as

$$u_j = u\dot{\xi}_j , \qquad j=0,1,2 ,$$

the linear Riemann solver converts to a characterstic flux average. In this case the quantities R_1 and R_2 can be multiplied out. They assume the surprisingly simple form

$$R_1 = \frac{\rho}{2\gamma}(\dot{\xi}_1 - \dot{\xi}_0) , \qquad R_2 = \frac{\rho}{2\gamma}(\dot{\xi}_2 - \dot{\xi}_0) , \tag{6.95}$$

where $\dot{\xi}$ is the non-normalized eigenvalue

$$\xi = \lambda A = \lambda\ (\xi_x^2 + \xi_y^2 + \xi_z^2)^{0.5} .$$

The Steger-Warming fluxes are obtained by implementing the eigenvalue sign interrogation.

$$\dot{\xi}^+ = \frac{1}{2}(\dot{\xi} + |\dot{\xi}|) , \qquad \xi^- = \frac{1}{2}(\dot{\xi} - |\dot{\xi}|) ,$$
$$R_j^+ = \frac{\rho}{2\gamma}(\dot{\xi}_j^+ - \dot{\xi}_0^+) , \qquad R_j^- = \frac{\rho}{2\gamma}(\dot{\xi}_j^- - \dot{\xi}_0^-) , \qquad j=1,2 . \tag{6.96}$$

The positive fluxes are formed with the left state

$$\rho^+ = [\rho\xi_o^+ + R_1^+ + R_2^+]_\ell \ ,$$

$$\ell^+ = [\ell\dot{\xi}_o^+ + (u+sx)R_1^+ + (u-sx)R_2^+]_\ell \ ,$$

$$m^+ = [m\dot{\xi}_o^+ + (v+sy)R_1^+ + (v-sy)R_2^+]_\ell \ , \tag{6.97}$$

$$n^+ = [n\dot{\xi}_o^+ + (w+sz)R_1^+ + (w-sz)R_2^+]_\ell \ ,$$

$$e^+ = [e\dot{\xi}_o^+ + (H+s\lambda_o)R_1^+ + (H-s\lambda_o)R_2^+]_\ell \ .$$

The negative fluxes are formed with the right state

$$\rho^- = [\rho\dot{\xi}_o^- + R_1^- + R_2^-]_r \ ,$$

$$\ell^- = [\ell\dot{\xi}_o^- + (u+sx)R_1^- + (u-sx)R_2^-]_r \ ,$$

$$m^- = [m\dot{\xi}_o^- + (v+sy)R_1^- + (v-sy)R_2^-]_r \ , \tag{6.98}$$

$$n^- = [n\dot{\xi}_o^- + (w+sz)R_1^- + (w-sz)R_2^-]_r \ ,$$

$$e^- = [e\dot{\xi}_o^- + (H+s\lambda_o)R_1^- + (H-s\lambda_o)R_2^-]_r \ .$$

Adding the negative and positive parts together gives the cell face fluxes

$$\tilde{\rho} = \rho^+ + \rho^- \ , \quad \tilde{\ell} = \ell^+ + \ell^- \ , \quad \tilde{m} = m^+ + m^- \ , \quad \tilde{n} = n^+ + n^- \ , \quad \tilde{e} = e^+ + e^- \ .$$

6.27 Generalized Steger-Warming Fluxes

The original Steger-Warming fluxes can be changed to a set of another system of fluxes by introducing a parameter, which allows some control on the strength of diffusion of these fluxes. This has been proven to be a valuable tool for hypersonic flow calculations where shocks may have a considerable strength, generating ratios of the pressures ahead and aft of the shock, which may be very large. The parameter mentioned above can be defined using the formulation of the previous paragraph. We see that the definition of the quantities R_1 and R_2 contains a constant coefficient $1/(2\gamma)$.

The coefficient is a measure of the size of the split mass fluxes at vanishing normal speed and can be used as a parameter for controlling this size:

$$M = \frac{1}{2\gamma} \ .$$

The coefficient M can be taken as a particular Mach number in the range

$$0 < M \leq 1 ,$$

which could be an input parameter. In order to guarantee that positive and negative fluxes do sum up to the original Euler fluxes the speed of sound occuring in the coefficients multiplying the quantities R_1 and R_2 must be replaced by the modified speed of sound

$$\bar{s} = \frac{s}{2\gamma M} .$$

Since the degree of the polynomial representation is not changed by the above manipulations, the eigenvalues remain those of the original Euler equations. Two particular values of the parameter M have a specific property. If M is very small the flux splitting is governed only by the zeroth eigenvalue, and the present formulation can be used for extreme low speed flow. If the parameter M is put one half then the flux splitting is governed by the extreme eigenvalues one and two only. this parameter setting is well suited for extreme high speed flow with very strong shocks. In addition this particular parameter setting allows a simplified formulation of the fluxes since many items of the flux calculation do cancel out and need not be calculated explicity therefore.

6.28 Einfeldt-Type Fluxes

Einfeldt discovered in his work[12] that at least for moderate to high speed flows the splitting due to the zeroth eigenvalue is of neglegible importance. So the splitting needs only the two extreme eigenvalues labeled "one" and "two" in our definition. This finding leads to the formulation of particularly simple positive and negative fluxes:

$$E^+ = \frac{1}{2}(1+\xi)E_\ell , \qquad E^- = \frac{1}{2}(1-\xi)E_r . \tag{6.99}$$

Here E means the conventional Euler flux vector, and the interpolation coordinate ξ can be specified for example as

$$\xi = \frac{\lambda_1 + \lambda_2}{|\lambda_1|+|\lambda_2|} . \tag{6.100}$$

The eigenvalues occuring in the definition of ξ can be calculated from a suitable mean of the left and the right state, say, the arithmetic mean. This formulation is the simplest of the Einfeld-type fluxes. If the left and the right state are identical, the sum of the negative and the positive flux recovers the original Euler flux. A somewhat more elaborate interpolation with better continuity and less diffusion is

$$
\begin{aligned}
E^+ &= \frac{1}{4}(2+\bar{\xi})E_\ell \; , \\
E^- &= \frac{1}{4}(2-\bar{\xi})E_r \; ,
\end{aligned}
\tag{6.101}
$$

where the new interpolation coordinate is expressed by

$$\bar{\xi} = \xi(3-\xi^3) \; .$$

The positive and negative fluxes are differentiable everywhere using this formulation. Finally we suggest a version of Einfeld-type fluxes, which is very much aligned with the Steger-Warming flux formulation:

$$
\begin{aligned}
E^+ &= \frac{1}{2}(1+\xi_\ell)E_\ell \; , \\
E^- &= \frac{1}{2}(1-\xi_r)E_r \; ,
\end{aligned}
\tag{6.102}
$$

where the interpolation coordinates are now formed with the left and right state variables

$$
\begin{aligned}
\xi_\ell &= \left(\frac{\lambda_1 + \lambda_2}{|\lambda_1|+|\lambda_2|}\right)_\ell \; , \\
\xi_r &= \left(\frac{\lambda_1 + \lambda_2}{|\lambda_1|+|\lambda_2|}\right)_r \; .
\end{aligned}
\tag{6.103}
$$

The summation rule is also fulfilled with these latter fluxes since, if the left and the right state is identical, it turns out that in addition to the fluxes also the eigenvalues of the left and right state are both the same.

Therefore in this case the identities hold

$$E_\ell = E_r = E \quad \text{and} \quad \xi_\ell = \xi_r = \xi \; ,$$

from which we conclude that the positive and the negative fluxes sum up to the original Euler fluxes. Whether the new split fluxes do possess the same eigenvalues as the Euler fluxes is not evident, since they are now formed by relatively complicated rational functions of the flow vairables in contrast to the polynomials the conventional Euler fluxes are formed of. Numerical tests, however, seem to work well, thus indicating that the positive fluxes contain positive eigenvalues only and vice versa.

6.29 Van Leer-Type Fluxes[13]

If the Steger-Warming flux of, say, the continuity equation (mass flux) is rewritten for the one-dimensional case, we obtain for the eigenvalue constellation

$$\lambda_o > 0, \qquad \lambda_1 > 0, \qquad \lambda_2 < 0 ,$$

the formula

$$E^+ = A\rho\left[u + \frac{1}{2\gamma}(s-u)\right] = \frac{A\rho s}{2\gamma}\left[M(2\gamma-1)+1\right] , \tag{6.104}$$

with M being the Mach number $M = u/s$, and A the area of the stream-tube cross section.

The negative flux for the same eigenvalue constellation is

$$E^- = A\,\frac{\rho}{2\gamma}(u-s) = A\,\frac{\rho s}{2\gamma}(M-1) .$$

This is valid for $0<M<1$ while for $M>1$ we simply have $E^+=E$, $E^-=0$.

The sum of the negative and the positive flux results, of course, in the unsplit conventional mass flux

$$E = E^+ + E^- = A\rho s M = E_M M = A\rho u . \tag{6.105}$$

For the other mixed-sign eigenvalue constellation

$$\lambda_o < 0 , \qquad \lambda_1 > 0 , \qquad \lambda_2 < 0 ,$$

the following formulas are valid:

$$E^+ = A\,\frac{\rho}{2\gamma}(u+s) = A\,\frac{\rho s}{2\gamma}(M+1) , \qquad E^- = A\rho\left[u - \frac{1}{2\gamma}(u+s)\right] = A\,\frac{\rho s}{2\gamma}\left[M(2\gamma-1)-1\right] . \tag{6.106}$$

This is valid for $-1<M<0$ while for $M<-1$ we simply have $E^+=0$, $E^-= E$.

The sum of the negative and the positive flux again does recover, of course, the conventional mass flux

$$E = E^+ + E^- = A\rho s M = A\rho u .$$

We summarize the results in a plot, Fig. 6.9, showing the normalized mass fluxes as a function of the Mach number.

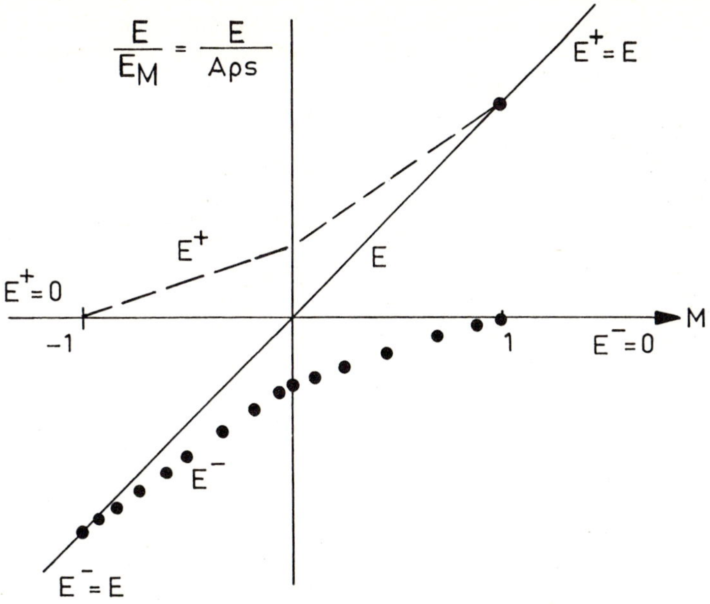

Fig. 6.9 Normalized Steger-Warming mass flux

As with the Steger-Warming fluxes, the approach of van Leer is also based on the formulation that the cell-face flux is the sum of the negative and the positive flux part $E = E^- + E^+$.

From the Fig. 6.9 we see immediately that the slope of the normalized positive and negative fluxes with respect to the Mach number does have jumps at $M=-1$, $M=0$ and $M=1$. Also the other fluxes in the Steger-Warming approach do exhibit this property. The derivation of van Leer type fluxes is now governed by the desire to make the positive and negative fluxes differentiable with respect to the Mach number everywhere. This includes that the positive and negative fluxes should be represented by one and the same analytical function within the mixed eigenvalue sign interval limited by

$$-1 < M < 1 \quad .$$

Van Leer postulates that this analytic function should be a polynomial of the Mach number of possibly lowest order. The rules for constructing such polynomials, which are differentiable everywhere, are evident from Fig. 6.9 and can be expressed by the following formulation:

1) The polynomial using the Mach number for the representation of the positive and the negative flux must be one order higher than the Mach number polynomial of the conventional normalized Euler flux.

2) The function value and the slope of the Mach number polynomial representing the positive and the negative fluxes should coincide with the function value and the slope of the equivalent unsplit fluxes at M=1 and M=-1.

3) Positive fluxes should not have negative eigenvalues.

4) Negative fluxes should not have positive eigenvalues.

The summary of all these rules is expressed mathematically by

$$\begin{aligned} &E^+ = E\ , && E_M^+ = E_M\ , && \text{for } M \geq 1\ , \\ &E^+ = 0\ , && E_M^+ = 0\ , && \text{for } M \leq -1\ , \\ &E^- = E\ , && E_M^- = E_M\ , && \text{for } M \leq -1\ , \\ &E^- = 0\ , && E_M^- = 0\ , && \text{for } M \geq 1\ . \end{aligned} \tag{6.107}$$

We start now with the construction of the positive and negative fluxes of the van Leer type.

6.30 1D-Mass Flux of van Leer Type

The general flux-relation reads:

$$E = A\rho u = A\rho s\, M\ , \quad E_M = A\rho s\ . \tag{6.108}$$

For the positive flux we need a quadratic polynomial in M:

$$E^+ = A\rho s(a + bM + cM^2)\ .$$

The coefficients a,b,c are defined by

$$\begin{aligned} &E^+(-1) = A\rho s(a-b+c) = 0\ , \\ &E_M^+(-1) = A\rho s(b-2c) = 0\ , \\ &E^+(1) = A\rho s(a+b+c) = E(1) = A\rho s\ , \\ &E_M^+(1) = A\rho s(b+2c) = E_M(1) = A\rho s\ . \end{aligned} \tag{6.109}$$

Adding the second to the last equation gives

$$b = \frac{1}{2}\ .$$

Subtraction gives

$$c = \frac{1}{4} .$$

The third equation yields a=1/4. The first equation is fulfilled identically with these coefficients, so no overspecification happens. The positive mass flux is therefore

$$E^+ = A\rho \, \frac{s}{4}(M+1)^2 . \tag{6.110}$$

The negative mass flux is easily obtained as

$$E^- = E - E^+ = - A\rho \, \frac{s}{4}(M-1)^2 .$$

6.31 1D-Momentum Flux of van Leer Type

Here we get

$$\left.\begin{aligned} E &= A(\rho u^2+p) = A\rho(u^2+\frac{s^2}{\gamma}) = A\rho s^2(M^2+\frac{1}{\gamma}) , \\ E_M &= 2A\rho s^2 M , \\ E(1) &= E(-1) = A\rho s^2 \, \frac{\kappa+1}{\gamma} , \\ E_M(1) &= -E_M(-1) = 2A\rho s^2 . \end{aligned}\right\} \tag{6.111}$$

Using the mass flux derived so far, the ansatz for the positive momentum flux should be a third order polynomial in M

$$E^+ = A\rho \, \frac{s^2}{4}(M+1)^2(a+bM) .$$

This has the advantage that the function value and the slope are automatically zero at M=-1. The coefficients a and b are obtained from

$$\left.\begin{aligned} E^+(1) &= A\rho s^2(a+b) = E(1) = A\rho s^2 \, \frac{\gamma+1}{\gamma} , \\ E_M^+(1) &= A\rho s^2(a+2b) = E_M(1) = 2A\rho s^2 , \end{aligned}\right\} \tag{6.112}$$

and read:

$$b = \frac{\gamma-1}{\gamma} \quad ; \quad a = \frac{2}{\gamma} .$$

The positive momentum flux is therefore

$$E^+ = A\rho \, \frac{s^2}{4\gamma}(M+1)^2[2+(\gamma-1)M] .$$

The negative momentum flux can be obtained from

$$E^- = E - E^+ .$$

This leads to rather tedious algebra. The other choice is the ansatz with the negative mass flux being incorporated:

$$E^- = A\rho \frac{s^2}{4}(M-1)^2(a+bM) ,$$

such that the unknown coefficients a and b can be obtained by fulfilling the constraints

$$E^-(-1) = E(-1) , \qquad E^-_M(-1) = E_M(-1) .$$

The result is

$$E^- = A\rho \frac{s^2}{4\gamma}(M-1)^2[2-(\gamma-1)M] .$$

The sum of the negative and the positive flux does recover the unsplit flux indeed. Once the positive flux is available the negative flux can be obtained even without any algebra. Let us consider the unsplit flux for this purpose. The momentum flux is a parabola being symmetric with respect to the coordinate origin. This property also is reflected in the positive and the negative flux. So the negative flux could have been obtained simply from the formula expressing plain symmetry:

$$E^-(M) = E^+(-M) .$$

6.32 1D-Energy Flux of van Leer Type

This flux reads:

$$E = A\rho uH = A\rho u(\frac{s^2}{\gamma-1} + \frac{u^2}{2}) = A\rho s^3 M(\frac{1}{\gamma-1} + \frac{M^2}{2}) . \qquad (6.113)$$

For the construction of the positive flux only the function value and the slope at M=1 is needed:

$$E(1) = A\rho s^3 \frac{\gamma+1}{2(\gamma-1)} = -E(-1) , \qquad E_M(1) = A\rho s^3 \frac{3\gamma-1}{2(\gamma-1)} = E_M(-1) .$$

The ansatz for the 1D-flux polynomial is not unique. We could for example augment the momentum flux by a linear function. Another choice is closer to the original energy flux and is composed of the mass flux multiplied by a parabolic function:

$$E^+ = A\rho \frac{s^2}{4}(M+1)^2(a+bM^2) .$$

Using the constraints mentioned previously the coefficients a and b can be fixed

$$E^{+}(1)= A\rho s^{3}(a+b)= E(1)= A\rho s^{3}\ \frac{\gamma+1}{2(\gamma-1)}\ ,$$

$$E_{M}^{-}(1)= A\rho s^{3}(a+3b)= E_{M}(1)= A\rho s^{3}\ \frac{3\gamma-1}{2(\gamma-1)}\ . \tag{6.114}$$

The solution for the coefficients is easily obtained

$$b = \frac{1}{2}\ , \qquad a = \frac{1}{\gamma-1}\ .$$

The initial ansatz becomes

$$E^{+}= A\rho\ \frac{s^{3}}{4}(M+1)^{2}\ (\frac{1}{\gamma-1} + \frac{M^{2}}{2})\ .$$

Here we see that the ansatz using a plain parabola was the preferable choice since the second bracket is exactly that of the unsplit energy flux. So no difficulties should arise if the 1D-flux is transformed to three dimensions.

The negative flux can be easily obtained from the observation, that the unsplit energy flux is a point symmetric function with respect to the coordinate origin. This should be reflected in the formula for the negative flux with the help of the formula expressing point symmetry

$$E^{-}(M)= -E^{+}(-M)\ . \tag{6.115}$$

So the negative energy flux is

$$E^{-}= -A\rho\ \frac{s^{3}}{4}(M-1)^{2}\ (\frac{1}{\gamma-1} + \frac{M^{2}}{2})\ .$$

Clearly the sum of the negative and the positive flux recovers the original flux.

There remains to transform the one-dimensional fluxes to three dimensions. This requires the introduction of the velocity normal to the cell face and component-wise decomposition of the one-dimensional momentum flux.

6.33 3D-Mass Flux

Since the unsplit mass flux in 3D is simply

$$E = A\rho\lambda_{o}\ , \tag{6.116}$$

with λ_o being the zeroth eigenvalue or the velocity normal to the cell face, the extension of the plus/minus fluxes to three dimensions is straight forward:

$$E^+ = A\rho s\left(\frac{\lambda_o}{s} + 1\right)^2 , \qquad E^- = -A\rho s\left(\frac{\lambda_o}{s} - 1\right)^2 . \qquad (6.117)$$

In other words, the Mach number needs only to be replaced by the Mach number normal to the cell face.

6.34 3D-Momentum Fluxes

The extension of the one-dimensional momentum flux to the three components of the multi-dimensional case is not quite trivial. Since it is desirable to retain the split mass fluxes as coefficients multiplying the linear function in M, we tentatively make a new ansatz for the plus/minus momentum fluxes retaining the structure of the one dimensional split fluxes:

$$E^+ = A\,\frac{\rho}{4\gamma}(\lambda_o + s)^2\left[2a + (\gamma-1)M\right] , \qquad E^- = A\,\frac{\rho}{4\gamma}(\lambda_o - s)^2\left[2a - (\gamma-1)M\right] , \qquad (6.118)$$

where a parameter "a" has been introduced. We first consider the x-momentum flux, the unsplit form of which is

$$E = A\rho\left(u\lambda_o + \frac{s^2 x}{\gamma}\right) .$$

The sum of the negative and the positive flux is

$$E^- + E^+ = A\,\frac{\rho}{\gamma}\,\{\lambda_o\left[a\lambda_o + s(\gamma-1)M\right] + as^2\} . \qquad (6.119)$$

By comparison with the original flux we see immediately that the following two conditions should hold:

$$a = x , \qquad \text{and} \qquad (\gamma-1)M = \frac{\gamma u - x\lambda_o}{s} .$$

The split x-momentum fluxes are therefore:

$$E^+ = A\,\frac{\rho s}{4\gamma}\left(\frac{\lambda_o}{s} + 1\right)^2\left[x(2s-\lambda_o) + \gamma u\right] , \quad E^- = A\,\frac{\rho s}{4\gamma}\left(\frac{\lambda_o}{s} - 1\right)^2\left[x(2s+\lambda_o) - \gamma u\right] . \qquad (6.120)$$

The other two momentum fluxes are evident:

y-momentum fluxes:

$$E^\pm = A\,\frac{\rho s}{4\gamma}\left(\frac{\lambda_o}{s} \pm 1\right)^2\left[y(2s \mp \lambda_o) \pm \gamma v\right] , \qquad (6.121)$$

z-momentum fluxes:

$$E^{\pm}= A \frac{\rho s}{4\gamma} \left(\frac{\lambda_o}{s} \pm 1\right)^2 \left[z(s\mp\lambda_o)\pm\gamma w\right] .$$

6.35 3D-Energy Flux

The energy flux in three-dimensional formulation is also evident. The first bracket is nothing but the mass flux and should therefore be formulated with the Mach number normal to the cell face, while the Mach number in the second bracket is the conventional Mach number formed with the total velocity:

$$E^{+}= A\rho \frac{s^3}{4}\left(\frac{\lambda_o}{s} +1\right)^2 \left(\frac{1}{\gamma-1} + \frac{M^2}{2}\right) = A\rho \frac{s}{4}\left(\frac{\lambda_o}{s} +1\right)^2 \left(\frac{s^2}{\gamma-1} + \frac{q^2}{2}\right) = \tag{6.122}$$

$$= A\rho \frac{s}{4}\left(\frac{\lambda_o}{s} +1\right)^2 H , \qquad E^{-}= -A\rho \frac{s}{4}\left(\frac{\lambda_o}{s} -1\right)^2 H .$$

While the splitting of the energy flux in one dimension is not unique, it is shown here for completeness that the mass flux should be multiplied by an incomplete quadratic function. The energy flux is then unique in the 3D case. Let us make the ansatzes

$$E^{+}= A\rho \frac{s}{4}(\lambda_o+s)^2(a^{+}-b^{+}M+c^{+}M^2) , \quad E^{-}= A\rho \frac{s}{4}(\lambda_o-s)^2(a^{-}+b^{-}M+c^{-}M^2) . \tag{6.123}$$

The sum of both fluxes produces a term

$$\lambda_o^2\left[a^{+}-a^{-}+M(b^{+}-b^{-})+ M^2(c^{+}-c^{-})\right] .$$

Therefore it is desireable to put

$$a^{-}= a^{+} , \qquad b^{-}= b^{+} , \qquad c^{-}= c^{+} ,$$

since the unsplit flux does not contain a square of the normal velocity. On the other hand we obtain from the condition of point symmetry applied to the 1D-case the equations

$$\begin{aligned} E_M^{-}(-1)= E_M^{+}(1)&: \quad a^{+}+2b^{+}+3c^{+}= a^{-}-2b^{-}+2c^{-} , \\ E^{-}(-1)= -E^{+}(1)&: \quad a^{+}+b^{+}+c^{+}= a^{-}-b^{-}+c^{-} . \end{aligned} \tag{6.124}$$

We see immediately that the constraints can easily been fulfilled by putting $b^{\pm}$ to zero:

$$b^{-}= b^{+}=0 .$$

At this stage it should be noticed that the energy flux is no more the original energy flux suggested by van Leer in 1982, Ref. 13. It is clear that our formulation with the total enthalpy as a coefficent in the energy flux is superior to the original flux, since the total temperature, which is a conserved quantity, remains unsplit. There remains to check the eigenvalue sign pattern of the new fluxes whether it coincides with that of the original fluxes. Since this has been proved elsewhere for the van Leer fluxes this exercise is not repeated in this context.

6.36 The Use of the Conservative Riemann Solver for Splitting Flux Differences

From the definition of the truth function

$$\begin{aligned} u &= \frac{1}{2}[(1+\text{sign}\lambda)u_\ell + (1-\text{sign}\lambda)u_r] = \\ &= \frac{1}{2}[u_\ell+u_r - \text{sign}\lambda(u_r-u_\ell)] , \end{aligned} \tag{6.126}$$

we conclude immediately, that any split flux at the cell face can be represented as the sum of the arithmetic mean and a difference between the left and right flux with a suitable eigenvalue interrogation being incorporated. The key point is to decompose the flux differences each into three parts being aligned with the three characteristics the signs of which decide which part of the flux difference is assigned to the left or the right cell. The conservative Riemann solver can be used for this purpose. Let us recall the definition of the initial values labeled 0, 1 and 2. They where introduced by the characteristic derivative:

$$U_{in} = U - (\dot{U}+U_\xi \dot{\xi}_{in})\Delta t , \qquad in=0,1,2 . \tag{6.125}$$

If we insert this formulation into the Riemann solver, say, for the density, we get

$$\rho = \rho - (\dot{\rho}+\rho_\xi \dot{\xi}_o)\Delta t - (\tilde{R}_1+\tilde{R}_2)\Delta t .$$

We see immediately that the cell face density cancels out. The tilde indicates that the characteristic derivatives are also inserted into the quantities R_1 and R_2. Rearrangement yields

$$-\dot{\rho} = \rho_\xi \dot{\xi}_o + \tilde{R}_1 + \tilde{R}_2 .$$

Without any further algebra we see that the Riemann solver can be used as a flux splitting rule simply by redefining the meaning of the left hand side. Instead of the cell face flow variable we understand that the left-hand side is now the negative time derivative of the appropriate flow variable, while the right hand side, provided a suitable eigenvalue sign interrogation is incorporated, decides which piece of the difference is assigned to either the left cell or to the right cell. The present formulation with the differences of the flow variables can be taken for the construction of split artificial viscosity terms added to the cell-face flux of the form

$$E = \frac{1}{2}[E_\ell + E_r + |\dot{\xi}|(U_\ell - U_r)] \ . \tag{6.127}$$

So far we considered the difference splitting of the conservative flow variables. Subsequent multiplication with the associated eigenvalue makes out of the split differences of the flow variables a quasilinear flux-split algorithm. If a splitting of the original conventional Euler fluxes differences is desired, also the linear Riemann solver can be used for this purpose. Let us consider again the density equation.

$$-\dot{\rho} = \rho_\xi \dot{\xi}_o + \tilde{R}_1 + \tilde{R}_2 \ . \tag{6.128}$$

Now we can add to the space derivatives $\partial/\partial\xi$ the one-dimensional version of the conservative Euler equations being of the form

$$\dot{\rho} + \overset{\rho}{E}_\xi = 0, \quad \dot{\ell} + \overset{\ell}{E}_\xi = 0, \quad \dot{m} + \overset{m}{E}_\xi = 0, \quad \dot{n} + \overset{n}{E}_\xi = 0, \quad \dot{e} + \overset{e}{E}_\xi = 0, \tag{6.129}$$

where the E's are the one-dimensional mass, momentum and energy fluxes. The result for, say, the density equation, turns out to be

$$-2\dot{\rho} = \rho_\xi \dot{\xi}_o + \overset{\rho}{E}_\xi + \approx{R}_1 + \approx{R}_2 \ ,$$

where the double tilde indicates that the conservative flux differences are also included in the quantities R_1 and R_2. Upon subtracting the initial equation, we easily obtain the flux difference splitting rule for the conservative Euler fluxes

$$-\dot{\rho} = \overset{\rho}{E}_{\xi o} + \bar{R}_1 + \bar{R}_2 \ ,$$

where the bars indicate that the quantities R_1 and R_2 contain plain differences of the conservative fluxes. The accompanying eigenvalue sign convention rules, to which side of the cell face – left or right – the flux difference contribution is directed.

We see easily, that the linear Riemann solver can be used in a manyfold of ways simply by redefining the left-hand side being either a state or the time derivative of the latter, while the right-hand side is thought of as being either composed of plain flow variable differences, or of flux differences weighted with a prescribed matrix of coefficients, which is the same regardless what kind of definitions – state differences or flux differences – are used.

6.37 The Use of the Conservative Riemann Solver for Splitting Flux Differences by Projection

The technique of projection is based on the following problem. Provided initial data are given at the cell centers, then over each cell face a definite flux difference

$$E_\xi = E_r - E_\ell$$

exists. The question is now, which fraction of the total flux difference is assigned to the zeroth, first and second eigenvalue for deciding to which side of the cell face – left or right – that signal should be attributed to. Again the linear Riemann solver can be used for the calculation of the three fractions of the total flux differences each being aligned with one of the three characteristic directions. Let us rewrite for example the density equation:

$$- \dot{\rho} = \overset{\rho}{E}_{\xi 0} + \bar{R}_1 + \bar{R}_2 \ .$$

By addition of the one-dimensional continuity equation to the left hand side we get

$$\overset{\rho}{E}_{\xi tot} = \overset{\rho}{E}_{\xi 0} + \bar{R}_1 + \bar{R}_2 \ ,$$

or for simpler short-hand notation

$$\rho = \rho_0 + R_1 + R_2 \ ,$$

with ρ being the total mass flux difference and the quantities with subscripts (j=0,1,2) being the differences of the mass-, momentum- and energy fluxes on the right-hand side. There is now an implicit formalism by means of which the total left-hand side flux differences can be broken up into three parts each being assigned to one of the three eigenvalues. This needs regrouping of the right-hand side, such that the first group belongs to the zeroth eigenvalue, the second group belongs to the first eigenvalue and the third group belongs to the second eigenvalue. Rearrangement of the linear conservative Riemann solver gives the result

$$\begin{aligned}
\rho &= \rho_o + \frac{\gamma-1}{s^2}(u\ell_o + vm_o + wn_o - \frac{q^2}{2}\rho_o - e_o) + r_1 + r_2 , \\
\ell &= \ell_o + u\frac{\gamma-1}{s^2}(u\ell_o + vm_o + wn_o - \frac{q^2}{2}\rho_o - e_o) + \\
&\quad + x(\lambda_o\rho_o - x\ell_o - ym_o - zn_o) + (u+sx)r_1 + (u-sx)r_2 , \\
m &= m_o + v\frac{\gamma-1}{s^2}(u\ell_o + vm_o + wn_o - \frac{q^2}{2}\rho_o - e_o) + \\
&\quad + y(\lambda_o\rho_o - x\ell_o - ym_o - zn_o) + (v+sy)r_1 + (v-sy)r_2 , \\
n &= n_o + w\frac{\gamma-1}{s^2}(u\ell_o + vm_o + wn_o - \frac{q^2}{2}\rho_o - e_o) + \\
&\quad + z(\lambda_o\rho_o - x\ell_o - ym_o - zn_o) + (w+sz)r_1 + (w-sz)r_2 , \\
e &= e_o + H\frac{\gamma-1}{s^2}(u\ell_o + vm_o + wn_o + \frac{q^2}{2}\rho_o - e_o) + \\
&\quad + \lambda_o(\lambda_o\rho_o - x\ell_o - ym_o - zn_o) + (H+s\lambda_o)r_1 + (H-s\lambda_o)r_2 .
\end{aligned} \tag{6.130}$$

Let us recall that the left-hand sides are nothing but the flux differences of the continuity-, the three momentum- and the energy equation written for one dimension. The quantities with the subscripts are taken to be unknowns, while the coefficients are thought of being calculated from a suitable mean state, say Roe's average or the arithmetic mean. So the latter are taken to be known as initial values. The unknown quantities aligned with the first and second eigenvalue r_1 and r_2 can be found from two operations applied to the above system of equations:

$$x\ell + ym + zn - \lambda_o\rho = s(r_1 - r_2) , \tag{6.131}$$

and

$$e + \frac{q^2}{2}\rho - u\ell - mw - wn = \frac{s^2}{\gamma-1}(r_1 + r_2) , \tag{6.132}$$

where use has been made of the definition of the total enthalpy

$$H = \frac{s^2}{\gamma-1} + \frac{q^2}{2} .$$

By addition and subtraction the two previous equations can be solved for the unknowns r_1 and r_2:

$$r_1 = \frac{1}{2s^2}\{(\frac{\gamma-1}{2}q^2 - s\lambda_o)\rho + [sx-(\gamma-1)u]\ell + $$
$$+ [sy-(\gamma-1)v]m + [sz-(\gamma-1)w]n + (\gamma-1)e\} ,$$
$$(6.133)$$
$$r_2 = \frac{1}{2s^2}\{(\frac{\gamma-1}{2}q^2 + s\lambda_o)\rho - [sx+(\gamma-1)u]\ell - $$
$$- [sy+(\gamma-1)v]m - [sz+(\gamma-1)w]n + (\gamma-1)e\} .$$

Note the formal similarity with the definitions of R_1 and R_2. There remains to solve the system of equations for the flux difference pieces assigned to the zeroth eigenvalue ρ_o, ℓ_o, m_o, n_o, e_o. This is easily achieved by the observation that the remainder of the total flux difference can only be the quantities with the subscript zero after the variables r_1 and r_2 are known:

$$\rho_o = \rho - r_1 - r_2 ,$$
$$\ell_o = \ell - (u+sx)r_1 - (u-sx)r_2 ,$$
$$m_o = m - (v+sy)r_1 - (v-sy)r_2 , \qquad (6.134)$$
$$n_o = n - (w+sz)r_1 - (w-sz)r_2 ,$$
$$e_o = e - (H+s\lambda_o)r_1 - (H-s\lambda_o)r_2 .$$

We see that we have obtained again the form of the conservative linear Riemann solver although we started from quite another concept.

Of course the method of projection can also be applied to the differences of the flow variables rather than the differences of the conservative fluxes. The conceptual idea is that any flux difference can be represented by the difference of the flow variables multiplied by the normal speed of the discontinuity. This speed is the same for the different elements of the flow variable difference vector:

$$\overset{\rho}{E}_\xi = s\rho_\xi \ ,$$

$$\overset{\ell}{E}_\xi = s\ell_\xi \ ,$$

$$\overset{m}{E}_\xi = sm_\xi \ , \qquad (6.135)$$

$$\overset{n}{E}_\xi = sn_\xi \ ,$$

$$\overset{e}{E}_\xi = se_\xi \ ,$$

where the operator $\partial/\partial\xi$ can replaced by

$$\frac{\partial}{\partial\xi} = \cdot \text{ right} - \cdot \text{ left} \ .$$

Upon dividing each line of the Riemann solver by the speed of the discontinuity s, conceptually the same formalism as for the plain flux differences can be applied. Subsequent multiplication of the quantities labeled zero and the quantities r_1 and r_2 by the non dimensionalized eigenvalues

$$\dot{\xi}_j = \lambda_j A = \lambda_j(\xi_x^2 + \xi_y^2 + \xi_z^2)^{0.5}, \quad j=0,1,2 \qquad (6.136)$$

results in the quasilinear flux difference pieces aligned with each of the three characteristics.

6.38 Evaluation of Eigenvalues by Projection

There are two techniques for the evaluation of eigenvalues. The first is based on the calculation of the three characteristic speeds from the coefficients formed from an average of the left and right flow quantities (arithmetic mean, Roe's average). There is also another possibility available now. After the strengths of the waves associated with the different wavespeeds are known from the projection technique and also the split parts of the flux difference are found by projection these two informations can be used for the implicit determination of the eigenvalues. Let us consider the following analogy:

$$
\begin{aligned}
r_o &= \dot{\xi}_o \; r_{oo} \, , \\
\ell_o &= \dot{\xi}_o \; \ell_{oo} \, , \\
m_o &= \dot{\xi}_o \; m_{oo} \, , \\
n_o &= \dot{\xi}_o \; n_{oo} \, , \\
e_o &= \dot{\xi}_o \; e_{oo} \, , \\
r_1 &= \dot{\xi}_1 \; r_{11} \, , \\
r_2 &= \dot{\xi}_2 \; r_{22} \, .
\end{aligned}
\tag{6.137}
$$

The left hand side are the pieces of the flux difference each being assigned to the appropriate eigenvalues. The quantities with the double subscripts are the strengths of the different waves obtained from the differences of the conservative flow quantities. Both types of quantities – the flux splitting pieces and the wave strengths – are assumed being known form the technique of projection. Then the first and the second eigenvalue are an immediate result of the projection technique:

$$
\begin{aligned}
\dot{\xi}_1 &= \frac{r_1}{r_{11}} \, , \\
\dot{\xi}_2 &= \frac{r_2}{r_{22}} \, .
\end{aligned}
\tag{6.138}
$$

The zeroth eigenvalue is obviously overspecified by five equations for one unknown. It can be found therefore, for instance, from the least square minimum of these five equations:

$$
\begin{aligned}
&(r_o-\dot{\xi}_o r_{oo})^2 + (\ell_o-\dot{\xi}_o \ell_{oo})^2 + (m_o-\dot{\xi}_o m_{oo})^2 + \\
&+(n_o-\dot{\xi}_o n_{oo}) + (e_o-\dot{\xi}_o e_{oo})^2 = \text{Min} \, .
\end{aligned}
\tag{6.139}
$$

From this condition the zeroth eigenvalue can be obtained with ease:

$$
\dot{\xi}_o = \frac{r_o r_{oo} + \ell_o \ell_{oo} + m_o m_{oo} + n_o n_{oo} + e_o e_{oo}}{r_{oo}^2 + \ell_{oo}^2 + m_{oo}^2 + n_{oo}^2 + e_{oo}^2} \, . \tag{6.140}
$$

There are other implicit techniques for the evaluation of the eigenvalues based on the assumption that the combinations of the wavestrength with particular coefficients should vanish analytically. Whether this is also true for the numerical counterpart cannot be said in advance. The present eigenvalue determination, however, is the most faithful alignment with the numerics.

6.39 Non-Oscillating Interpolation: Introduction

Up to now only first-order accurate schemes were considered. They are all based on directly inserting the flow values of either the left or the right cell into the Riemann solvers formed with the coefficients calculated at the cell face using either the arithmetic mean or Roe's average. Such schemes are three-point schemes, since in one dimension the total flux difference calls for the flow quantities of three points:

i-1	i	i+1

First-order schemes are exceptional stable schemes since they do not allow the generation of new extrema of the flow variables distribution at least in the linear case as shown previously. So it would be desireable to generate higher-order accurate schemes at least in the space derivatives by adding somehow higher-order differences for the calculation of the flow in smooth regions. If, however, non-monotonic distributions of the flow variables do occur, then the scheme should be switched down to first order accuracy locally in order to prevent the solution to grow unbounded at these places. The techniques how to make the scheme know unwanted extrema are manifold and their incorporation into a particular code is quite heuristic. It can be said even that inventing such interpolation procedures is the key point the program developer has to consider. At the present stage of technics for solving the Euler equations, research on non-oscillating interpolation is by far not at its end. Therefore in the following sections only those methods can be mentioned which where tested successfully up to now.

At this stage several remarks are in order: First, as was stated already, it has been accepted by the numerical community that the order of accuracy is checked by Taylor series expansions in the computational space with an orthogonal equidistant grid rather than in the physical space with an usually non-equidistant point distribution. Second, in principle a multi-dimensional finite-volume method cannot exceed second-order accuracy, because the flux integration over a cell-face is performed by the trapezoidal rule. Higher order cell-face integration such as, say, Simpson's rule or Gauss' integration do have a destabilizing effect on the numerical scheme. Therefore, no previous attempts are known to increase the accuracy by a cell face integration higher than second order. The analysis to follow takes only into account the order of the interpolation function in the computational space. Extensive studies incorporating the individual distances between grid points into the interpolation did not lead to the desired success and are therefore mentioned only marginally.

6.40 Five-Point Schemes

The one-dimensional difference star layout is clear from the following sketch:

$i-2$	$i-1$	i	$i+1$	$i+2$	cell
$i-\frac{5}{2}$	$i-\frac{3}{2}$	$i-\frac{1}{2}$	$i+\frac{1}{2}$	$i+\frac{3}{2}$	$i+\frac{5}{2}$ cell face

Let us consider for the moment the case with the eigenvalue being positive ($\lambda>0$). If the cell labeled i is to be updated then for the left state at the right cell face labeled i+1/2 we can make the following most general ansatz using differences of the flow variables:

$$U_{\ell,i+1/2} = U_i + a(U_{i+2}-U_{i+1}) + b(U_{i+1}-U_i) + c(U_i-U_{i-1}) \; . \tag{6.141}$$

The analoguous ansatz for the left state at the left cell face labeled i-1/2 is

$$U_{\ell,i-1/2} = U_{i-1} + a(U_{i+1}-U_i) + b(U_i-U_{i-1}) + c(U_{i-1}-U_{i-2}) \; .$$

Keeping in mind the homogeneous property of the Euler equations, we easily recognize that the flux difference formed with the states defined above between the left and the right cell face has the same order of accuracy as the difference of the conservative flow variables and vice versa. Therefore, for an estimate of the order of accuracy of the flux difference it is sufficient to consider the difference of the states of the conservative flow quantities at the right and the left cell face. For positive eigenvalues these differences do read

$$U_\xi = U_{\ell,i+1/2}-U_{\ell,1-1/2} = U_i(1+a-2b+c) + U_{i-1}(b-1-2c) + \\ + U_{i-2}c+U_{i+1}(b-2a) + aU_{i+2} \; . \tag{6.142}$$

A first rule for constructing upwind interpolation functions can be easily seen: From the derivation of the primitive first-order schemes we see immediately, that a stable scheme needs the coefficient of the update value labeled i be positive for positive eigenvalues. Only in this case the contribution to the matrix diagonal is such that it may be well conditioned. If this coefficient is zero or negative (for positive eigenvalues) we can be sure that the scheme is unstable. Now we try to adjust the coefficients a,b,c such that the scheme gets the desired order of accuracy. For this purpose the individual function values are expressed by a Taylor series expansion with respect to the update point labeled i:

$$U_{\xi num}=(1+a-2b+c)U_i+(b-1-2c)(U_i-\overset{1}{U}+\overset{2}{U}-\overset{3}{U}+\overset{4}{U})+c(U_i-2\overset{1}{U}+4\overset{2}{U}-8\overset{3}{U}+16\overset{4}{U})+$$

$$+\ (b-2a)(U_i+\overset{1}{U}+\overset{2}{U}+\overset{3}{U}+\overset{4}{U})+a(U_i+2\overset{1}{U}+4\overset{2}{U}+8\overset{3}{U}+16\overset{4}{U})\ . \qquad (6.143)$$

The superscripts 1 through 4 indicate the order of the derivative. The denominators are omitted and can be thought of as being included in the yet unknown coefficients. Since we are working in the computational space, the spacing between the cell centers can be taken to be unity without loss of generality. Rearrangement yields the numerical difference of the flow variables in terms of derivatives of increasing order:

$$U_{\xi num}=\overset{1}{U}+\left[2(a+b+c)-1\right]\overset{2}{U}+\left[1+6(a-c)\right]\overset{3}{U}+\left[14(a+c)+2b-1\right]\overset{4}{U}\ . \qquad (6.144)$$

The last formula is the most general scheme from which stable schemes up to third-order accuracy can be derived. We start with the classic upwind and biased upwind schemes. The first one is the classic first-order scheme.

6.41 First-Order Upwind Scheme

Here we have

$$a = b = c = 0\ ,$$

$$U_{\ell,i+1/2} = u_i\ , \qquad (6.145)$$

$$U_{r,i-1/2} = u_{i+1}\ .$$

The left and right state at the left cell face is obtained by replacing the subscript i by i-1 and needs not to be repeated here explicitly. This scheme is capable of capturing relatively strong shoks without oscillations.

6.42 Second-Order Upwind Scheme

Here only

$$a = b = 0 .$$

Since the coefficient of the second-order derivative must vanish, we get one single equation for the coefficient c:

$$c = 1/2 ,$$

$$U_{\ell,i-1/2} = U_i + 1/2(U_i - U_{i-1}) , \qquad (6.146)$$

$$U_{r,i+1/2} = U_{i+1} + 1/2(U_{i+1} - U_{i+2}) .$$

This is simply a linear extrapolation from either the left side or the right side to the cell face i+1/2. This scheme is capable of capturing shocks of moderate strength. But during the transient shock motion to its final steady state position it may generate considerable pre-shock spikes in all flow quantitites until the shock finds its final steady state position. The matrix conditioning check gives a desired positive value for the coefficient of the update value labeled i:

$$1 + a - 2b + c = 3/2 > 0 .$$

6.43 Third-Order Biased Upwind Scheme

In this case only a vanishes:

$$a = 0.$$

In order to make the coefficients of the second and third derivative vanish, the two equations have to be satisfied

$$\begin{aligned} 2(b+c) - 1 &= 0 , \\ 1 - 6c \quad &= 0 . \end{aligned} \qquad (6.147)$$

The solution is:

$$c = 1/6 , \qquad b = 1/3 .$$

The state vectors are then

$$\begin{aligned} U_{\ell,i+1/2} &= U_i + \frac{U_{i+1} - U_i}{3} + \frac{U_i + U_{i-1}}{6} , \\ U_{r,i+1/2} &= U_{i+1} + \frac{U_i - U_{i+1}}{3} + \frac{U_{i+1} - U_{i+2}}{6} . \end{aligned} \qquad (6.148)$$

The matrix conditioning check gives a positive number

$$1 + a - 2b + c = 1/2 > 0 .$$

This scheme produces spurious oscillations at moderate shocks and makes any code fail for strong shocks. The results, however, for smooth flows are excellent, particularly in three-dimensional low to moderate speed flow.

In Chapter XI this statement will be verified in conjunction with automobile aerodynamics.

6.44 Fourth-Order Centered Scheme

All derivatives from second to third order must vanish ,which yields the system of equations

$$\begin{aligned} 2(a+b+c) &= 1 , \\ 6(a-c) &= -1 , \\ 14(a+c)+2b &= 1 . \end{aligned} \qquad (6.149)$$

The solution is

$$\begin{aligned} c &= - a = 1/12 , \\ b &= 1/2 . \end{aligned} \qquad (6.150)$$

The matrix conditioning check reveals because of

$$1 + a - 2b + c = 0 ,$$

that this scheme certainly is unstable. It is a centered scheme, which gets no directional preference from the eigenvalue sign interrogation, since to the flow quantities to the left and the right of the cell face the same weighting coefficients are asigned. Still this scheme can be implemented in an upwind scheme for controlling the fourth-order diffusion. Therefore, also for this scheme the states are given:

$$U_{\ell,i+1/2} = U_{r,i+1/2} = U_i + 1/12(U_i - U_{i-1} + U_{i+1} - U_{i+2}) + 1/2(U_{i+1} - U_i) . \qquad (6.151)$$

This concludes the development of the classic schemes covering the orders of accuracy from first-order up to fourth-order. As mentioned previously, only the first-order scheme has a clear stability criterion, since for the linear model equation no new extrema of the solution function may be generated provided the CFL-restriction is fulfilled. The other schemes presented do not exhibit such a clear stable behaviour.

6.45 The von Neumann Stability Test for Upwind Schemes

Out of many considerations to test the stability of a particular numerical scheme, the von Neumann test (see e.g. Ref. 14) is certainly the most popular. It is based on the question whether the error of the numerical scheme relative to the exact solution will grow or decay with the development of the solution during the iterative process. Again we consider the model equation used so often in this text. The von Neumann stability test requires the coefficent in the model equation, that is the eigenvalue $\dot{\xi}$, not to vary with respect to time. So the von Neumann test is only to apply to a strictly linear problem. Provided the time step is sufficiently small, this assumption also may hold for the Euler equations which are certainly non-linear. Now we think about an intermediate state during the course of iterations of our numerical solution U_{num} of the model equation. It can be taken as a superposition of the exact solution U and an error distribution ε which somehow creeped into our numerical solution:

$$U_{num} = U + \varepsilon \ .$$

Upon inserting this expression into the model equation, we arrive at

$$\dot{U}_{num} + \dot{\xi} U_{num\xi} = (U+\varepsilon)^{\bullet} + \dot{\xi}(U+\varepsilon)_{\xi} = 0 \ . \tag{6.152}$$

Since U is the exact solution fulfilling the model equation at all instances, we obtain easily the error equation

$$\dot{\varepsilon} + \dot{\xi}\, \varepsilon_{\xi} = 0 \ .$$

It can be solved by the ansatz of separated variables

$$\varepsilon = XT \ ,$$

where X is a function of ξ only and T is a function of time only. The model equation assumes with this ansatz the form

$$X\dot{T} + \dot{\xi} T X_{\xi} = 0 \ ,$$

or upon division by XT

$$\frac{\dot{T}}{T} + \dot{\xi}\, \frac{X_{\xi}}{X} = 0 \ .$$

Since the first group depends only on time, and the second group depends only on the abscissa, any solution can be composed of

$$\frac{\dot{T}}{T} = b\dot{\xi} \text{ and } \frac{X_\xi}{X} = -b \ , \tag{6.153}$$

which trivially fulfills the last equation. The constant b can be any arbitrary number. The solution of the last two ordinary differential equations, which also can be written as

$$\frac{dT}{T} = b\dot{\xi}dt, \text{ and } \frac{dX}{X} = -bd\xi \ , \tag{6.154}$$

is

$$\ell nT = A + b\dot{\xi}t, \text{ or } T = e^{A+b\dot{\xi}t} \tag{6.155}$$

and

$$\ell nX = B - b\xi, \text{ or } X = e^{B-b\xi} \ , \tag{6.156}$$

where A and B are constants of integration.

The error of the numerical scheme can therefore be expanded as a function of the time and the abscissa:

$$\varepsilon = XT = e^{A+B+b(\dot{\xi}t-\xi)} \ .$$

We are going now to fix the constants of integration A and B. Provided at the time t=0 the error at the update point $\xi=0$ is ε_o we obtain the error function to be

$$\varepsilon = \varepsilon_o e^{b(\dot{\xi}t-\xi)} \ .$$

The constant b is an arbitrary number as stated above. So we may also assign a complex number to it:

$$b = c + i\theta \ ,$$

where i is the imaginary unity. This allows the error distribution in the vicinity of the update point labeled "o" to have wave character rather than a monotonuous functional behaviour:

$$\varepsilon = \varepsilon_o e^{(c+i\theta)(\dot{\xi}t-\xi)} = \varepsilon_o e^{c(\dot{\xi}t-\xi)} e^{i\theta(\dot{\xi}t-\xi)} .$$

The von Neumann stability criterion says that the absolute ratio of the new error ε relative to the old error ε_o at ξ=t=0 (our definition) should be less than unity:

$$\left|\frac{\varepsilon}{\varepsilon_o}\right| < 1 .$$

This introduces another restriction from which we can evaluate the constant c. The absolute value of the third coefficient is always unity:

$$\left|e^{i\theta(\dot{\xi}t-\xi)}\right| \equiv 1 .$$

In order to make the amplitude of the error growth indifferent with the sign and size of the exponent $\dot{\xi}t-\xi$, where the eigenvalue may be both, positive or negative, we simply put the constant c to zero:

$$\varepsilon = \varepsilon_o e^{i\theta(\dot{\xi}t-\xi)} .$$

Let us check now whether the first-order scheme of the previous Sub-Chapter 6.40 is stable or not in the case of the simple forward update (Euler integration), which is expressed by:

$$\frac{u_i - u_o}{\Delta t} + \dot{\xi}(u_{oi} - u_{oi-1}) = 0 , \qquad \dot{\xi}>0 .$$

The associated error equation is, with the definition of the CFL-number $\dot{\xi}\Delta t$=CFL,

$$\varepsilon = \varepsilon_{oi} - \mathrm{CFL}(\varepsilon_{oi} - \varepsilon_{oi-1}) ,$$

where "o" indicates the initial time level $t=t_o=0$.

With the previous ansatz we obtain

$$\varepsilon_{oi} = \varepsilon(t=\xi=0) = \varepsilon_o , \qquad \varepsilon_{oi-1} = \varepsilon(t=0, \xi= -\Delta\xi= -1) = \varepsilon_o e^{i\theta} . \tag{6.157}$$

The new error is

$$\varepsilon = \varepsilon_o[1 - \mathrm{CFL}(1-e^{i\theta})] = \varepsilon_o[1 - \mathrm{CFL}(1-\cos\theta - i\sin\theta)] . \tag{6.158}$$

Therefore the complex growth coefficient is

$$G \equiv \frac{\varepsilon}{\varepsilon_o} = 1-CFL(1 - \cos\theta - i\sin\theta) \ .$$

In the complex plane the number G are circles with their origin at x=1-CFL and z=0, Fig. 6.10. They have a radius equal to the CFL-number. Now we can plot all the G-functions depending on the phase angle θ with the CFL-number being a fixed parameter.

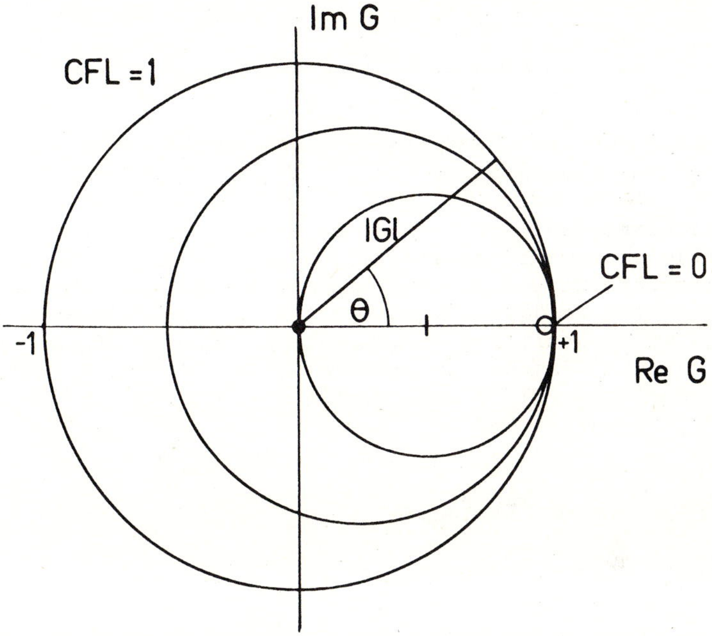

Fig. 6.10 Plot of the complex growth coefficient for the first-order scheme

The von Neumann stability test says that for stability the G-curves must lie inside the curve $|G|=1$, which is the unit circle around the coordinate origin. This in fact is the case for the classic first-order scheme up to the limiting CFL-number being unity. From the plot we see that obviously the vicinity of vanishing phase angle θ=0 is the critical place, where the stability should be analyzed with particular care. It is clear that the amplitude $|G|$ should decrease with increasing phase angle in the vicinity of θ=0. A sketch of the region θ=0 makes things clear, Fig. 6.11.

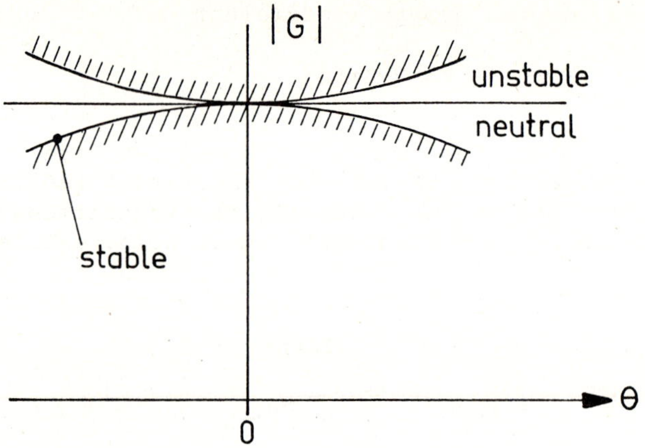

Fig. 6.11 Growth coefficient in the vicinity of vanishing phase angle θ

We see immediately by it that a stable scheme should have negative curvature there. This does not exclude, however, that the amplification factor G penetrates the unit circle at some other place indicating instability. If the curvature at vanishing phase angle $\theta=0$ is positive, then the scheme is for sure unstable in the von Neumann sense. Let us do this check for the classic first order scheme:

$$\left|G\right|^2 = (1 - CFL + CFL\cos\theta)^2 + CFL^2\sin^2\theta\ ,$$

$$\left|G\right|^2_\theta = 2CFL(CFL - 1)\sin\theta\ , \tag{6.159}$$

$$\left|G\right|^2_{\theta\theta}\ (\theta=0) = 2CFL(CFL - 1)\ .$$

The last line gives the curvature, which is negative for CFL-numbers less than unity as it should be for a stable scheme.

Now we turn our attention to the classic second-order upwind scheme. Again we consider the simple one-step update or forward-in-time integration.

The error equation reads for this case

$$\varepsilon = \varepsilon_o - \frac{CFL}{2}\ (3\varepsilon_o - 4\varepsilon_{i-1} + \varepsilon_{i-2})\ . \tag{6.160}$$

The growth factor is therefore

$$G = 1 - \frac{CFL}{2}\ (3 - e^{i\theta} + e^{2i\theta})\ .$$

After forming the square of the amplitude

$$|G|^2 = [1 - \frac{3FL}{2} + \frac{CFL}{2}(4\cos\theta - \cos 2\theta)]^2 + \frac{CFL^2}{4}(4\sin\theta - \text{ain}2\theta)^2 , \tag{6.161}$$

we obtain the curvature of the squared amplitude at θ=0 to be

$$|G|^2_{\theta\theta}\ (\theta=0) = 2CFL^2 > 0 .$$

Therefore this scheme is for sure unstable in the von Neumann sense.

A plot reveals the fact that for small θ there is always a portion of the growth factor outside of the unit circle regardless of the size of the CFL-number.

The third-order scheme of the previous Sub-Chapter 6.42 also shows this property. Its error equation is with the formulas of the previous paragraph (λ>0):

$$\varepsilon = \varepsilon_o - CFL[\varepsilon_o + \frac{\varepsilon_{i+1} - \varepsilon_o}{3} + \frac{\varepsilon_o - \varepsilon_{i-1}}{6} - (\varepsilon_{i-1} + \frac{\varepsilon_o - \varepsilon_{i-1}}{3} + \frac{\varepsilon_{i-1} - \varepsilon_{i-2}}{6})] =$$

$$= \varepsilon_o - \frac{CFL}{6}(3\varepsilon_o - 6\varepsilon_{i-1} + \varepsilon_{i-2} + 2\varepsilon_{i+1}) , \tag{6.162}$$

where the subscript 'o' indicates the place i.

The amplification factor is therefore

$$G = 1 - \frac{CFL}{6}(3 - 6e^{i\theta} + e^{2i\theta} + 2e^{-i\theta}) .$$

Also this scheme has a growth coefficient the amplitude of which is greater than unity in the vicinity of zero phase angle θ=0. In the von Neumann sense the scheme is unstable for any CFL-number.

Finally the fourth-order centered scheme is considered. From the previous Sub-Chapter 6.43 we find the error equation to be

$$\varepsilon = \varepsilon_o - \frac{CFL}{12}[8(\varepsilon_{i+1} - \varepsilon_{i-1}) + \varepsilon_{i-2} - \varepsilon_{i+2}] . \tag{6.163}$$

The growth factor is

$$G = 1 - \frac{CFL}{12}[8(e^{-i\theta} - e^{i\theta}) + e^{i2\theta} - e^{-2i\theta}] =$$

$$= 1 - i\frac{CFL}{12}(\sin 2\theta - 4\sin\theta) .$$

We see immediately from Fig. 6.10 that this is the equation of a vertical line through the point ReG=1. Therefore, the amplitude is larger than one for any CFL-number. That is why the scheme is clearly unstable.

6.46 Criticism on the von Neumann Stability Test for Upwind Schemes

Although the von Neumann Stability Test is a simple tool widely used for checking numerical stability, some critical remarks are justified. It is clear from practical applications that the classic second-order upwind scheme and also the third-order biased upwind scheme proved to be stable in hundreds of Euler calculations. Even the fifth-order scheme, which will be described later, does a very nice job without any tendency to instability at least in subsonic flow calculations, although the measure of instability is even worse than that of the third-order scheme. So the von Neumann test is pretty useless for upwind schemes. Upwind schemes, though unstable in the von Neumann sense, often are superior to centered schemes with clear von Neumann stability. Another fact is that the von Neumann test says nothing about the formation of wiggles or short wave oscillations in the profiles of the solution. Once they occur, instability is on its way in the numerical calculation regardless of the fact that the von Neumann test says that the scheme is stable. A much better approach to check stability therefore is the concept of monotonicity preservation and that of the generation of new extrema. An older principle is the check of the scheme to remain globally bounded within the extrema of the intermediate solution profiles. This is the idea of total variation diminishing (TVD) difference schemes. So at least in the context with upwind schemes we simply can forget about the von Neumann test, since it is useless for all of the latter. Another feature makes the von Neumann test doubtful for schemes solving the Euler equations: The von Neumann test is only valid for constant coefficients. The Euler equations, however, do exhibit a non-linear behaviour of the coefficients.

6.47 Extremum Principles for Upwind Schemes

As we have seen, the von Neumann stability test is an useless tool for upwind schemes. A more rigorous justification of the stability of upwind schemes can be obtained from the question whether the function value at an extremum, which may be a relative or an absolute maximum or minimum, grows or decays. For the classic first-order scheme the situation is easy to analyze in the case of the one-step update. The new updated value is found simply from linear interpolation. Let us consider the case with a positive eigenvalue. Furthermore, we apply the pseudo-unsteady update process with locally varying time step, but constant CFL-number, since we are interested only in the steady state solution. The new update value of the flow quantity of the model equation at a point i is then simply

$$U_{new} = U_i(1 - CFL) + CFL\ U_{i-1} \ .$$

Provided the CFL-number is less than unity, the updated value is simply an asymmetric weight of the point to be updated and its left neighbour. The new value is therefore bounded by the interval given by U_i and U_{i-1}:

$$\min(U_i, U_{i-1}) < U_{new} < \max(U_i, U_{i-1}) \ .$$

Therefore, if U_i is the function value of a maximum or a minimum, that extremum is degraded under all circumstances. It never may grow nor fall over, respectively below, the value given initially at the extremum point i. So the classic first-order scheme must be stable for CFL-numbers less than unity at least in the linear scalar case.

The situation for the classic second-order upwind scheme is slightly different. The update reads

$$U_{new} = U_i - \frac{CFL}{2}(3U_i - 4U_{1-1} + U_{i-2}) = U_i(1-CFL) + CFL(U_{i-1} - \frac{U_i + U_{i-2}}{2}) \ . \tag{6.164}$$

The first part guarantees boundedness of the updated new value while the second part is sort of a perturbation to it, expressed by the curvature of the second-order one-sided polynomial. If the function value at the point to be updated is a maximum and the curvature term is negative, no difficulty arises from the latter. The new value will be less than the maximum initial value at the point i. On the other hand, if the initial value at the point i is a minimum and the curvature term is positive, the updated value will be greater than the initial minimum value at the point i. For both cases the extrema will be degraded and the scheme will be stable therefore. Difficulties do arise if the curvature term at a maximum is positive, or if the curvature term at a minimum is negative. Since this may make maxima grow or minima fall. There seems to be, however, still some boundedness also in the latter cases. Again we consider the case of a maximum, and do relax the requirement that maxima may not grow at all by the statement that maxima may not grow unbounded. Expressing the increment by which a maximum may grow – which only may happen if the curvature term is positive – by a fraction ε of the curvature term, we obtain the condition of boundedness to be

$$U_{new} < U_i + \varepsilon \, (U_{i-1} - \frac{U_i+U_{i-2}}{2}) \, , \qquad (6.165)$$

where ε is a positive arbitrarily small number different from zero. Inserting the expression for the updated value we find

$$U_i(1-CFL)+CFL\ U_{i-1}+(CFL-\varepsilon)(U_{i-1}- \frac{U_i+U_{i-2}}{2}) \leq U_i \, . \qquad (6.166)$$

If the CFL-number is less than ε and if the curvature term is positive, we see easily that the new updated value is bounded and may not grow to an arbitrarily large value. This means that the classic second-order upwind scheme is always stable provided the CFL-number is sufficiently small. The fact that new extrema may be generated by this scheme does not mean that it tends to instability. Many caluclations with this scheme proved always successful for smooth flow problems. It is even capable to capture shocks up to moderate strength without producing wiggles or oscillations. Problems do arise only with strong shocks. In this case the pseudo-unsteady shock motion may generate isolated spikes in the flow variables at intermediate shock locations. If the steady state is approached, usually the spikes do disappear.

The situation for the third-order biased upwind scheme is quite different from that of the second-order upwind scheme. Its pseudo-unsteady update reads

$$U_{new} = U_i - \frac{CFL}{6} \, (3U_i-6U_{1-1}+ U_{1-2}+ 2U_{i+2}) =$$

$$= U_i(1-CFL) + CFL\ U_{i-1}+ \frac{CFL}{6} \, [U_i-U_{i-1}+ 2(U_i-U_{i+1})] \, . \qquad (6.167)$$

If U_i is a maximum, the curvature like term is always positive and is added to the value constituting safe boundedness. Still in most smooth flow situations it happens that the differences inside the square bracket are very small, besides the fact that only one sixth of them is added to the interpolated value stated by the first group of the formula. But we see that the stability bound is very narrow. In practical applications it turned out that this scheme is only well suited for subsonic flow calculations. Captured shocks cannot be treated by this scheme, not only because of the relatively low stability limit, but because of the upward influence introduced by the flow quantities at the point i+1.

We have now listed some heuristic arguments for the previous schemes to be stable. They are no mathematical proofs. Their suitability for solving the Euler equations is proved at best numerically. Regardless whether the von Neumann test is positive or not, or the heuristic arguments do hold or not, the schemes considered did run successfully for the Euler equations. The fourth-order centered scheme is, however, unstable for sure. On the other hand, neither the von Neumann stability test nor any extremum principles give a hint that any centered scheme may run stable in the one-step update mode.

6.48 Foundation of Flux Limiting

All the previous arguments and the von Neumann stability test did not care about the non-linearity of the Euler equations and the development of shock wave discontinuities in flows at transonic and higher Mach numbers. All we know is that the classic first-order scheme is the one with the best properties with respect to stability and boundedness. All higher-order schemes do, indeed, exhibit sort of a "grey" region, where we do not exactly know whether they might work or not. One of those situations is the occurence of an extremum of the flow variables in the flow field or a discontinuity. To make sure that in these situations an upwind scheme (fully upwinded or biased upwinded) works without exhibiting a tendency for instability or wiggles or oscillations in the distribution of the flow variables, we always can switch the higher-order scheme down to the first-order upwind scheme at those particular places. Another alternative is to interpolate a new state somewhere in between the higher-order state and the first-order state at these places, maybe parameter controlled, but such, that the overall convergence and the local cosmetics – no wiggles, no oscillations – is guaranteed.

As expected, an interpolation as just mentioned or a switch from high order to first order is a non-linear process which hardly can be analyzed by pure mathematics. It is rather the vast number of successful numerical experiments, using non-linear switches, which justifies this approach. In the literature the process of switching a difference scheme from a high order of accuracy down to a low order of accuracy at places where this is needed is called flux limiting. There is no unique theory about flux limiting. There are numerous ways to achieve flux limiting and practically each author uses that limiter he trusts in most. The design of limiters is one of the most important ingredients of a characteristics-based Euler code. Their effectiveness can be judged only by numerical experiments.

6.49 Flux Limiting by Sensing Functions

A very simple approach of flux limiting can be obtained by a formally linear superposition of the schemes of different orders of accuracy in the following way:

$$U_{i+1/2} = \{[\overset{4}{U}(1-C) + C\overset{3}{U}](1-B) + B\overset{2}{U}\}(1-A) + A\overset{1}{U} \; . \qquad (6.168)$$

The superscripts 1 through 4 indicate here the order of accuracy of the schemes derived so far. For brevity, the subscript "left" or "right" is dropped. We see easily, that we can recover the original schemes by putting the parameters A,B,C to either one or zero.

Another choice allows the scheme being of formal high-order accuracy everywhere, if we make A being a function proportional to the square of a characteristic distance between gridpoints

$$A \sim \Delta x^2 \; ,$$

and if B is made a function of at least the distance of two adjacent grid points. Of course any higher order of the functions A and B than indicated above is admissible. The choice of the functions A and B should be such that wiggles or oscillations in any of the flow variables are detected immediately as well as, of course, shock waves. This suggests that the local Mach number should be used as entry for the two functions, since it is built up from all the primitive flow variables available and thus will detect irregularities in the distribution of any of them.

With

$$M^2{}_{\xi\xi i} = M^2{}_{i-1} + M^2{}_{i+1} - 2M^2{}_{i} \; , \qquad (6.169)$$

and

$$M^2{}_{\xi\xi i+1} = M^2{}_{i} + M^2{}_{i+2} - 2M^2{}_{i+1} \; ,$$

we specify the function A to be

$$A_{i+1/2} = \min(1, \alpha \left| M^2{}_{\xi\xi i+1} - M^2{}_{\xi\xi i} \right|) \; , \qquad (6.170)$$

and the function B to be

$$B_{i+1/2} = \min[1, \beta(\left|M^2{}_{\xi\xi i+1}\right| + \left|M^2{}_{\xi\xi i}\right|)] \; ,$$

where α and β are dimensionless user specified constants of the order unity.

The third parameter C is a constant. If it is taken greater than one, the smoothing effect of the fourth derivative may be considerably increased. In the explicit mode a number of 2 or 3 for C allows a doubling of the admissible CFL-number. Going beyound those values makes the scheme tending to instability.

There are advantages of the present approach which consist of the easy parameter controlled adaptation of the scheme to all kind of flow situations, whether it is extreme low speed or highly compressible flow with strong shocks and the simple calculation of the sensing functions. There are, however, also drawbacks: in case the scheme should be extended to the solution of the Navier-Stokes equations, the sensing functions are nearly zero exactly in the boundary layer where high solution gradients do occur. What comes to our mind is using the pressure as an entry to the sensing functions. Thus at least in the flow direction of a boundary layer the sensing function is not identical to zero. In the direction normal to the body, however, the sensing functions would be almost zero again in a boundary layer, since the pressure variation in normal direction is very small. If still the sensing functions are formed with the pressure a normalization procedure has become familiar which will be briefly considered:

$$A = \min\left[1, \alpha\left(\frac{|p_{i+2}-p_{i-1}-3(p_{i+1}-p_i)|}{p_{i+2}+p_{i-1}+3(p_{i+1}+p_i)}\right]\right. ,$$

$$B = \min\left[1, \beta\left(\frac{|p_{i+1}+p_{i-1}-2p_i|}{p_{i+1}+p_{i-1}+2p_i} + \frac{|p_{i+2}+p_i-2p_{i+1}|}{p_{i+2}+p_i+2p_{i+1}}\right)\right] . \qquad (6.171)$$

An undesirable property of the sensing functions using the Mach number square, is that it is non-differentiable because of the minimum option. The same holds also for the sensing functions using the normalized pressure formulas except for the parameters α and β being less than unity.

The sensing functions specified so far do not include the total variation diminishing restriction, since they distinguish between smooth and abrupt extrema. Smooth extrema are treated in the high-order mode. Only at wiggles or shock waves the scheme returns to first-order accuracy. Finally it is remarked that one and the same sensing function acts on the interpolation of all five flow quantities. This is in very contrast to the more elaborate (and computationally more expensive) flux limiters discussed later in this text.

6.50 Flux Limiting by Biased Differences

Again only five-point schemes are considered here. Using directionally biased differences for the left and right state of variables is in principle nothing new. It is only a formal recast of the superposition of interpolations of different order of accuracy derived in the last sub-chapter. The formalism is easily obtained by inserting the difference expressions of the various schemes into the full interpolation formula of the last paragraph. We first consider only the left-state formula:

$$U_{i+1/2,\ell} = \{[(U_i + \frac{U_i - U_{i-1} + U_{i+1} - U_{i+2}}{12} + \frac{U_{i+1} - U_i}{2})(1-C) +$$

$$+ C(U_i + \frac{U_{i+1} - U_i}{3} + \frac{U_i - U_{i-1}}{6})](1-B) + B(U_i + \frac{U_i - U_{i-1}}{2})\}(1-A) + U_i A. \tag{6.172}$$

After rearrangement we obtain:

$$U_{i+1/2,\ell} = U_i + \{(U_i - U_{i-1}) \frac{1+5B+C(1-B)}{12} +$$

$$+(1-B)[(U_{i+1} - U_i) \frac{3-C}{6} + (U_{i+1} - U_{i+2}) \frac{1-C}{12}]\}(1-A) . \tag{6.173}$$

Most authors drop the option of increased fourth-order smoothing and make the coefficient of the leading difference vanish by putting the constant parameter C unity. The formula for the interpolation then reduces to

$$U_{i+1/2,\ell} = U_i + [(U_i - U_{i-1})(1+2B) + (U_{i+1} - U_i)2(1-B)] \frac{1-A}{6} . \tag{6.174}$$

A clearer structure of the formula has become popular by the linear transformation of the parameter B

$$B = \frac{1 - 3C}{4} ,$$

where C is a new parameter. The interpolation is then

$$U_{i+1/2,\ell} = U_i + \frac{1-A}{4} [(U_i - U_{i-1})(1-C) + (U_{i+1} - U_i)(1+C)] . \tag{6.175}$$

Introducing a switching function s which is unity in smooth flow regions and tends to zero in regions of large flow gradients a slight modification of the previous formula has also become very popular:

$$U_{i+1/2,\ell} = U_i + \frac{s}{4}[(U_i - U_{l-1})(1-\phi s) + (U_{i+1} - U_i)(1+\phi s)] \; . \tag{6.176}$$

This formula is used by many authors, e.g. Refs. 15, 16.

It contains all classes of second-order schemes the truncation error of which is controlled basically by the parameter ϕ. The classic second-order upwind scheme is easily identified by putting $\phi=-1$. The second-order biased upwind scheme is recovered with $\phi=0$. The second-order centered scheme is obtained with $\phi=1$. The third-order biased upwind scheme is found if we put $\phi=1/3$. Thus the scheme contains practically all options a five-point scheme offers, except the possibility of increasing the fourth-order smoothing term in the third-order mode. For completeness also the right-state interpolation function is given:

$$U_{i+1/2,r} = U_{i+1} + \frac{s}{4}[(U_{i+1} - U_{i+2})(1-\phi s) + (U_i - U_{i+1})(1+\phi s)] \; . \tag{6.177}$$

6.51 Flux Limiting with Minimum Dispersion

A finite-difference scheme with as many uneven derivatives of the truncation error as possible being missing, is called a minimum-dispersion scheme. For the five-point schemes considered in the present context this means that the coefficient of the third derivative of the Taylor-series expansion should be made vanish. The Taylor expansion for the general five-point ansatz is repeated here for the sake of convenience:

$$U_{\xi num} = \overset{1}{U} + [2(a+b+c)-1]\overset{2}{U} + [1+6(a-c)]\overset{3}{U} + [14(a+c)+2b-1]\overset{4}{U} \; . \tag{6.178}$$

Again only the case of a positive eigenvalue is considered. For completeness it is repeated that a,b,c are yet free constant coefficients multiplying the forward difference and the backward difference. The superscripts indicate the order of the derivatives of the Taylor-series expansion. In order to exploit the second derivative for controlled diffusion its coefficient should be negative. The fourth derivative also provides diffusion, if its coefficent is positive. Minimum dispersion requires the coefficient of the third derivative be zero. Summarizing these statements we obtain a system of equations

$$\begin{aligned} 2(a+b+c) - 1 &= -A \, , \quad A > 0 \, , \\ 1 + 6(a-c) &= 0 \, , \\ 14(a+c) + 2b - 1 &= B \, , \quad B > 0 \, , \end{aligned} \tag{6.179}$$

where A and B are positive numbers. After inversion we obtain

$$a = \frac{A + B - 2}{24} ,$$

$$b = \frac{6 - 7A - B}{12} , \qquad (6.180)$$

$$c = \frac{A + B + 2}{24} .$$

The matrix conditioning check requires

$$1 + a + c - 2b = \frac{5A + B}{4}$$

be positive, which is in fact true.

In order to make the scheme third-order accurate, the coefficient A should be a function at least proportional to Δx^2, which is zero in smooth regions of the flow and assumes a maximum value at shocks A_s. How this maximum value can be found requires the consideration, that for a positive eigenvalue the pre-shock-point should not be affected by the flow quantities of the post-shock-points. This means, that at a shock wave the coefficients a and b multiplying the forward difference and the cell face difference should become simultaneously zero at this place:

$$a = \frac{A + B - 2}{24} = 0 ,$$

$$b = \frac{6 - 7A - B}{12} = 0 . \qquad (6.181)$$

It is now easy to see that the shock value must be

$$A_s = \frac{2}{3} , \qquad B_s = 2 - A_s = \frac{4}{3} . \qquad (6.182)$$

This suggests that A can be one of the sensing functions mentioned above, say

$$A = \min(\tfrac{2}{3}, \alpha \, |M^2{}_{\xi\xi\xi}|) .$$

The coefficient B may be constructed as

$$B = \beta(\tfrac{2}{3} - A) + 2A ,$$

where α and β are user-specified constants. A particular choice for β makes the coefficient of the forward difference vanish:

$$a = \frac{A + B - 2}{12} = \frac{3A + \beta\big((2/3)-A\big)- 2}{12} = 0 .$$

Therefore $\beta=3$ and the coefficient B becomes

$$B = 2 - A .$$

For this particular one-parameter formulation the left state is

$$U_{i+1/2,\ell} = U_i + \frac{1}{6}\left[(U_{i+1} - U_i)(2-3A) + U_i - U_{i-1}\right] . \tag{6.183}$$

The right state follows from the reflection principle:

$$U_{i+1/2,r} = U_{i+1} + \frac{1}{6}\left[(U_i - U_{i+1})(2-3A) + U_{i+1} - U_{i+2}\right] . \tag{6.184}$$

6.52 Limiters

As we have seen, an important ingredient of the higher-order interpolation functions is a numerical device to switch the scheme back to low order at places in the flow field where shocks occur or spurious oscillations tend to evolve. One out of a big choice of such devices are sensing functions, which use higher-order differences of a sensitive flow variable in order to detect those pathologic spots in the flow as mentioned above.

A limiter in this sense is nothing but a sensing function. It has become, however, familiar to use the term limiter for sensing functions, which test each of the flow variables individually for non-smoothness. Thus each flow variable is the entry of its own limiter.

In the literature we find a large number of limiters, some of which are named after their authors, e.g. Refs. 16 to 18. Most of them do have, however, a drawback as serious as that of the sensing functions. Since they are based usually on minimum options or on "if-then" structures, they are not differentiable. That means, that these limiters are not smooth functions of their arguments. To our knowledge out of a vast variety there is only <u>one differentiable</u> limiter which is suggested by van Albada, see e.g. Ref. 19. It is particularly well suited for the interpolation by directionally biased differences. Let us consider the left state at the cell face i+1/2. It is calculated from the flow values given at the points i-1, i, i+1. So the limiter also should be extended over these three points (for positive eigenvalues). Van Albada's limiter reads for this constellation:

$$L_\ell = L_i = \frac{2(U_{i+1} - U_i) \cdot (U_i - U_{i-1}) + \varepsilon}{(U_{i+1} - U_i)^2 + (U_i - U_{i-1})^2 + \varepsilon} ,$$

where ε is a small number preventing a division by zero. The same formula can be used of course for the right state interpolation function if the operator is shifted to i+1:

$$L_r = L_{i+1} .$$

The number ε, which for simplicity is chosen equal for all five flow variables, has quite an impact on the efficiency of the limiter. If it is taken a great value no limiting is achieved at all. If it is taken as a small value then – with low diffusion schemes – it may have some influence on the entropy error distribution particularly on solid body boundaries. The rule of thumb for fixing the parameter ε is: as large as possible, but still preventing wiggles. For the following analytical investigations we put ε to zero and define an independent variable η as the ratio of the slopes:

$$U_{i+1} - U_i = \eta(U_i - U_{i-1}) . \tag{6.185}$$

Therefore

$$L = \frac{2\eta}{1+\eta^2} .$$

A graph reveals some properties of the limiter, Fig. 6.12.

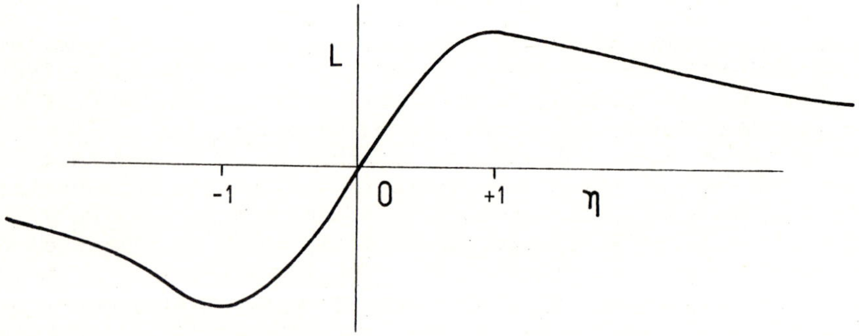

Fig. 6.12 The van Albada limiter.

Smooth flow is represented by $\eta \approx 1$. At this place the limiter has a smooth maximum which is a desirable property.

Extrema are represented by $\eta<0$. A symmetric extremum generates $\eta=-1$ which is the hardest case for a scheme to overcome. Limiting η to non-negative values – as van Leer did, see Ref. 20 – leads to a non-differentiable function, since the slope at the origin is non-vanishing. The biased differencing is for $\eta=-1$:

$$U_{i+1/2,\ell} = U_i - \frac{1}{4}\left[(U_i - U_{i-1})(1+\phi) + (U_{i+1} - U_i)(1-\phi)\right] , \tag{6.186}$$

$$U_{i+1/2,r} = U_{i+1} - \frac{1}{4}\left[(U_{i-1} - U_{i-2})(1+\phi) + (U_i - U_{i-1})(1-\phi)\right] .$$

The numerical difference

$$U_{\xi num} = U_{i+1/2,\ell} - U_{i-1/2,\ell} \tag{6.187}$$

produces a coefficient of the update value U_i which is

$$\frac{5 - 3\phi}{4} .$$

Now ϕ may vary from -1 to 1. We see immediately that the coefficient contributing to the matrix diagonal is always positive, even in the case of a formally second-order centered scheme ($\phi=1$).

Finally we check the order of accuracy of the von Albada limiter by a Taylor-series expansion up to second order:

$$U_{i+1} = U_i + U_x \Delta x + \frac{U_{xx}}{2} \Delta x^2 , \quad U_{i-1} = U_i - U_x \Delta x + \frac{U_{xx}}{2} \Delta x^2 .$$

After some trivial algebra we find

$$L = \frac{4U_x{}^2 - 2(U_{xx}\Delta x)^2 + (U_{xx}\Delta x)^2}{4U_x{}^2 + (U_{xx}\Delta x)^2} , \tag{6.188}$$

where a dummy term has been incorporated in the numerator in order to split the truncation error from the pure function value:

$$L = 1 - \frac{2(U_{xx}\Delta x)^2}{4U_x{}^2 + (U_{xx}\Delta x)^2} \approx 1 - O(\Delta x^2) . \tag{6.189}$$

We see that the von Albada limiter is of sufficient formal accuracy for schemes up to third-order accuracy. Recent developments by Eberle led to a new limiter derived from that of van Albada, which can be adapted to formally arbitrarily high-order schemes by an additional parameter p. It reads:

$$L = \frac{2\, d_\ell d_r \cdot [\max(|d_\ell|, |d_r|)]^p + \varepsilon}{|d_\ell|^{p+2} + |d_r|^{p+2} + \varepsilon} , \qquad (6.190)$$

where

$$d_r = U_{i+1} - U_i ,$$

$$d_\ell = U_i - U_{i-1} .$$

Assigning a large positive number to the exponent p makes this limiter a very sharp sensor, which vanishes only at extrema such as wiggles but otherwise is practically the base value retaining the high-order formulation throughout the smooth part of the flow field.

6.53 Seven-Point Schemes

Some attempts have been made to increase the accuracy of the scheme by extending the support of the left and right state being entry to the Riemann-Solver under consideration, to seven points at a whole. The strategy again is to adapt the coefficients of the derivatives in the Taylor-series expansion such that the low-order derivatives are multiplied by zero, while a high-order even derivative, multiplied with a coefficient of appropriate sign, acts as background smoothing device stabilizing the scheme. At shocks and places where spurious oscillations do appear, the scheme has to be switched back to low order.

How this is achieved is a matter of much ingenuity of the program designer. The general strategy certainly is to switch off at first the leading differences (for positive eigenvalues these are the differences on the right side of the cell face under consideration) and then the trailing differences. In any case, the scheme returns to first-order accuracy, and among all first-order modes the classic three-point scheme is the most robust. It also should be mentioned that either the sensing functions or limiters should be extended over the same difference star as the basic high-order scheme. This way a seven-point scheme can be at best of formal fifth-order accuracy. It should be mentioned, however, that the real accuracy cannot exceed second order because of the cell face integration, which is at most only second-order accurate. Also the usually non-equidistant grid spacing prohibits an order of accuracy being higher than at most two. Trials to compensate non-equidistant grid errors by incorporating the geometric distances in between the grid points did not lead to the desired success and are extremely hard to incorporate into conservative difference schemes which are essentially based on simple differences of fluxes. This is why there are no obvious rules how to enter the geometry into the interpolation functions in order to achieve a desired order of accuracy in the physical space rather than in the computational space.

Another unresolved problem is the extrapolation of flow values at dummy points outside of the computational space at boundaries. Safe stability is usually generated by simple linear extrapolations in the computational space. But clearly the order of accuracy is abruptly lowered at the cell layers adjacent to boundaries, thus introducing errors of uncontrollable size. High-order extrapolations tend to make the scheme unstable, since they produce difference representations opposite to those required by the sign of eigenvalues at the boundaries. So this is no worked-out theory of numerical boundary conditions concerning the dummy-point flow, which is needed, however, by high-order schemes at the boundaries, regardless whether they are farfield boundaries or those representing a solid body. Thus any boundary formulation can be only a compromise between accuracy and stability. Maintaining both requirements with exactly the same priority is not possible at the present state of the art.

6.54 Truth Functions

All tools have now been provided so far in order to generate an at least second-order accurate solution to the Riemann problem at the cell face from which the total flux can be calculated at that place. A key point for this purpose is the eigenvalue-controlled weighting of the left and right states from which the averaged flow variables at the cell face are obtained by using one of the many methods described above. Since the left and the right state are assumed being independent of the normal coordinate ξ, the interpolation problem boils down to a left/right decision, the simplest form of which was mentioned previously:

$$U_j = \frac{1}{2}[(1+\mathrm{sign}\lambda_j)U_\ell + (1-\mathrm{sign}\lambda_j)U_r] , \qquad (6.191)$$

$$j = 0,\ 1,\ 2 .$$

We see that a problem may arise if one of the eigenvalues calculated with either the left or the right state variables has different signs. So the sign of this averaged eigenvalue may greatly depend on the averaging procedure adopted. Moreover, if an iterative Riemann solver is used and the eigenvalues are calculated from the newest available averaged set of flow variables, it may happen that one of the eigenvalues changes sign from one iteration to another, such that no convergence can be expected.

Quite evidently this happens often at shockwaves which are anyway a source of numerical difficulties, while the zeroth eigenvalue sign change gives no rise to trouble, since the items belonging to the zeroth eigenvalue are multiplied by small to vanishing quantities, when the flux is formed. In order to circumvent these difficulties, truth functions can be developed depending only on the stationary set of the initial values given by the left and right state. We got to know similar truth functions when we worked out the Einfeldt-type fluxes. The first one is a linear interpolation:

$$U_j = \frac{1}{2}[(1+\xi_j)U_\ell + (1-\xi_j)U_r] , \tag{6.192}$$

with ξ_j being

$$U_j = \frac{\lambda_{\ell_j} + \lambda_{r_j}}{|\lambda_{\ell_j}|+|\lambda_{r_j}|+\varepsilon[1- \frac{j}{2}(3-j)]} , \quad j = 0,\ 1,\ 2 . \tag{6.193}$$

The introduction of a small positive number ε in the denominator for the zeroth eigenvalue is necessary, since the denominator may be zero at many places in the flow field otherwise. This formulation switches on linear interpolation at all places where an eigenvalue changes sign. The interpolation may be even made more continuous in the slopes if the following formula is used

$$U_j = \frac{1}{4}[(2+\bar{\xi}_j)U_\ell + (2-\bar{\xi}_j)U_r] , \tag{6.194}$$

where the new interpolation coordinate depends on the previous one by

$$\bar{\xi} = (3-\xi^2)\xi . \tag{6.195}$$

6.55 Non-Oscillating Interpolation and Riemann Solvers

Up to now nothing has been said about the variables which should be entries to the non-oscillating interpolation prior to the calculation of the averaged flow variables at the cell face from one of the Riemann solvers suggested in this book. If the scheme is classic first order, the present discussion is irrelevant, since in this case the left and the right state are independent of the choice of the set of flow variables, say for example conservative or non-conservative. In both cases the first-order evaluation of, for example, the left state and the right state speed of sound is identical. Things change dramatically, if a higher-order interpolation is used for the left and the right state. Now it turns out that, for example, the state "speed of sound" depends very much on the choice of the variables, which are used for the interpolation.

Usually for a conservative code the conservative quantities are stored and updated during the course of iteration. But some of the Riemann solvers do require the left and the right state as a set of non-conservative variables, two of which have been used in the present context:

s, u, v, w, S ,

and ρ, u, v, w, p .

Now the serious question arises: Shall we first interpolate the conservative variables for the left and the right state and then convert the latter to those sets of variables, which are direct entries to the Riemann solver, or shall we first convert the conservative variables to those required as entries to the Riemann solver and then apply the high-order interpolation for finding the left and right state? Some authors even suggest to first convert the update variables to characteristic form upon which the high-order interpolation is applied, and then the characteristic variables are decoded to those variables needed for the particular Riemann solver chosen for the cell-face variable evaluation. This is by far the most expensive way to form a flux with maximum inaccuracy, as we found out in a test code. Still this method is seriously suggested in Ref. 21. We see that the choice of the sequence of the operations:

- conversion of variables,
- high-order interpolation,
- conversion of the interpolated states to the entries of the Riemann/Solver,
- conversion of the result of the Riemann solver to fluxes,

plays an important role and may lead us to dozens of different methods. There is no uniqueness in this field.

By far the outstanding best choice concerning accuracy and low diffusion is the following high-resolution method for a conservative scheme. It exploits rigorously the increased accuracy, which can be obtained by applying the homogeneous property of the Euler equations to the numerical formulation. It is characterized by the following items:

- store and update the conservative variables,

- apply one of the high-order interpolations of the conservative variables to evaluate the left and the right state,

- use the conservative Riemann solver,

- use the characteristic-averaged conservative flow-variable vector for calculating the conventional Euler fluxes.

By this method non-uniqueness is reduced to the evaluation of the coefficients of the conservative Riemann solver, which contain only the quantities u, v, w, s. For low to moderate Mach-number flows they can be calculated from some mean value of the high-order interpolated left and right state. For high Mach-number flows, however, it is recommended to use the left and right cell variables for this purpose, since it may happen that the high-order interpolated flow variables generate either a negative density and/or a negative pressure. So the choice to calculate the coefficients from the adjacent cell variables rather than from the left and right state is another piece of uniqueness. Some latitude is given for the averaging procedure for the coefficients of the conservative Riemann solver. Popular choices are

- the arithmetic mean applied to the conservative variables and subsequent conversion to u, v, w, s,

- the arithmetic mean applied to u, v, w, s directly,

- Roe's average,

- use of a first-order non-conservative Riemann solver for u, v, w, s.

Numerical tests with all of these choices showed no visible differences between them, where the entropy error was used as a quality check. So we dare say that the cheapest way may be used for the average: the arithmetic mean of the conservative variables.

6.56 Riemann Solvers and Strong Shocks

Since some years the flow prediction at hypersonic speeds has become increasingly important in the design phase of high speed aerodynamic vehicles, be it a shuttle, a rapid transport or an orbital vehicle. In the context of inviscid flow, two outstanding problems do accompany the numerical treatment of such air flows:

- shocks of considerable strength.
- near vacuum on the leeward side.

The second item poses a problem to numerical methods in so far as errors creeping into the calculations may cause the pressure or the density or both becoming negative. In all Riemann solvers we employ the speed of sound. The argument entering the square root for calculating the latter must be positive. Otherwise the numerical scheme blows up. Attempts to assign a small positive number to the speed of sound in such cases obviously cannot lead to the desired success, since for the Riemann solvers that small positive number occurs in the denominator, and the question arises how large this number should be chosen to be.

The first item named above also suffers from a similar problem. If the shock is captured rather than fitted it may be a source of generating negative pressures and/or densities by an isolated spike usually placed upstream of the shock. For smooth geometries this problem can be circumvented by bow-shock fitting. But embedded shocks then still pose the problem mentioned above. Rigorous flux limiting usually is not a sufficient tool to prevent negative pressures and/or densities. This is because even a first-order scheme places a more or less centered difference exactly across the shock jump for the flux difference aligned with that extreme eigenvalue, which changes sign, when the flow changes from supersonic to subsonic flow. The undesirable property of a centered difference across a shock is its lack of diffusion, and it is exactly this numerical phenomenon, which may cause the scheme to form a wiggle at this place, causing the trouble mentioned.

The numerical mechanism of the shock treatment by one of the Riemann solvers can be estimated at best by the scalar model equation.

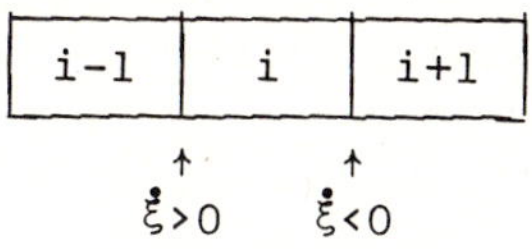

In the above sketch we indicate the shock by the sign change of one of the extreme eigenvalues (if the flow is directed from left to right, it is the second eigenvalue which has different signs at the two cell faces). We see at once that conventional Riemann solvers give rise to a conceptual difficulty: The eigenvalue sign constellation indicates that the shock must be in between the two cell faces of the cell labeled 'i'.

This is in very contrast to the finite-volume concept, which allows steps in the distribution of the flow variables only at the cell faces. The scalar flux difference obtained from a conventional Riemann solver for the shock point i is:

$$\frac{1}{2}[(\dot{\xi}_i+\dot{\xi}_{i+1})U_{i+1} - (\dot{\xi}_i+\dot{\xi}_{i-1})U_{i-1}] , \tag{6.196}$$

which is essentially a central difference in the U_is. We find easily that no positive contribution of the matrix is assigned to the update point i. So no numerical diffusion can be expected. This gives rise to spikes at the shock, even for an upwinding scheme.

The situation is quite different for a Riemann solver which uses the conservative fluxes rather than the flow variables, which technique also is called flux-vector splitting. A good example are the Steger-Warming fluxes, or the generalized form of them. In these techniques the eigenvalues are not averaged, but calculated with the flow variables of either the left cell for the positive sign, or the right cell for the negative sign. The eigenvalue sign constellation for a shock can be seen in the following sketch.

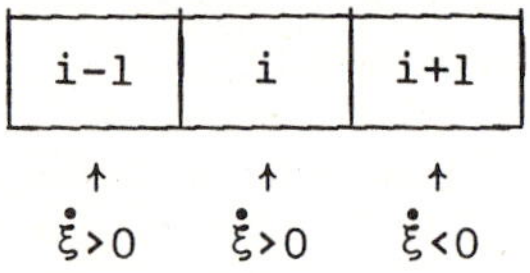

We see clearly that the shock position coincides with the right cell face, which is in accordance with the finite volume concept, since the flow variable distribution forms a step at this place anyway. The scalar difference representation for the shockpoint update cell i reads:

$$\dot{\xi}_i U_i + \dot{\xi}_{i+1} U_{i+1} - \dot{\xi}_{i-1} U_{i-1} . \tag{6.197}$$

Clearly the update point i receives a positive matrix element $\dot{\xi}>0$, which may have a considerable value, if the flow speed ahead of the shock is large (the flow is assumed from left to right).

This is the desired property of numerical shock capturing. From the discussion of the conventional conservative Riemann solver we know that it is the most accurate. But it has the drawback of poor shock-capturing capabilities. To combine both features - high accuracy, robust shock capturing - two methods for forming fluxes can be combined by interpolation:

$$E = E_{average}\,\omega + (1-\omega)E_{split} \; . \qquad (6.198)$$

Here "average" means the conventional Euler flux formed from the result of the highly accurate conservative Riemann solver. The label "split" means a flux obtained from flux-vector splitting, say the Steger-Warming flux, or the cheapest alternative being the generalized Steger-Warming flux with the particular parameter M=0.5. The interpolation coordinate ω stands for a sensing function which is near unity in smooth flow regions and vanishes at shocks. The van Albada limiter used for the density, required anyway by the Riemann solver, is a good choice for the definition of the interpolation coordinate ω. It can be incorporated in the following way:

$$\omega_{i+\frac{1}{2}} = \max(L_i, L_{i+1}) \; .$$

More elaborate is the construction

$$\omega_{i+1} = \min\left[1, \; \delta \frac{\max(L_i, L_{i+1})}{\left| \left|\frac{\lambda_i}{L_i}\right| - \left|\frac{\lambda_{i+1}}{L_{i+1}}\right| \right|}\right] ,$$

where δ is an input parameter, and the denominator is the difference of the normal cell face Machnumbers. If δ is taken a large number, the scheme runs in the high-accuracy mode only. The denominator guarantees that only at shocks flux-vector splitting is switched on, avoiding in this way the amplification of the entropy error in the vicinity of stagnation points. It is evident that the interpolation can be refined in many aspects, which cannot be given all in the context of this book. But the basic principles of strong-shock capturing are laid out.

6.57 Riemann Solvers and Boundary Conditions

All the Riemann solvers developed so far are written in structured form using the quantities r_1 and r_2 or R_1 and R_2. This has a good reason, since these entities play an important role in the formulation of boundary conditions. The most important condition is the kinematic condition, that a solid body surface cannot be penetrated by the flow, or, in other words, the normal velocity must be zero at the skin of a configuration:

$$ux + vy + wz = 0 \; .$$

All Riemann solvers derived so far are well suited for the incorporation of this boundary condition, because the velocity components can be written the following way:

$$u = u_o + xa \ ,$$

$$v = v_o + ya \ , \qquad (6.200)$$

$$w = w_o + za \ .$$

The solid body boundary condition reads therefore

$$ux + vy + wz = u_o x + v_o y + w_o z + a = 0 \ , \qquad (6.201)$$

where use has been made of the fact

$$x^2 + y^2 + z^2 = 1 \ .$$

The auxiliary quantity a then can be evaluated easily at solid body boundaries:

$$a = -u_o x - v_o y - w_o z \ .$$

By inspection of the Riemann solvers we see that the coefficient a is composed of the quantities r_1 and r_2 or R_1 and R_2. One of these quantities is always known from the flow field variables, but the other one would require flow data from inside the solid body, where of course no data are available. So the quantity a is solved either for r_1 or r_2, depending on which of these quantities is known by the flow field data. For the particular case of the solid body boundary condition it is easy to decide for which of both quantities the auxiliary entity "a" should be solved. If the solid body is placed to the right side of the cell face, then the second eigenvalue points inside the body, where no data are available. So the missing quantity r_2 is calculated form the boundary condition rather than from extrapolated values.

The definition of right and left is clear by inspecting the direction of the normal coordinate adopted. "Right" means on that side of the cell face, where the surface normal increases, and "left" means on the opposite side of the cell face. From this discussion it is now clear that the quantity to be solved for is r_1 or R_1, if the solid body is placed on the left side of the cell face under consideration.

6.58 Other Updates

Explicit updates of the flow variables are usually simple to program and do allow the fast incorporation of the different split-flux formulations mentioned above, since the update process is in principle an isolated part of the computer program being almost independent of the flux-difference representation. A simple update for characteristics-based schemes is the simple one-step update being first-order accurate in time. Since we are interested only in steady state solutions, the spatially maximum allowable time step for a specified CFL-number may be used for that purpose. This is a simple tool for increasing the speed of wave propagation and thus the convergence speed. It can be seen easily, however, that a wave front is propagated in a one-step scheme by at most the width of one cell. An example enlightens this process: If we have a grid which contains for example 100 cells in the x-direction, then at least 100 time steps are necessary, until the last cell of the grid "feels" a disturbance, which is entered at the first left-most cell. Because of the scheme being diffusive the wave strength, however, decays enormously on its way from the first to the last cell, such that we need many more iterations in order to settle the pseudo-unsteady perturbation to the steady state. So it is desirable to invent update schemes allowing the increase of the CFL-number in order to converge to the steady state with less overall computational work. This is the starting point to consider update mechanisms being different from the one-step scheme provided above. There are three requirements for those new update schemes, which we will retain in the following considerations:

- the scheme should allow for spatially varying time steps for fast wave propagation,

- the steady state solution must be independent of the spatially distributed varying time steps,

- the CFL-number should increase more than proportional to the number of sweeps, otherwise no advantage over the simple one-step scheme is achieved.

If only the steady-state solution is of interest, no emphasis needs be put on the order of accuracy of the time integration.

6.59 Lax-Wendroff (L-W) Type Updates[22]

A popular update, which exists now for more than two decades, is the L-W-update. It is based on the explicit calculation of the first two members of the Taylor-expansion in time of the flow variables:

$$U^{n+1} = U^n + (\dot{U} + \ddot{U}\,\frac{\Delta t}{2})\,\Delta t \ . \qquad (6.202)$$

Taking the time derivatives from the unsteady Euler equations we find for the model equation

$$\dot{U} + E_\xi = 0 \qquad (6.203)$$

the first time derivative of the flow-variable vector to be

$$\ddot{U} = -\dot{E}_\xi = -(E_u \dot{U})_\xi = (E_u E_\xi)_\xi \ , \qquad (6.204)$$

where use has been made of the homogeneous property of the Euler equations. Now, calculating the second derivative in time of the flow variables this way is rather expensive because of the evaluation of the Jacobian matrix E_u. Therefore L-W-schemes are usually split into two steps, a predictor step and a corrector step, which are both simpler to calculate than the original formulation. A general update of this kind can be constructed the following way:

The first solution (predictor) is calculated almost the same way as the simple classic one-step update, however, with a big difference. The finite differences are downwinded rather than upwinded. So, for, say, a positive eigenvalue, the right-side weighted state at a cell face is taken for calculating the flux at this place. The update is performed with half the time step of the overall update. It is evident that this solution, if taken as isolated final update, is unstable. In the second solution, however, stability and the Lax-Wendroff type diffusion is entered into the update procedure. The second solution (corrector) takes the flow values obtained in the predictor and forms the fluxes the proper way, using the upwind formulae. The update is performed with the full time step. The result of the corrector is formally second-order accurate in time, provided a constant global time step is taken for the update (not the largest allowable local time step!). The accuracy of the space-difference representation is that of the one-step base code. An example clears up the update process. For simplicity the second-order one-sided difference representation of the scalar model equation is taken for this purpose:

$$\dot{U} + E_\xi = 0 \qquad \text{(conservative)} ,$$
$$\dot{U} + \dot{\xi} U_\xi = 0 \qquad \text{(non-conservative for accuracy checks)} . \tag{6.205}$$

Following the rules described above the predictor first gives the solution which is indicated by a star, and then the corrector the final solution (the eigenvalue is taken to be positive).

Predictor:

$$\begin{aligned} \overset{*}{U}_i &= U_i + \frac{\Delta t}{4} \dot{\xi}(3U_i - 4U_{i+1} + U_{i+2}) , \\ \overset{*}{U}_{i-1} &= U_{i-1} + \frac{\Delta t}{4} \dot{\xi}(3U_{i-1} - 4U_i + U_{i+1}) , \\ \overset{*}{U}_{i-2} &= U_{i-2} + \frac{\Delta t}{4} \dot{\xi}(3U_{i-2} - 4U_{i-1} + U_i) . \end{aligned} \tag{6.206}$$

Corrector:

$$U_i^{n+1} = U_i + \frac{\Delta t}{2} \dot{\xi}(3\overset{*}{U}_i - 4\overset{*}{U}_{i-1} + \overset{*}{U}_{i-2}) . \tag{6.207}$$

After inserting the predictor into the corrector the following result is obtained:

$$\begin{aligned} U_i^{n+1} &= U_i - \frac{\Delta t}{2} \dot{\xi}(3U_i - 4U_{i-1} + U_{i-2}) - \\ &- \frac{(\Delta t \dot{\xi})^2}{8}(26U_i - 16U_{i+1} + 3U_{i+2} - 16U_{i-1} + 3U_{i-2}) . \end{aligned} \tag{6.208}$$

We see immediately that the coefficient multiplying the update value U_i is smaller than unity provided that the CFL-number is given a positive value below a maximum upper bound. This can be found from the condition that the coefficient at U_i may be at most zero:

$$1 - \frac{3}{2} \Delta t \dot{\xi} - \frac{13}{4}(\Delta t \dot{\xi})^2 > 0 . \tag{6.209}$$

This means in terms of the CFL-number:

$$\text{CFL} < \frac{1}{13}(\sqrt{61} - 3) .$$

There remains the accuracy of this particular L-W type update to be checked. A Taylor-series expansion up to third order reveals the result to be second-order accurate:

$$U_i^{n+1} = U_i - \Delta t \dot{\xi}(U_\xi - \frac{2}{3} U_{\xi\xi\xi}) + \frac{(\Delta t \dot{\xi})^2}{2} U_{\xi\xi} . \tag{6.210}$$

Note, that the third derivative in the first bracket representing the first derivative does not spoil second-order accuracy. Finally we check whether the present update is a correct representation of the L-W update. It is evident that the first bracket is the approximation of the plain flux difference E_ξ multiplied by the time step. From

$$\ddot{U} = (E_u E_\xi)_U \tag{6.211}$$

we find:

$$\ddot{U} = E_u^2 \, U_{\xi\xi} = \dot{\xi}^2 \, U_{\xi\xi} \,, \tag{6.212}$$

where multiple use has been made type of the homogeneous property of the Euler equations, which may be expressed by

$$E_{uu} U \equiv 0 \,. \tag{6.213}$$

A last remark is necessary concerning the switch to low order at shock waves or at places where wiggles do appear. Since the second derivative needs the information at points on both sides of the shock wave, a switching procedure must be entered by sensing the flow field with respect to spurious oscillations. If the sensor detects a place with oscillations, then the time step in the predictor should be set zero, while in the corrector step the usual flux limiting can be applied. This means that with the present update the method switches back to first order in time and space at shock waves. Although the present L-W type update is valid for all types of base point interpolations regardless of its order of accuracy this certainly is a certain drawback. For the particular case of second-order pure upwinding, however, a very popular update technique developed by Moretti for the so called λ-scheme is available, which is also based on the Lax-Wendroff second-order formula, Ref. 23. It reads for positive eigenvalues:

Predictor 1:

$$\begin{aligned} \overset{*}{U}_i &= U_i - \Delta t \dot{\xi} (U_i - U_{i-1}) \,, \\ \overset{*}{U}_{i-1} &= U_{i-1} - \Delta t \dot{\xi} (U_{i-1} - U_{i-2}) \,, \\ \overset{*}{U}_{i-2} &= U_{i-2} - \Delta t \dot{\xi} (U_{i-2} - U_{i-3}) \,. \end{aligned} \tag{6.214}$$

Predictor 2:

$$\bar{U}_i = U_i - \Delta t \dot{\xi} (2\overset{*}{U}_i - 3\overset{*}{U}_{i-1} + \overset{*}{U}_{i-2}) \,, \tag{6.215}$$

Corrector:

$$U_i^{n+1} = \frac{1}{2} (\bar{U}_i + \overset{*}{U}_i) \ . \tag{6.216}$$

With this update also the second derivative is upwinded. A Taylor-series expansion reveals the proper second-order representation of the Lax-Wendroff formulation. For weak shocks no flux limiting needs be incorporated. If limiting is still necessary, it needs only be implemented in the second predictor, which is monotonicity preserving per se.

Besides Lax-Wendroff two-step explicit updates another class of multi-step updates is popular, which is derived from the Runge-Kutta schemes solving numerically ordinary differential equations. It consists of a series of numerical solutions of the govering equations, such that the time integration of the flow variables may assume any desired order of accuracy in time depending only on the number of time levels, on which we calculate intermediate solutions. Since we are not interested in improving the time accuracy, Runge-Kutta updates can be invented which drive the iterations with less overall computer work to the steady state solution than one-step schemes. A member of this class of schemes can be defined as (see also Chapter V):

$$\begin{aligned} U^1 &= U^n + \alpha_1 \Delta t \dot{U}^n \ , \\ U^2 &= U^n + \alpha_2 \Delta t \dot{U}^1 \ , \\ U^3 &= U^n + \alpha_3 \Delta t \dot{U}^2 \ , \\ U^{n+1} &= U^n + \Delta t \dot{U}^3 \ . \end{aligned} \tag{6.217}$$

Here the dot derivative of the flow-variable vector stands for the non-linear flux difference operator. The time step may be the largest allowable in each cell. The present example is a four-step scheme. In order to make the scheme at least formally second-order accurate in time, the coefficient multiplying the time step of the update previous to the last update must be one half, in our example this means:

$$\alpha_3 = \frac{1}{2} \ .$$

There is much latitude in the choice of the other fractions of the time step, in our example α_1, α_2. But one choice can be said to be for sure unstable for an upwinding scheme. This choice is putting all α's unity (Laasonen scheme). This is in very contrast to central-difference schemes, which would exhibit improved stability with this particular choice. A rational choice for the present example, which works quite well, is the sequence of coefficients:

$$\alpha_1 = \frac{1}{8}, \quad \alpha_2 = \frac{1}{4}, \quad \alpha_3 = \frac{1}{2}. \tag{6.218}$$

An N-step scheme could be constructed with the coefficients

$$\alpha_i = \frac{2^{i-1}}{2^{N-1}}, \qquad i = 1, \ldots, N.$$

Finally some remarks are added to multi-step explicit updates. If a variable local time step is used convergence is speeded up on one hand, but on the other hand the steady state solution inevitably depends on the time-step distribution in the flow field. The error due to this phenomenon is hard to estimate, since it depends on the cell size variation and the local Mach number. Anyway, it is a source of uncertainty. The gain in the overall computer time is not dramatic compared to the simple one-step update, usually only a small percentage of the latter, in some cases even no gain at all could be realized.

Inherent in the Runge-Kutta updates there is a tendency to generate small amplitude wiggles if the coefficients α are not carefully chosen. In addition, all multi-step updates described in this paragraph exhibited a larger total pressure error than the simple one-step update.

6.60 Implicit Updates

An extremely powerful tool for characteristic-based schemes is the implicit solution of the Euler equations, because the left-hand splitmatrix, being the linearized counterpart of the non linear right-hand side fluxes, can be transformed to sufficient diagonalization suited to drive the residuals very fast to vanish. An overall computer time gain of one or two orders of magnitude compared with a one-step explicit update scheme has been observed with the EUFLEX scheme, Ref. 24, as an example. Some low-speed problems at free-stream Mach numbers less than 0.05 even could not be converged with explicit updating, while the implicit residual driver did manage these problems. The overall computation speed depends of course on the flow case considered.Extreme low-speed cases such as micro-meteorological aerodynamics past buildings may need one thousand iterations (compared to several ten thousand explicit updates). Moderate supersonic flows past wing - pointed fuselage - external store configurations could be driven to convergence within ten iterations with the EUFLEX code. For supersonic flows without subsonic portions implicit relaxation schemes with an infinite time step convert to improved space marching schemes and are usually superior to them concerning computational speed because of the implicitness of the crossflow surface update in which the characteristics are subsonic.

Before going into detail it should be mentioned, that Alternating-Direction-Implicit (ADI) schemes will not be treated. They do have the same amount of arithmetic work as any two-step relaxation scheme, but have a very bad convergence rate in three-dimensional flow because the CFL-number is restricted to nearly explicit values.

From the Riemann solvers using fluxes instead of the flow variables we see that the positive fluxes are formed with the left-state and the negative fluxes are calculated from the right-state variables. So any linearization of the total flux requires the evaluation of two fully populated Jacobian matrices for each cell face, which are then added together to give the total Jacobian matrix. By inspection of the conventional Riemann solvers we see that only one Jacobian matrix has to be calculated, which means of course that flux vector plus/minus splitting requires roughly twice as much computer time for the left-hand side matrix evaluation. That is why we also discard the dicussion of the implicit solution of flux-vector splitting methods in the following.

Which method is left over after these remarks is now clear. It only can be the fully homogeneous approach using the conservative Riemann solver prior to forming the conventional Euler fluxes. Thus we see that the choice of the method adopted has an enormous influence on the simplicity of an implicit Euler solver.

The consequent exploitation of the homogeneous property has severeal advantages:

- highest possible accuracy,
- cheapest possible left-hand side for the implicit solution.

6.61 Implicit Formulation

The starting point is the update

$$\frac{U^{n+1} - U^n}{\Delta t} + E_\xi{}^{n+1} + F_\eta{}^{n+1} + G_\zeta{}^{n+1} = 0 \ . \qquad (6.219)$$

Since the flux differences at the new time level are not known, they can be estimated form a time-like linear Taylor series expansion about an iteration level ν somewhere in between the old and the new time level:

$$\frac{U^{n+1} - U^n}{\Delta t} + \left[E^\nu + E_U^\nu (U^{n+1} - U^\nu)\right]_\xi + \\ + \left[F^\nu + F_U^\nu (U^{n+1} - U^\nu)\right]_\eta + \\ + \left[G^\nu + G_U^\nu (U^{n+1} - U^\nu)\right]_\zeta = 0 \; . \tag{6.220}$$

After rearrangement we obtain

$$\frac{\Delta U}{\Delta t} + (E_u^\nu \; \Delta U)_\xi + (F_u^\nu \; \Delta U)_\eta + (G_u^\nu \; \Delta U)_\zeta = \frac{U^n - U^\nu}{\Delta t} + \dot{U}^\nu , \tag{6.221}$$

where the abbreviation

$$\Delta U = U^{n+1} - U^\nu$$

has been adopted. $\dot{U}$ stands for the non-linear flux difference representation

$$\dot{U}^\nu = -E_\xi^\nu - F_\eta^\nu - G_\zeta^\nu \; . \tag{6.222}$$

If the left and right-hand side are iterated to convergence, the right-hand side becomes the conventional explicit operator, however, with the difference operator being driven to the new time level. From the present formulation any higher-order time-accurate method can be derived by a suitable iteration strategy. Since we are only interested in the steady state solution, we can put the iteration level to that of the present time level. This is justified by two reasons. The method converts to Newton's first order in time method. Second, since we try to put the time step as large as possible, the first right-hand side term can be neglected. Under these circumstances the scheme boils down to

$$\frac{\Delta U}{\Delta t} + (E_U \; \Delta U)_\xi + (F_U \Delta U)_\eta + (G_U \Delta U)_\zeta = \dot{U} \; , \tag{6.223}$$

where the superscript has been omitted with the understanding that all these quantities are computed form the solution vector available at the present time level. The implicit delta solution is followed by the simple update

$$U^{n+1} = U^n + \omega \Delta U \; , \tag{6.224}$$

where ω is a relaxation factor being less than unity. This coefficient is necessary, if the left-hand side differences are first-order accurate and the right-hand side differences are of higher-order accuracy. To use first-order differences on the left hand side has good reasons. It is simple and enhances stability, since low-order upwind differencing enters more diffusion acting on the residuals, than a high-order difference representation (more smoothing of the delta distribution).

6.62 The Split Matrix

Here we turn our attention to the evaluation of the left-hand side Jacobian matrix in upwinded form. A good starting point for this purpose is the homogeneous Riemann solver with the conservative flow variables used as entries. Since the result of this Riemann solver is the conservative flow variable vector at the cell face, the left-hand side matrix is based fully on the homogeneous property of the Euler equations. Let us consider at first the increment of the flow-variable vector at the cell face to the right of the update cell i:

$$\Delta U_{i+\frac{1}{2}} = f(\Delta U_i, \Delta U_{i+1}) \; . \qquad (6.225)$$

The function $f(\Delta U)$ can be easily obtained from the conservative Riemann solver by the understanding that the conservative quantities there are replaced by their time like increments. The matrix contribution, to the update cell named "i" would then be

$$E_{U,i+\frac{1}{2}} \cdot \bar{f}(\Delta U_i) \; , \qquad (6.226)$$

while the off-diagonal contribution is

$$E_{U,i+\frac{1}{2}} \cdot \bar{f}(\Delta U_{i+1}) \; , \qquad (6.227)$$

where the left/right truth function depending on the eigenvalue sign has been incorporated. This is indicated by the bar over the function $\bar{f}$, which directly can be read from the conservative Riemann solver. In order to form the total matrix contribution to the update cell the above operation must be performed for each of the six (four) cell faces for the three- (two-) dimensional implicit left-hand side matrix evaluation incorporating the flux differencing with the appropriate sign. We see that at each cell face first the coefficients of the conservative Riemann solver must be calculated upon which the evaluation of the Jacobian matrix E_u follows. For the homogeneous approach this arithmetic work can be drastically reduced, as we will see in the next sub-chapter.

6.63 The Homogeneous Implicit Solution

In principle the conservative Riemann solver in delta form is nothing but the operation

$$\Delta U_{i+\frac{1}{2}} = T(I^+ + I^-)\, T^{-1}\, (\Delta U_{i+1}, \Delta U_i) \; ,$$

where the identity matrices I contain only one's and zero's depending on the sign of the eigenvalues. The T's are eigenvectors calculated from the averaged components of the velocity vector and the speed of sound. The whole calculation of the right-hand side matrix is comprised by the evaluation of twenty-five numbers in 3D. The true mapping of the right-hand side flux to the linearized matrix form of the left-hand side implicit operator consists in multiplying the result of the Riemann solver by the fully populated Jacobian matrix of the conventional Euler fluxes, which is comprised again by twenty-five numbers in 3D. So the evaluation of the true left-hand side is pretty expensive:

$$E_{U,i+\frac{1}{2}}\, \Delta U_{i+\frac{1}{2}} = E_{U,i+\frac{1}{2}}\, T(I^+ + I^-)T^{-1}\, (\Delta U_{i+1}, \Delta U_i). \qquad (6.228)$$

It is obvious that the unsplit operation

$$E_u \equiv T\, \Lambda\, T^{-1} \; , \qquad (6.229)$$

with the diagonal matrix Λ containing the eigenvalues λ_j (j=0,0,0,1,2), is in fact an identity, since it is only a similarity transformation.

This is a trivial statement. Not so trivial is that the identity also holds for the split version, where the diagonal contains now negative and positive parts of the eigenvalues of the form

$$\lambda \pm |\lambda| \; .$$

Eberle, see Ref. 24, found out by inspecting computer printouts that obviously the same numbers are generated by the two operations, the first of which is called true Jacobian and the second false Jacobian:

$$E_u^{\pm} = E_u\, T\, \Lambda^{\pm} T^{-1} \overset{!}{\equiv} T\, \Lambda^{\pm} T^{-1} A \; , \qquad (6.230)$$

where $\bar{\Lambda}$ is a diagonal matrix containing only one's and zero's depending on the sign of the eigenvalues and Λ is a diagonal matrix containing only zero's and eigenvalues depending on the contributions of the form

$$\lambda \pm |\lambda| \ .$$

The latter operation is multiplied by A, the cell face area, and is called false Jacobian. The use of the false Jacobian is much more desirable than the use of the true Jacobian, since it is much cheaper to calculate. While the true Jacobian requires the evaluation of another 25 numbers (in 3D) the false Jacobian needs no extra number calculation, since the eigenvalues have to be available in both cases.

Recently Brenneis, Ref. 26, gave a detailed analytical proof that the false and the true plus/minus-splitted Jacobian matrix are in fact identical.

At this stage it also should be mentioned that the arithmetic count of the false Jacobian for the conservative variables is the lowest possibly attainable out of all choices of the fluxes mentioned in this volume. All plus/minus flux vector splittings, for example the Steger-Warming and van Leer fluxes do require the calculation of two fully populated Jacobian matrices, one on each side of the cell faces.

All split-flux versions written in artificial viscosity form, such as Harten's, even require three Jacobian matrices for each cell face: two unsplit Jacobians for the left and the right state, and a split Jacobian calculated from an average of both states.

6.64 Matrix Conditioning

There is a big difference in the diagonal dominance of the one-dimensional formulation and the multi-dimensional approach. For moderate time steps diagonal dominance is always guaranteed since each line of the left-hand side contains a positive contribution of arbitrary size to the matrix diagonal, depending on the CFL-number. Our goal, however, is to assign a large number, possibly infinity, to the CFL-number, in order to obtain the fastest possible convergence rate. The situation for an infinite time step can be analyzed at best in two dimensions. Let us consider a rectangular cell the lateral sides of it being aligned with the flow direction, which is from left to right, Fig. 6.13.

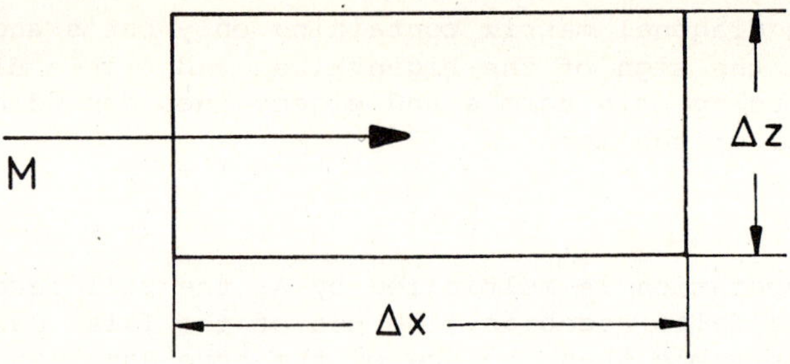

Fig. 6.13 Layout of a 2D-finite volume cell in parallel flow

The aspect ratio of the cell is

$$\Lambda = \frac{\Delta z}{\Delta x} \ ,$$

and the local flow Mach number is M. Now we check through some of the matrix diagonal elements whether they are larger then zero or not. Starting with the matrix diagonal element of the continuity equation we find directly from the conservative Riemann solver for the subsonic Mach number $M<1$ the following result:

$$\begin{aligned} A_{11} = u\Delta z + \frac{1}{2s^2}\Big[&\Big(\frac{\gamma-1}{2}u^2 - us\Big)\Delta z(u+s-u) - \\ &- \Big(\frac{\gamma-1}{2}u^2 - us\Big)\Delta z u - \\ &- \Big(\frac{\gamma-1}{2}u^2 - us\Big)\Delta z(u-s) + \\ &+ \frac{\gamma-1}{2}u^2\,\Delta x s - \\ &- \frac{\gamma-1}{2}u^2\,\Delta x(-s)\Big] \ . \end{aligned} \qquad (6.231)$$

All contributions multiplied by the height of the cell Δz come from the right and the left cell face. The remainder is the result of the flux difference of the upper and the lower face.

Using the Mach number and the aspect ratio, the formula is abbreviated to:

$$A_{11} = s\Delta x M\Big(\Lambda\Big\{1 + \frac{M}{2}\big[\gamma-3-(\gamma-1)M\big]\Big\} + \frac{\gamma-1}{2}M\Big) \ . \qquad (6.232)$$

One sees immediately that the density matrix diagonal vanishes for incompressible flow $M \to 0$. The coefficient multiplying the aspect ratio is always positive for the Mach number range under consideration. Seen as a whole the matrix diagonal of the density equation is larger than zero at subsonic speeds.

For supersonic speeds $M>1$, the matrix diagonal of the continuity equation is

$$A_{11} = s\Delta x \frac{\gamma-1}{2} M^2 , \quad (6.233)$$

which is always positive. Now we turn our attention to the x-momentum equation. The subsonic ($M<1$) result is

$$A_{22} = s\Delta x\left(\Lambda\{1+M^2[3-2\gamma + (\gamma-1)M]\} - (\gamma-1)M^2\right) . \quad (6.234)$$

Here we see clearly the exceptional result that the matrix diagonal can vanish for all aspect ratios fulfilling the equation

$$\Lambda = \frac{(\gamma-1)M^2}{1 + M^2[3-2\gamma+(\gamma-1)M]} . \quad (6.235)$$

This incidence often occurs in Navier-Stokes solutions at the cell layers close to the solid body, where both, the aspect ratio and the local Mach number, are small. But also for supersonic speeds desaster is on its way. Here the matrix diagonal element for the x-momentum equation reads

$$A_{22} = s\Delta x M[\Lambda(3-\gamma) - (\gamma-1)M] , \quad (6.236)$$

which vanishes for all aspect ratios following the equation

$$\Lambda = \frac{(\gamma-1)M}{3-\gamma} .$$

We omit now the calculation of the two remaining matrix diagonal elements, which are always safely positive. We suspect that the non-positiveness of the split matrix in multi-dimensions seems not to have been recognized by numerical analysts, since much literature about time step strategies is available to avoid the loss of diagonal dominance, but no rigorous analysis of the split matrix for the conservative variables. The problem of matrix illconditioning can be solved relatively easily. For this purpose we inspect the matrix diagonal elements which we would get if we solved the quasi-linear non-conservative equations by matrix splitting.

The result for the density equation is for all Mach numbers

$$A_{11} = s\Delta x M\Lambda .$$

We see that the vanishing of the diagonal element for incompressible flow obvious cannot be avoided for the continuity equation, which is a consequence of the decoupling of the continuity equation in the limit of constant density flow.

The result for the x-momentum equation which updates now the u-velocity component rather than the x-momentum is

$$A_{22} = s\Delta x\Lambda \max(1, M) ,$$

which is always clearly positive. We need not care about the two remaining diagonal elements since they ar clearly positive.

Summarizing the results on matrix conditioning obtained so far we arrive at the following statements:

- the split matrix for the conservative deltas of the flow variables is not diagonal dominant,
- the split matrix for the conservative deltas is differentiable everywhere,
- the split matrix for the non-conservative variables is diagonal dominant,
- the split matrix for the non-conservative variables has slope discontinuities at all places, where the eigenvalues change sign.

With this knowledge it is easy to construct a perfect implicit point Gauss-Seidel or point Jacobi scheme for vector computers. The five × five system of equations for an update cell can be written as

$$A\Delta U = \dot{U} + ODIAG ,$$

where A stands for the conservative split matrix and ODIAG samples all contributions from the adjacent cells. Now we have seen that the matrix A may be not well diagonalized under certain circumstances. So we do not solve for the conservative deltas, but for the non-conservative deltas which are connected to the latter by the differential representation

$$\Delta U = \frac{\delta U}{\delta \bar{U}} \Delta \bar{U} , \qquad (6.238)$$

where the bar indicates the non-conservative variables. Written in full length we have

$$\begin{aligned} \Delta\rho &\equiv \Delta\rho , \\ \Delta\ell &= u\Delta\rho + \rho\Delta u , \\ \Delta m &= v\Delta\rho + \rho\Delta v , \\ \Delta n &= w\Delta\rho + \rho\Delta w , \\ \Delta e &= \frac{\Delta p}{\gamma-1} + \frac{q^2}{2}\Delta\rho + \rho(u\Delta u+v\Delta v+w\Delta w) . \end{aligned} \qquad (6.239)$$

The new matrix is then

$$B = A \frac{\delta U}{\delta \bar{U}} ,$$

which is always well diagonalized and, as a by-pass result, also differentiable everywhere.

Upon solving the five × five system for the update cell, the non-conservative deltas are immediatly back-substituted by the previous transformation to the conservative deltas, which are then taken as off-diagonal contributions to the next update cell. This way the left-hand side remains a true mapping of the right-hand side homogeneous flux-difference representation and still is well conditioned.

Other tools for improving matrix conditioning are local selective time steps and local residual smoothing. Both can be incorporated easily in the left-hand side operator. First we consider local residual smoothing which needs be applied if a mixture of homogeneous fluxes and flux vector splitting is used on the right-hand side. For this purpose the sensing or limiter function L controlling the switch to an non-homogeneous flux must be stored. By inspecting the split matrix we see that each of is elements is multiplied by a quantity of the form

$$\lambda \pm |\lambda| .$$

Local residual smoothing can be easily incorporated by the redefinition of the absolute eigenvalue

$$|\lambda|_{new} = \max(|\lambda|, \; \varepsilon L \; |\lambda|_{max}) , \tag{6.240}$$

where ε is an input parameter.

This way the absolute value of an eigenvalue is limited to a lower bound given by some fraction of the maximum absolute eigenvalue out of the three being available. This measure imposes an additional Laplace smoother on the deltas as can be easily studied using the scalar model equation.

The local selective time step strategy may be incorporated the following way. Each line of the equations has the form

$$(\frac{1}{\Delta t} + A)\Delta U = (\frac{|\lambda|_{max}}{CFL} + A)\Delta U = RHS . \tag{6.241}$$

The maximum eigenvalue is the maximum absolute eigenvalue of all those eigenvalues, which have to be calculated anyway at each of the six cell faces of the update cell under consideration. Selective time stepping can now be incorporated by an IF-statement:

$$\begin{aligned} \frac{1}{\Delta t} &= 0 \,, \quad \text{if Diag (A)} > \alpha|\lambda|_{max} \,, \\ \frac{1}{\Delta t} &= \alpha|\lambda|_{max} \quad \text{otherwise} \,. \end{aligned} \qquad (6.242)$$

Here, α is an input parameter and λ means the eigenvalue being multiplied by the associated cell-face area.

6.65 References

1. Courant, R., Isaacson, E., Rees, M.: "On the Solution of Nonlinear Hyperbolic Differential Equations by Finite Differences". Comm. Pure & Appl. Math., Vol. 52, 1952, pp. 243-255.

2. Steger, J.L., Warming, R.F.: "Flux Vector Splitting of the Inviscid Gas Dynamic Equations with Application to Finite Difference Methods". J. Comp. Phys., Vol. 40, No. 2, 1981, pp. 263-293.

3. Harten, A.: "High Resolution Schemes for Hyperbolic Conservation Laws". J. Comp. Phys., Vol. 49, 1983, pp. 357-393.

4. Yee, H.C., Warming, R.F., Harten, A.: "Implicit Total Variation Diminishing (TVD) Schemes for Steady State Calculations". J. Comp. Phys., Vol. 57, 1985, pp. 327-360.

5. Chakravarthy, S.R.: "The Versatility and Reliability of Euler Solvers Based on High Accuracy". AIAA-Paper 86-0243, 1986.

6. Godunov, S.K.: "A Finite Difference Method for the Numerical Computation of Discontinuous Solutions of the Equations of Fluid Dynamics". Math. Sbornik, Vol. 47, 1959, pp. 357-398.

7. Courant, R., Hilbert, D.: "Methoden der mathematischen Physik". Springer Verlag, Berlin/Heidelberg/New York, 1968.

8. Osher, S., Solomon, F.: "Upwind Difference Schemes for Hyperbolic Systems of Conservation Laws". Math. Comp., Vol. 38, No. 158, 1982, pp. 339-374.

9. Engqvist, B., Osher, S.: " One-Sided Difference Approximations for Non-Linear Conservation Laws". Math. Comp., Vol. 36, No. 154, 1981, pp. 321-351.

10. Pandolfi, M.: "On the 'Flux-Difference Splitting' Formulation". In "Non-Linear Hyperbolic Equations - Theory, Computation Methods and Applications", Ballmann, J., Jeltsch, R. (eds.), Vol. 24 of Notes on Numerical Fluid Mechanics, Vieweg, Braunschweig/Wiesbaden, 1989, pp. 466-481.

11. Roe, P.L.: " Approximate Riemann Solvers, Parameter Vectors and Difference Schemes". J. Comp. Phys., Vol. 43, No. 2, 1981, pp. 357-372.

12. Einfeldt, B.: "Zur Numerik der stossauflösenden Verfahren". Doctoral thesis, RWTH Aachen, 1988.

13. Van Leer, B.: " Flux-Vector Splitting for the Euler Equations". Lecture Notes in Physics, Vol. 170, Springer, Berlin/Heidelberg/New York, 1982, pp. 507-512.

14. Anderson, D.A., Tannehill, J.C., Fletcher, R.H.: "Computational Fluid Mechanics and Heat Transfer". Hemisphere Publishing Corporation, New York, 1984.

15. Chakravarthy, S.R,: "Relaxation Methods for Unfactored Implicit Upwind Schemes". AIAA-Paper 84-0165, 1984.

16. Yee, H.C.: "A Class of High Resolution Explicit and Implicit Shock-Capturing Methods". NASA Technical Memorandum 101088, 1988.

17. Roe, P.L.: "A Survey of Upwind Differencing Techniques". Proceedings of the 11th Conference on Numerical Methods in Fluid Dynamics, Williamsburg, VA, USA, 1988.

18. Sweby, P.K.: "High Resolution Schemes Using Flux Limiters for Hyperbolic Conservation Laws". SIAM J. Num. Analy., Vol. 21, 1984, pp. 995-1101.

19. Van Albada, G.D., van Leer, B., Roberts, W.W.: "A Comparative Study of Computational Methods in Cosmic Gas Dynamics". Astron. Astrophys. 108, 1982, pp. 76-84.

20. Hänel, D.: "On the Accuracy of Upwind Schemes in Solutions of Navier-Stokes Equations". AIAA-Paper 87-1105 CP, 1987.

21. Thomas, J.L., Walters, R.W., van Leer, B., Rumsey, C.L.: "Implicit Flux-Split Algorithm for the Compressible Navier-Stokes Equations". In "Numerical Simulation of Compressible Navier-Stokes Flows", a GAMM-Workshop, Bristeau, M.O., Glowinski, R., Periaux, J., Viviand, H. (eds.), Vol. 18 of Notes on Numerical Fluid Mechnics, Vieweg, Braunschweig/Wiesbaden, 1986, pp. 326-341.

22. Richtmeyer, R.D., Morton, K.W.: "Difference Methods for Initial Value Problems". Wiley & Sons, New York/London/ Sidney, 1967.

23. Moretti, G.: "The λ-Scheme". Computer and Fluids, Vol. 7, 1979, pp. 191-205.

24. Eberle, A.: "Characteristic Flux Averaging Approach to the Solution of Euler's Equations". VKI-Lecture Series 1987-04.

25. Eberle, A., Schmatz, M.A., Schäfer, O.: "High-Order Solutions of the Euler Equations by Characteristic Flux Averaging". ICAS-Paper 86-1.3.1, 1986.

26. Brenneis, A.: "Berechnung instationärer zwei- und dreidimensionaler Strömungen um Tragflügel mittels eines impliziten Relaxationsverfahrens zur Lösung der Euler-Gleichungen". Doctoral thesis, University of the Armed Forces, München. VDI-Fortschrittsberichte, Reihe 7: Strömungstechnik, Nr. 165, 1989.

VII CONVERGENCE TO STEADY STATE

The time-dependent formulation is most often used to compute steady state solutions to the Euler equations. There are several mechanisms that drive the solution to a steady state. Here we shall concentrate on the dissipation effect due to the boundary conditions, and not to the effect of artificial viscosity. Therefore we shall study hyperbolic partial differential equations where the boundary effects are dominant. The results are also valid for more general classes of differential equations of essentially hyperbolic character, as for example the Navier-Stokes equations for high Reynolds numbers. The study is mathematical, much of it repeated from Ref. 1.

7.1 Introduction

We begin to analyze the convergence properties to steady state first for the continuous problem, and then for the corresponding discrete approximation. This is the approach taken by Engqvist and Gustafsson[1]. They studied the behaviour of the spectrum of the differential and the difference operators in the context of two model problems, the scalar advection problem and the isentropic Euler equations problem. They found that the choice of boundary conditions radically affects the convergence rate to steady state as time increases, even to the extent that the asymptotic rate may change from exponential to algebraic convergence. For certain sets of boundary conditions, there may even be no convergence at all.

The usual way to study the convergence rate for time-marching procedures is to investigate the eigenmodes of the solution. Giles[2] did this for the one-dimensional Euler equations under various boundary conditions, and Eriksson and Rizzi[3] carried out such a study numerically for their centered finite-volume method applied to the two-dimensional Euler equations. This involved the approximate eigensystem analysis of the linearized Euler equations around a computed solution of an airfoil flow. Details are given in Section 5.5.2 of this book.

One can select a pseudo-time path for better convergence to steady state. Many methods use a modified form of the system, as for example when a scaling matrix multiplies the time-derivatives, or in calculations using the common practice of local time stepping where different time steps are used in different parts of the domain.

An interesting result[1] is that in many cases upwind differencing gives much faster convergence than centered differencing. This effect is not only due to the numerical dissipation. In fact, centered differencing with arbitrary large artificial viscosity added, does not give as good a convergence rate as upwind differencing. Engqvist and Gustafsson[1] give a simple example showing that prescribing the characteristic quantities at the boundaries may actually produce a solution with faster convergence to steady state than prescribing boundary conditions which are equivalent to the pure initial value problem. Thus, in this case the simpler characteristic boundary conditions are more effective than higher-order radiation boundary conditions.

7.2 Mathematical Understanding of Convergence[1]

7.2.1 The Continuous Problem

We are concerned with the convergence to steady state for the Euler equations and start by considering the quasi-linear system of partial differential equations

$$\frac{\partial q}{\partial t} + A\frac{\partial q}{\partial x} + Qq = 0\ , \tag{7.1a}$$

satisfying the homogeneous boundary conditions

$$B(x_b,t)\ q(x_b,t) = 0\ , \tag{7.1b}$$

and initial conditions

$$q(x,0) = f(x)\ . \tag{7.1c}$$

A necessary condition for convergence to steady state is that the solution to Eq.(7.1) converges to zero as $t \to \infty$.

There are essentially two phenomena causing the energy decay required for reaching a steady state (see also Ref. 4).

(i) Internal Decay Mechanism

Variable coefficients $A = A(x,t)$ and the presence of a lower-order term Q create the mechanism of internal decay. One example of this type of mechanism is enthalpy damping. This is to be distinguished from the dissipation mechanism in parabolic problems, where the decay rate depends on the frequency content of the solution. The decay rate for our type of hyperbolic problem is exponential, which can be shown by using the energy method (introduced in Sub-Chapter 3.18). Let x,t vary over the unit square, assume periodic boundary conditions, and let A be symmetric. The scalar product and the norm are defined by

$$(u,v) = \int_\Omega u^* v \, dx, \quad \|u\|^2 = (u,u) .$$

Integration of Eq.(7.1) by parts then gives[1]

$$\frac{d}{dt} \|q\|^2 = -2\mathrm{Re}(q,Dq), \quad D = Q - \frac{1}{2}\frac{\partial A}{\partial x} ,$$

which implies that the decay rate is determined by D. We get

$$e^{-d^* t}\|q(x,0)\| \leq \|q(x,t)\| \leq e^{-d_* t}\|q(x,0)\| ,$$

where the constants d_*, d^* satisfy

$$2d_* I \leq D + D^* \leq 2d^* I \text{ and } d_* > 0 .$$

(ii) Boundary Decay Mechanism

This mechanism is determined by the boundary matrix B together with the matrix A. When the integration by parts technique used above is applied to the non-periodic case, boundary terms will remain. Then the extra boundary terms are given by the boundary integral $\int_{\partial\Omega} q^* A q \, dS$, and Engqvist and Gustafsson describe two possibilities:

(1) q^*Aq is positive for some vectors q satisfying the boundary conditions. In this case the energy method fails to show even well-posedness of the problem.

(2) q^*Aq is non-positive for all vectors q satisfying the boundary conditions. This is sufficient for well-posedness of the problem. However, not even the stronger condition

$$q^*Aq \leq -\alpha q^* q , \quad \alpha > 0$$

is sufficient for proving convergence to steady state as $t \to \infty$, since the boundary terms may vanish even if the solution in the interior has not reached a steady state.

Clearly it is important for convergence to steady state that energy in the solution be transported to the boundary, such that the boundary conditions can cause the norm $\|q(x,t)\|$ to decay. This is one of the ways by which the multigrid technique accelerates convergence. This mechanism has been studied extensively in the following general form: Is there a decay law

$$\|q(\cdot,t)\|_{\Omega'} \leq f(t) ,$$

where $f(t) \to 0$ as $t \to \infty$ and Ω' is a subset of the boundary Ω?

Very often, Ω is an exterior domain and Ω' is bounded. Different conditions on the domain and the differential operator are derived in order to guarantee that the energy is not trapped in Ω' and that there exists a decay law. This has been the approach taken by Morawetz[5-7].

The Laplace transform is a useful tool for obtaining estimates of the decay rate of the solution to the homogeneous problem (7.1). Engqvist and Gustafsson use it in the following way. Let $\hat{q}(x,s)=Lq(x,t)$ be the Laplace transform of q, where L is defined by

$$\hat{q}(x,s)= \int_0^\infty e^{-st}q(x,t)dt \ . \tag{7.2}$$

The inverse Laplace operator L^{-1} is formally given by

$$q(x,t)=L^{-1}\hat{q}(x,s)= \frac{1}{2\pi i} \int_{\alpha-i\infty}^{\alpha+i\infty} e^{ts}\hat{q}(x,s)ds \ .$$

Parseval's relation

$$\int_0^\infty e^{2\alpha t}|u(x,t)|^2 dt = 2\pi \int_{-\infty}^{\infty} |\hat{u}(x,-\alpha+i\beta)|^2 d\beta$$

is also needed where the integrals are assumed to exist.

Consider the coefficients to be $A=A(x)$, $Q=Q(x)$, and $B=B(x)$ for the homogeneous problem (7.1). The Laplace-transformed problem (7.1), for which we assume a classical solution exists with uniformly bounded derivatives, is

$$s\hat{q} + P\hat{q} = f, \quad x \in \Omega \ ; \qquad B\hat{q} = 0, \ x \in \partial\Omega \ ; \quad P= A\frac{\partial}{\partial x} +Q \ . \tag{7.3}$$

Then the method for estimating the decay rate becomes based on the spectral properties of P. Assume that the problem (7.1) is well posed such that

$$\int_0^\infty e^{-2\alpha_0 t}\|q(x,t)\|^2 dt<\infty \text{ for some constant } \alpha_0 \ . \tag{7.4}$$

Parseval's relation then gives that the integral

$$\int_{-\infty}^{\infty} |\hat{q}(x,\alpha_0 + i\beta)|^2 d\beta$$

is also finite. Engqvist and Gustafsson have used this to reach the following result.

If the spectrum of the differential operator P is disjoint from the strip $-\alpha_1 < \text{Re } s < \alpha_0$, $\alpha_1 > 0$, such that for the solution to

(7.3) $L^{-1}\hat{q}$ is well defined in that strip, then q decreases exponentially, i.e. for any constant $\alpha, \alpha > \alpha_1$:

$$\| q(x,t) \| \leq c e^{-\alpha t} \| f(x) \| \; . \qquad (7.5)$$

Their result follows from the assumption of bounded time-derivatives (7.5).

Let us illustrate these concepts by an example.

Example: Linear Advection

The example of simple linear advection demonstrates that the solution in the typical case of a finite domain converges to a steady state after a finite time. The advection equation is

$$\begin{aligned} &(a) \quad \partial q/\partial t + \partial q/\partial x = 0, \qquad 0 \leq x \leq 1, \quad 0 \leq t \; , \qquad (7.6) \\ &(b) \quad q(0,t) = 0 \; , \\ &(c) \quad q(x,0) = f(x) \; , \end{aligned}$$

and the solution to this problem is given by

$$q(x,t) = \begin{cases} f(x-t), & x > t \; , \\ 0, & x < t \; . \end{cases}$$

The steady state $q \equiv 0$ is reached at $t=1$.

For analyzing the corresponding difference approximation, the Laplace transformation technique is applied to the continuous problem (7.6). The result of transforming (7.6) is

$$\begin{aligned} &(a) \quad s\hat{q} + \partial \hat{q}/\partial x = f, \qquad 0 \leq x \leq 1 \; , \qquad (7.7) \\ &(b) \quad \hat{q}(0,s) = 0 \; , \end{aligned}$$

which has the solution

$$\hat{q}(x,s) = \int_0^x e^{-s(x-\xi)} f(\xi) d\xi \; .$$

Engqvist and Gustafsson go on to show that this leads to an arbitrarily fast exponential decay because the differential operator $\partial/\partial x$ does not have a spectrum. But because of this it makes a clear example of the need to look at the discrete spectrum.

7.2.2 Linear Semi-Discrete Problem

We therefore turn now to the semi-discrete approximation of (7.6):

$$
\begin{aligned}
&(a)\quad \partial q_j/\partial t + Qq_j = 0\ , && j=1,2,..,N-1,\quad 0\le t\ ,\\
&(b)\quad q_o(t)=0\ ,\\
&(c)\quad B_r q_N(t)=0\ , && && (7.8)\\
&(d)\quad q_j(0)=f_j\ , && j=0,1,..,N,
\end{aligned}
$$

where $q_j(t)$ approximates $q(x_j,t)$, $x_j=j\Delta x$. The operator Q is the general consistent 3-point difference operator parametrized in the form

$$Q = D_o - \frac{c}{2}\,\Delta x D_+ D_-\ .$$

The standard notation for difference operators is used, i.e.

$$D_o u_j = \frac{1}{2\Delta x}(u_{j+1}-u_{j-1}),\qquad D_\pm u_j = \pm\frac{1}{\Delta x}(u_{j\pm1}-u_j)\ .$$

B_r is the rth-order extrapolation operator, i.e.

$$B_r u_j = (I-E^{-1})^r u_j,\qquad E^{-1}u_j = u_{j-1}\ .$$

The goal here is to investigate the rate of convergence to steady state. First Laplace transform the semidiscrete Equation (7.8) and treat the homogeneous problem $f_j=0$:

$$
\begin{aligned}
&(a)\quad s\hat{q}_j + \frac{1}{2\Delta x}(\hat{q}_{j+1}-\hat{q}_{j-1}) - \frac{c}{2\Delta x}(\hat{q}_{j+1}-2\hat{q}_j+\hat{q}_{j-1}) = 0,\quad j=1,2,..,N-1\ ,\\
&(b)\quad \hat{q}_o = 0\ ,\\
&(c)\quad B_r\hat{q}_N = 0\ . \qquad\qquad (7.9)
\end{aligned}
$$

In order to see that there exists no non-trivial solution for Re $s>0$ consider first the case $c=1$. Equation (7.9a) then becomes

$$s\,\hat{q}_j + \frac{1}{\Delta x}(\hat{q}_j-\hat{q}_{j-1}) = 0\ ,$$

which has the general solution

$\hat{q}_j = \alpha\kappa^j$, where κ is given by

$$s\,\kappa + \frac{1}{\Delta x}(\kappa-1) = 0\ .$$

The boundary condition (7.9b) shows that there is only the trivial solution $q_j\equiv 0$. For the general case $c\neq 1$ Engqvist and Gustafsson go on to prove that the system (7.8) has only exponentially decaying solutions.

One can see this more simply by computing the whole spectrum of Q using a standard eigenvalue routine. The linear extrapolation procedure for boundary condition at x=1 used for the case c=0, i.e. $Q=D_o$, yields the spectrum in Fig. 7.1. It indicates clearly how one end of the spectrum (for -Q) approaches the imaginary axis as Δx gets smaller. Actually, one can show that the distance between the imaginary axis and the right end of the spectrum is proportional to $(\Delta x)^2$, indicating a very poor convergence rate for steady state calculations using centered non-dissipative difference operators. In contrast, the upwind difference operator $-D_-$ obtained for c=1 has only one eigenvalue $-1/\Delta x$ (but with multiplicity N).

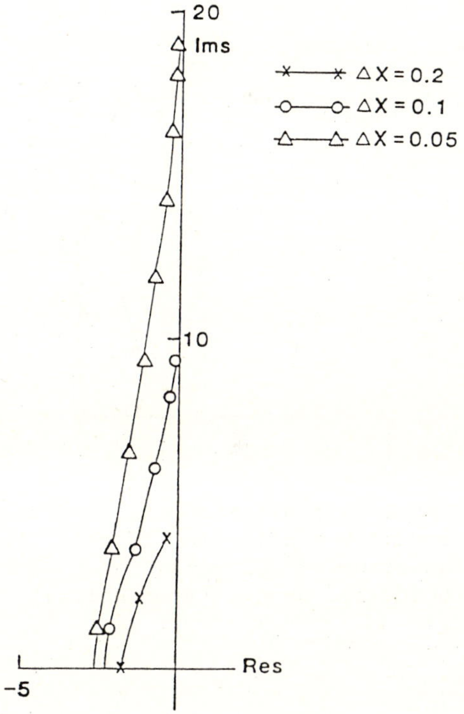

Fig. 7.1 Upper half spectrum of $-D_o$ with linear extrapolation at the boundary (from Ref. 1)

As we have seen, the introduction of boundary conditions has a dissipative effect on the non-dissipative operator $-D_o$. For comparison we shall show what effect the introduction of the dissipative term $c\Delta x D_+ D_-/2$ has in the periodic case. The eigenvalues for -Q are, in the periodic case:

$$s = \frac{1}{\Delta x}\,(-i\sin 2\pi\omega\Delta x - 2c\sin^2\pi\omega\Delta x), \quad \omega=0,1,..,N-1 .$$

Figure 7.2 shows the periodic case for c=0 and c=0.2, and the boundary condition case for $-D_o$ (shown also in Fig. 7.1). In all cases the step-length is $\Delta x=0.1$.

The general convergence rate is governed by the eigenvalue s with the largest real part. In Fig. 7.3, Res is presented as a function of c, and a very sharp peak is obtained at c=1 corresponding to the upwind operator D_-.

The use of second-order accurate upwind differencing requires an extra numerical inflow boundary condition for non-periodic problems. If that condition is given by first-order upwind differencing, s is again equal to $-1/\Delta x$. However, contrary to the previous case, this eigenvalue is simple.

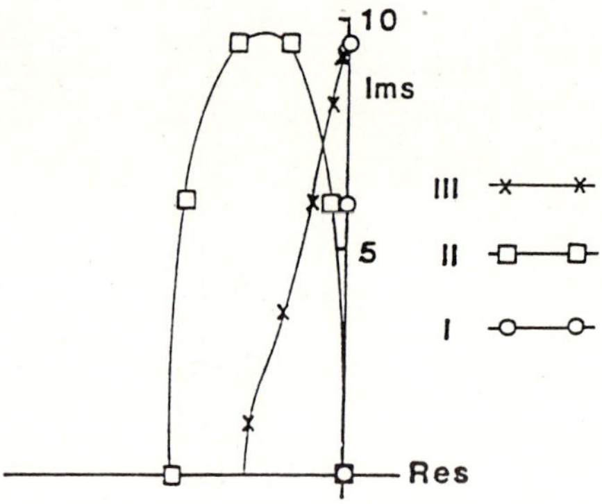

Fig. 7.2 Upper half spectrum of $-D_o+c\Delta x D_+ D_-/2$ with periodic boundary conditions. I: c=0, II: c=0.2, and III: spectrum of $-D_o$ with linear extrapolation at the boundary. $\Delta x=0.1$ (from Ref.1)

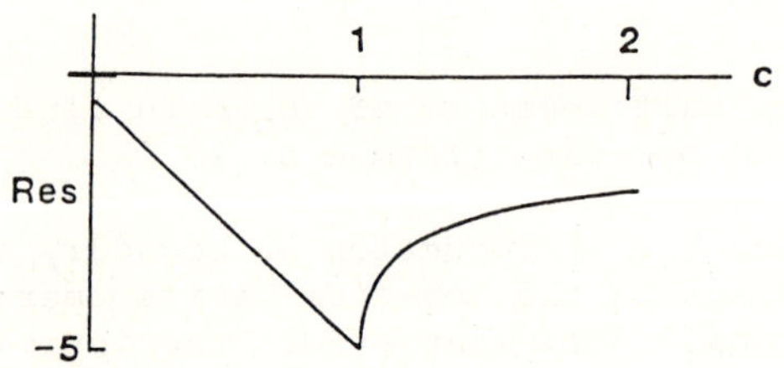

Fig. 7.3 Largest real part (Re s) of the eigenvalues to $-D_o+c\Delta x D_+ D_-/2$ as a function of c. Linear extrapolation at the boundary for $c\neq 1 (u_o=0, u_N=2u_{N-1}-u_{N-2})$ $\Delta x=0.2$ (from Ref. 1)

7.2.3 Linearized Euler Equations

Consider now inviscid isentropic flow in a channel. The Euler equations are

$$\frac{\partial q}{\partial t} + A(q)\,\frac{\partial q}{\partial x} + B(q)\,\frac{\partial q}{\partial y} = 0\ , \tag{7.10}$$

where

$$q = \begin{bmatrix} \rho \\ u \\ v \end{bmatrix},\quad A(q) = \begin{bmatrix} u & \rho & 0 \\ c^2/\rho & u & 0 \\ 0 & 0 & u \end{bmatrix},\quad B(q) = \begin{bmatrix} v & 0 & \rho \\ 0 & v & 0 \\ c^2/\rho & 0 & v \end{bmatrix}.$$

u and v are the velocities in the x- and y-directions, respectively.

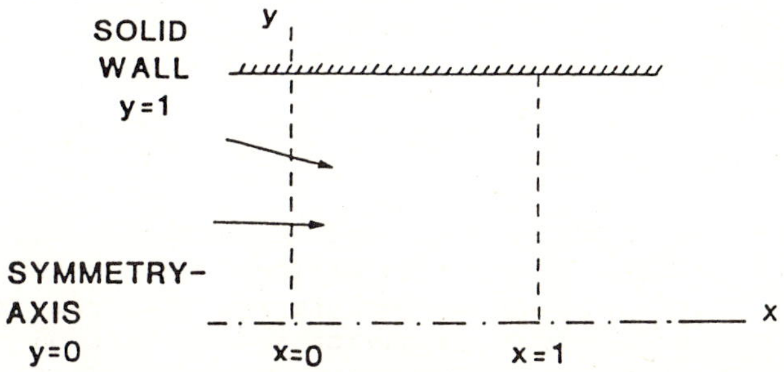

Fig. 7.4 Computational domain for the Euler equations (from Ref. 1)

It is assumed that the pressure p and the density ρ are related by an algebraic relation $p=p(\rho)$. The local speed of sound c is defined by

$$c^2 = dp/d\rho\ .$$

The equations (7.10) are defined on a domain corresponding to a channel with infinite extension in the x-direction according to Fig. 7.4.

The system (7.10) is linearized around some constant state $\bar{q}$ and the variables are made dimensionless. The normalizing velocity is the x-component u, so that the diagonal of A becomes the unit matrix. Since the y-component v is zero at both horizontal boundaries, it is assumed that $\bar{v}=0$ is the constant state which we are linearizing around. In order to get symmetric coefficient matrices A, B, the velocity components are scaled by the factor $\overline{\rho}/\overline{c}$. For convenience, the original notation $q=(\rho,u,v)^T$ is kept also for the new variables, and we get the linearized Euler equations with constant coefficients

$$\frac{\partial q}{\partial t} + A\frac{\partial q}{\partial x} + B\frac{\partial q}{\partial y} = 0 \,, \qquad 0\leq x,y\leq 1 \,, \quad 0\leq t \,, \tag{7.11}$$

where

$$q = \begin{bmatrix} \rho \\ u \\ v \end{bmatrix}, \qquad A = \begin{bmatrix} 1 & c & 0 \\ c & 1 & 0 \\ 0 & 0 & 1 \end{bmatrix}, \qquad B = \begin{bmatrix} 0 & 0 & c \\ 0 & 0 & 0 \\ c & 0 & 0 \end{bmatrix}.$$

For the case of subsonic flow $c>1$, two boundary conditions must be given at $x=0$, and one condition must be given at $x=1$. The standard procedure is to specify the Riemann invariants, which in this linear example corresponds to the specification of the characteristic variables

$$\begin{aligned} &\rho(0,y,t) + u(0,y,t) = g^{I}(y,t) \,, \\ &v(0,y,t) = g^{II}(y,t) \,, \\ &\rho(1,y,t) - u(1,y,t) = g^{III}(y,t) \,, \\ &v(X,1,t) = 0. \end{aligned} \tag{7.11a}$$

The energy method shows that these conditions lead to a well-posed problem. Now in order to estimate the convergence rate to steady state, we want to compute the spectrum of the differential space operator. Engqvist and Gustafsson make that computation easier by applying the Fourier transform

$$q(x,y,t) = \sum_{\omega=0}^{\infty} \hat{q}_{\omega}(x,t) \cos \pi\omega y \,. \tag{7.11b}$$

Introducing this in Eq.(7.11) gives for each ω

$$\frac{\partial \hat{q}}{\partial t} + A\frac{\partial \hat{q}}{\partial x} + \pi\omega\hat{B}\hat{q} = 0 \,, \tag{7.11c}$$

where

$$\hat{q} = \begin{bmatrix} \hat{\rho}_{\omega} \\ \hat{u}_{\omega} \\ \hat{v}_{\omega} \end{bmatrix}, \qquad \hat{B} = \begin{bmatrix} 0 & 0 & c \\ 0 & 0 & 0 \\ -c & 0 & 0 \end{bmatrix}. \tag{7.11d}$$

The spectrum is given by the set of s-values which satisfy

$$s\hat{q} + A\frac{\partial \hat{q}}{\partial x} + \pi\omega\hat{B}\hat{q} = 0 \tag{7.11e}$$

for non-trivial vectors $\hat{q}$, which fulfill the homogeneous boundary conditions. The general solution to (7.11e) can be written in the form

$$\hat{q} = \sigma_1 \begin{bmatrix} 0 \\ \pi\omega c \\ cs \end{bmatrix} e^{\kappa_1 x} + \sigma_2 \begin{bmatrix} s+\kappa_2 \\ -c\kappa_2 \\ \pi\omega c \end{bmatrix} e^{\kappa_2 x} + \sigma_3 \begin{bmatrix} s+\kappa_3 \\ -c\kappa_3 \\ \pi\omega c \end{bmatrix} e^{\kappa_3 x} . \qquad (7.11f)$$

Here the coefficients σ_1, σ_2, σ_3 are to be determined by the boundary conditions, and the exponential coefficients κ_1, κ_2, κ_3 are the distinct roots of the characteristic equation

$$\det \begin{vmatrix} s+\kappa & c\kappa & \pi\omega c \\ c\kappa & s+\kappa & 0 \\ -\pi\omega c & 0 & s+\kappa \end{vmatrix} = 0 . \qquad (7.11g)$$

Engqvist and Gustafsson then insert the general solution (7.11f) into the boundary conditions

$$\hat{\rho} + \hat{u} = \hat{v} = 0 , \quad \text{for} \quad x = 0 ,$$
$$\hat{\rho} - \hat{u} = 0 , \quad \text{for} \quad x = 1 .$$

There results a scalar equation in the complex variable s which they solve by Newton's method using different initial values such that all solutions within a certain domain around the origin were found. Their result is shown in Fig. 7.5 and for $c=2$, $\omega=1$, and $\omega=4$. The whole spectrum is for each ω contained in the left half-plane.

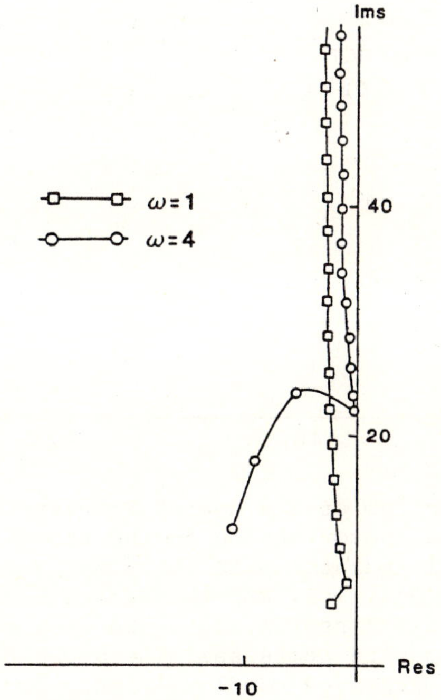

Fig. 7.5 Upper half spectrum for the Euler subsonic differential operator with characteristic boundary conditions (from Ref.1)

However, there is always one eigenvalue close to the imaginary axis whose distance from the axis goes to zero as $1/|\omega|$ with increasing ω.

Engqvist and Gustafsson have used this approach to study the effect of specifying the boundary conditions on the primitive variables q instead of the characteristic variables, and they also studied the case of supersonic flow. They found that the decay rate and thus the convergence rate as $t\to\infty$ may vary drastically depending on the flow field and the boundary conditions. For supersonic flow, the solutions vanish after finite time. For subsonic flow with boundary conditions on the primitive variables the energy does not decay at all, and with boundary conditions on the characteristic variables the decay is slower than any exponential and depends on the smoothness of the solution. The higher frequencies (large ω) decay at a slower rate than the lower frequencies (small ω).

7.2.4 Effect of Discrete Space Operator

Those are their conclusions from the study of the differential space operator. They went on to analyze the spectrum of the

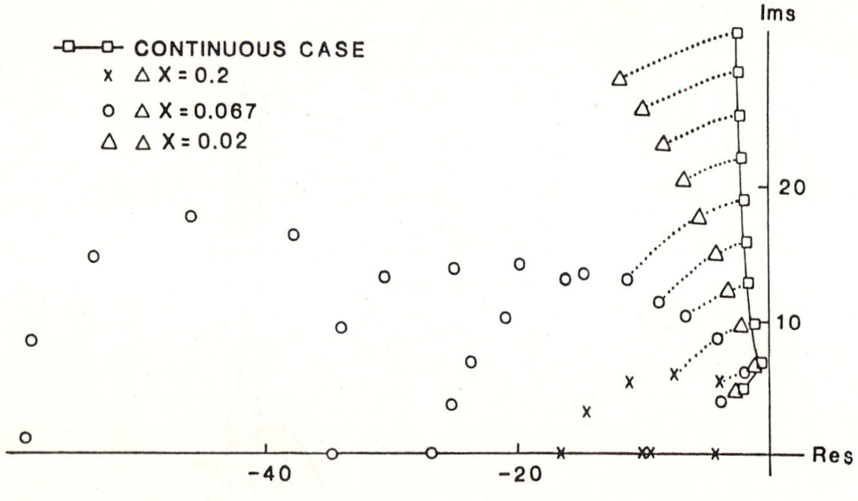

Fig. 7.6 Spectrum of the upwind Fourier-transformed difference approximation to the subsonic Euler differential operator with characteristic boundary conditions, $\omega=1$. For the discrete cases $\Delta x=\Delta y$. The dotted lines indicate the path of convergence as the step size decreases. The upper half of the spectrum is displayed completely only for $\Delta x=0.2$ and $\Delta x=0.067$ (from Ref. 1)

discrete operator -Q obtained by a first-order upwind scheme and characteristic boundary conditions. They computed the spectrum numerically, and the result is shown in Fig. 7.6 for $\omega=1$.

They reached the following conclusions. The spectrum for the continuous operator is "bent down" towards the negative real axis when the differential operator is discretized; the picture is similar also for all the higher frequencies ω. The consequence of this feature is that the use of upwind-differencing accelerates the convergence to steady state. We note, in particular, that the approximation of the lower part of the spectrum approaches the true values from the left, which is marked by dotted lines in the figure (These eigenvalues have the smoothest eigenvectors, and they are therefore better approximated than the others when a coarse mesh is used for the difference operator).

7.3 Multi-Grid Scheme

The technique of using multiple grids to obtain rapid convergence to the steady state of non-linear elliptic equations has met with large success, and a substantial body of theory is now in place so that there is a good understanding of how and why it works. That theory, however, is not strictly applicable to hyperbolic equations, but heuristic reasoning suggests that it ought to be possible to accelerate the evolution of a hyperbolic system to a steady state by using large time steps in coarse grids, because disturbances then will pass through the outer boundary more rapidly. The concept is based on the fact that the time step allowed with an explicit scheme is larger for a coarse mesh than for a fine mesh, and hence for a given number of time steps a wave travels further.

But one wants the accuracy of a fine mesh, which means interpolating corrections back to the fine grid. These interpolations necessarily introduce errors which cannot be rapidly expelled from the fine grid by propagation, and so are locally damped. All multigrid methods for hyperbolic problems tacitly assume that the errors are made up of high frequency modes only, and the driving scheme is constructed to have the property of rapidly damping these modes.

Jameson[8] has devised one of the most successful multigrid methods for hyperbolic problems by introducing auxiliary meshes into his explicit multistage scheme by simply doubling the mesh spacing. Values of the flow variables q are transferred to a coarser grid by the rule

$$q_{2h}^{(o)} = \Sigma\, V_h\, q_h / V_{2h} \;,$$

where the subscripts denote values of the mesh interval, V is the cell volume, and the sum is over the 4 cells on the fine grid composing each cell of the coarse grid. This rule conserves mass, momentum, and energy. Jameson then defines the forcing function, or projection operator

$$P_{2h} = \Sigma\, R_h(q_h) - R_{2h}(q_{2h}^{(0)}) \;,$$

where R is the residual of the difference scheme. In order to update the solution on a coarse grid, he reformulates the multistage scheme as

$$\begin{aligned} q_{2h}^{(1)} &= q_{2h}^{(0)} - \alpha_1 \Delta t[R_{2h}^{(0)} + P_{2h}] \;, \\ q_{2h}^{(m+1)} &= q_{2h}^{(0)} - \alpha_m \Delta t[R_{2h}^{(m)} + P_{2h}] \;, \end{aligned} \qquad (7.12)$$

where $R^{(m)}$ is the residual at the mth stage. In the first stage of the scheme, the addition of P_{2h} cancels $R_{2h}(q_{2h}^{(0)})$ and replaces it by $\Sigma\, R_h(q_n)$, with the result that the evolution of the coarse grid is driven by the residuals on the fine grid. This process is repeated on successively coarser grids. Finally the correction calculated on each grid is passed back to the next finer grid by bilinear interpolation.

Since the evolution on a coarse grid is driven by residuals collected from the next finer grid, the final solution on the fine grid is independent of the choice of boundary conditions on the coarse grids. The surface boundary condition is treated in the same way on every grid, by using the normal pressure gradient to extrapolate the surface pressure from the values in the cells adjacent to the wall. The far field conditions can either be transferred from the fine grid, or recalculated by the Riemann invariant procedure described in Sub-Chapter 3.4.

An effective damping of high frequency modes is crucially important in multigrid calculations, much more so than maximizing the CFL limit. Jameson finds that his three-stage scheme (7.12) used with only a single evaluation of the dissipative terms and the coefficients $\alpha_1=0.6$ and $\alpha_2=0.6$ is a good choice for a hyperbolic smoothing operator. The resulting stability region D for this scheme is shown in Fig. 7.7a. Figure 7.7b shows that with a Courant number $\nu=1.8$ and artificial viscosity coeffecients $\lambda=1.5$ and $\mu=0.04$, this scheme has an amplification modulus $|g|<0.35$ for all wave numbers in the range $1\leq\theta_1\leq\pi$.

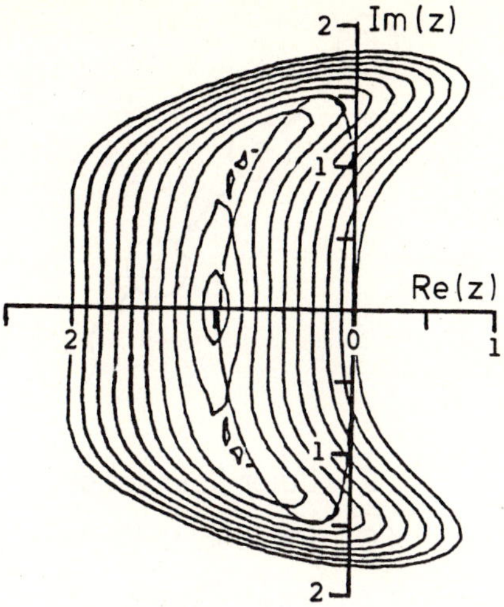

Figure 7.7a Stability region of 3 stage scheme with single evaluation of dissipation. Contour lines $|g|=1.$, $.9$, $.8, \ldots$ and locus of $z(\xi)$ for $\lambda=1.5$ and $\mu=0.04$. Coefficients $\alpha_1=0.6$, $\alpha_2=0.6$ (from Ref. 8)

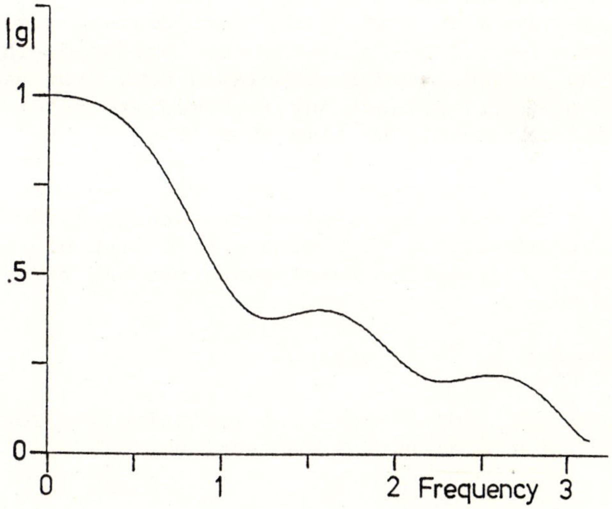

Figure 7.7b Amplification factor $|g|$ of 3 stage scheme with single evaluation of dissipation for $\lambda=1.5$ and $\mu=0.04$. Coefficients $\alpha_1=0.6$, $\alpha_2=0.6$ (from Ref. 8)

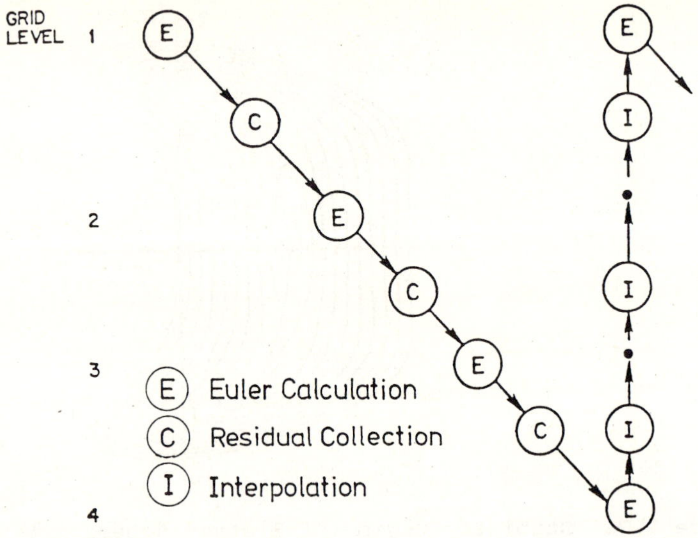

Fig.7.8 Saw tooth multigrid cycle (from Ref. 8)

Jameson has determined that an effective multigrid strategy is to use a simple saw-tooth cycle (Fig. 7.8) in which a transfer is made from each grid to the next coarser grid after a single time step. After reaching the coarsest grid, the corrections are then successively interpolated back from each grid to the next finer grid without any intermediate Euler calculations. On each grid the local time step limit is used and yields a constant Courant number through the field. The same Courant number is generally used on all grids, so that progressively larger time steps are used after each transfer to a coarser grid. In comparison to a single time step of the Euler scheme on the fine grid, the total computational effort in one multigrid cycle is

$$1 + \frac{1}{4} + \frac{1}{16} + \ldots < 4/3$$

plus the additional work of calculating the forcing functions P, and interpolating the corrections.

7.4 Enthalpy Damping

Enthalpy damping is a means Jameson[8] devised to accelerate convergence to steady state. He introduces a term into the Euler equations that is proportional to the difference between the total enthalpy H and its free stream value H_∞. In a steady flow with a uniform free stream the total enthalpy is a constant $H=H_\infty$ throughout the domain. The density and energy equations (2D)

$$\frac{\partial}{\partial x}(\rho u) + \frac{\partial}{\partial y}(\rho v) = 0, \quad \text{and} \quad \frac{\partial}{\partial x}(\rho u H) + \frac{\partial}{\partial y}(\rho v H) = 0$$

thus are consistent. This property is not preserved by the usual predictor - corrector difference schemes, e.g. the MacCormack scheme, but it is preserved by the centered Runge-Kutta schemes, provided that the dissipative operator is applied to ρH and not e in the energy equation. If this is done, then a forcing term proportional to $H-H_\infty$ does not alter the steady state.

Jameson introduced such a term, and thus altered the time path, because he wanted to provide additional damping to reach the steady state. He designed his term to mimic the role of ϕ_t in the telegraph equation. Although this is a second-order equation he argued that in the absence of strong shocks, the Euler equations are equivalent to the unsteady potential equation. Then the unsteady Bernoulli equation $\phi_t=-(H-H_\infty)$ led him to introduce the vector E to the Euler equations, where

$$E = \begin{bmatrix} \alpha\ \rho\ (H-H_\infty) \\ \alpha\ \rho u(H-H_\infty) \\ \alpha\ \rho v(H-H_\infty) \\ \alpha\ (H-H_\infty) \end{bmatrix} .$$

In the computational algorithm these non-differentiated terms are treated in a separate fractional step at the end of each time step. Numerical experiments have confirmed that the addition of these terms does enhance convergence.

7.5 Residual Averaging[8]

An obvious way to accelerate convergence to a steady state is to increase the time step. The time step of an explicit scheme is limited by the Courant-Friedrichs-Lewy condition, which requires that the region of dependence of the difference scheme must at least contain the region of dependence of the differential equation. This motivates the introduction of implicit schemes, which are discussed in Chapter VI.

The problem of slow convergence is not due to large changes in the time derivative, rather it results from the restricted CFL number caused by the fastest wave speeds of the problem. One idea to increase the convergence attempts to remove the fastest waves by smoothing the residuals. Consider the hyperbolic equation

$$q_t = f(q, q_x, x, t) . \tag{7.13}$$

It can be shown that if the time derivatives of q are small, then the space derivatives of f are also small. This property means that the right-hand side function is quite smooth in the space directions if the solution varies slowly in time. But this does not necessarily imply that q itself is smooth in space, but it does suggest that we could smooth the residuals, i.e. f, without altering the basic solution.

One way to smooth the residual is to replace it at each point by a weighted average of residuals at neighbouring points, i.e. $\bar{R}_i = SR_i$, where S is the averaging operator. In the one-dimensional case of the multi-stage scheme described by equations (7.12) one might replace the residual R_i by the average

$$\bar{R}_i = \varepsilon R_{i-1} + (1-2\varepsilon)R_i + \varepsilon R_{i+1}$$

at each stage of the scheme. This smoothes the residuals and also increases the support of the scheme, thus relaxing the restriction on the time step imposed by the Courant-Friedrichs-Lewy condition. If $\varepsilon > 1/4$, however, there are Fourier modes such that $\bar{R}_i = 0$, when $R_i \neq 0$. To avoid this restriction it is better to perform the averaging implicitly by setting

$$-\varepsilon\bar{R}_{i-1} + (1+2\varepsilon)\bar{R}_i - \varepsilon\bar{R}_{i+1} = R_i . \tag{7.14}$$

For an infinite interval this equation has the explicit solution

$$\bar{R}_i = \frac{1-r}{1+r} \sum_{q=-\infty}^{\infty} r^{|q|} R_{i+q} , \tag{7.15}$$

where

$$\varepsilon = \frac{r}{(1-r)^2} , \qquad r<1 . \tag{7.16}$$

Thus the effect of the implicit smoothing is to collect information from residuals at all points in the field, with an influence coefficient which decays by a factor r at each additional mesh interval from the point of interest.

Consider the linear advection model problem. According to equation (7.15) the Fourier symbol will be replaced by

$$z = -\lambda \frac{i \sin \xi + 4\mu(1-\cos\xi)^2}{1 + 2\xi(1-\cos\xi)} .$$

In the absence of dissipation one now finds that stability can be maintained for any Courant number λ, provided that the smoothing parameter satisfies

$$\varepsilon \geq \frac{1}{4} \{\frac{\lambda^2}{\lambda^{*2}} - 1\} ,$$

where λ^* is the stability limit of the unsmoothed scheme. In terms of the decay parameter r the stability condition becomes

$$\frac{1+r}{1-r} \geq \frac{\lambda}{\lambda^*} .$$

Suppose that the time step is increased to the limiting value permitted by a given value of r. Then according to equation (7.15) the Fourier symbol in the absence of dissipation is

$$z = -\lambda^* \{1+2 \sum_{q=-\infty}^{\infty} r^q \cos q\xi\} i\sin\xi = -\lambda^* \frac{1-r^2}{1-2r \cos\xi+r^2} i\sin\xi ,$$

and it may be verified that $|z| \leq \lambda^*$.

In the case of a finite interval with periodic conditions, Eq. (7-15) still holds, provided that the values R_{i+q} outside the interval are defined by periodicity. In the absence of periodicity one can also choose boundary conditions such that (7.15) is a solution of (7.14), with $R_{i+q}=0$, if i+q lies outside the interval. This solution can be realized by setting

$$\tilde{R}_1 = R_1 , \qquad \tilde{R}_i = R_i - r(R_i - \tilde{R}_{i-1}) \quad \text{for } 2 \leq i \leq n ,$$

and then

$$\bar{R}_n = \frac{1}{1+r} \tilde{R}_n ; \qquad \bar{R}_i = \tilde{R}_i - r(\tilde{R}_i - \bar{R}_{i+1}) \quad \text{for } 1 \leq i \leq n-1 .$$

The new modified residual field $\tilde{R}$ is used to advance the solution in time. This process alters the time-dependent solution without changing its steady state.

To bring out the essentials of the residual-averaging process, a simple wave equation is considered:

$$\phi_t + c\phi_x = 0 . \qquad (7.17)$$

The residual-averaging process as described is equivalent, to the lowest order, to adding an additional term to the original simple wave equation and converting it to the following equation:

$$\phi_t + c\phi_x - \varepsilon(\Delta x)^2 \phi_{txx} = 0 . \qquad (7.18)$$

The dispersion relation for this equation can be given as

$$\frac{\omega}{k} = \frac{c}{1 + \varepsilon k^2(\Delta x)^2} , \qquad (7.19)$$

where ω is the frequency and k is the wave number. By increasing the parameter ε, the wave speed for the high wave number component is substantially decreased. This decrease in wave speed for the restrictive short waves contributes to the substantial increase in the time step.

In the two-dimensional case smoothing is applied in product form:

$$(1 - \varepsilon_x\delta_x^2)(1 - \varepsilon_y\delta_y^2)\bar{R} = R ,$$

where δ_x^2 and δ_y^2 are second-difference operators in the x and y directions, and ε_x and ε_y are the corresponding smoothing parameters.

In the three-dimensional case smoothing is applied in product form too:

$$(1 - \varepsilon\delta_x^2)(1 - \varepsilon\delta_y^2)(1 - \varepsilon\delta_z^2)\ \bar{P}_{ijk} = P_{ijk} .$$

Thus it is only necessary to solve a sequence of tridiagonal equations for separate scalar variables, and in comparison with other implicit schemes this scheme has the advantage that it requires a relatively small amount of computational effort per time step.

Two other interesting formulations of implicit schemes have been proposed by MacCormack[9], and Lerat and Sidès[10]. MacCormack adds implicit stages requiring bidiagonal inversions to his explicit scheme. If the stages are reordered and the two implicit stages combined, it is equivalent in the scalar case to residual averaging. Lerat and Sidès consider a general class of implicit schemes using 9-point support in two dimensions. Their choice of parameters again leads to a form of residual averaging.

7.6 Mesh Sequencing

Convergence to the steady state of a hyperbolic problem is in general slow for both explicit as well as implicit methods. The reason for this is that the basic mechanism for eliminating transients is not diffusion, but the emission of these disturbances as travelling waves through the farfield boundaries. This mechanism is quite different from that in parabolic or elliptic problems, and is a consequence of the hyperbolic character of the equations. The process is faster naturally if one can propagate the waves faster, i.e. take as large a time step as possible by using an implicit technique.

Explicit schemes in this regard are restricted by the CFL condition, however. One technique that has proved helpful to explicit schemes is the one of mesh sequencing. The idea here is to begin the calculations, usually with initial conditions taken as the free stream values, with a very coarse mesh. Although the accuracy is not acceptable, the advantage is that transients travel very quickly to the boundaries and leave the domain. Remember that the step size varies linearly with the mesh length. It has also been observed in actual computations, that coarse mesh solutions like this do achieve a reasonable global accuracy. The lift and drag for example are usually within 10 per cent of their final values. This observation implies that most of the energy of the exact solution is carried in the very long wavelength modes. The short wavelength modes seem to be only a superposition on the long wavelengt modes. Mesh sequencing uses this fact by interpolating the converged coarse solution to a mesh refined by doubling the number of points in each direction. The initial conditions for this new calculation thus are very close to the global features of the exact solution. The transients that result from the interpolation are of high frequency and these tend to propagate rather quickly to the boundaries.

The overall result of this technique is to split the development of the solution. The low spatial frequencies, which travel rather slowly, are obtained on the coarse mesh. The high frequencies then are found in the fine-mesh solution as a superposition on the coarse mesh. High-frequency waves are seen to travel faster than low-frequency waves. The sequencing is a means to allow the boundary conditions in the farfield to work effectively on each component separately.

The sequencing can be repeated as often as one wishes. The most usual is a sequence of three meshes, a coarse, a medium, and a fine mesh. An overall round figure for the efficiency of this technique is a factor of three in reduced computation time.

7.7 References

1. Engqvist, B., Gustafsson, B.: "Steady State Computations for Wave Propagation Problems". Math. Comp., Vol. 49, No. 179, July 1987, pp. 39-64.

2. Giles, M.B.: "Eigenmode Analysis of Unsteady One-Dimensional Euler Equations". ICASE Report No. 83-47, 1983.

3. Eriksson, L.E., Rizzi, A.: " Computer-Aided Analysis of the Convergence to Steady State of Discrete Approximations to the Euler Equations". J. Comput. Phys., Vol. 57, 1985, pp. 90-128.

4. Ferm, L., Gustafsson, B.: "A Down-Stream Boundary Procedure for the Euler Equations". Comput. & Fluids. Vol. 10, 1982, pp. 261-276.

5. Morawetz, C.S.: "The Decay of Solutions to the Exterior Initial-Boundary Value Problem for the Wave Equation". Comm. Pure Appl. Math., Vol. 14, 1961, pp. 561-568.

6. Morawetz, C.S.: "Decay of Solutions of the Exterior Problem for the Wave Equation". Comm. Pure Appl. Math., Vol. 28, 1975, pp. 229-264.

7. Morawetz, C.S., Ralston, J.V., Strauss, W.A.: "Decay of Solutions of the Wave Equation Outside Nontrapping Obstacles". Comm. Pure Appl. Math., Vol. 30, 1977, pp. 447-508.

8. Jameson, A,: "Transonic Flow Calculations for Aircraft". Lecture Notes in Mathematics, Vol. 1127, Numerical Methods in Fluid Dynamics, F. Brezzi (ed.), Springer Verlag, 1985, pp. 156-242.

9. MacCormack, R.W.: "A Numerical Method for Solving the Equations of Compressible Viscous Flows". AIAA-Paper 81-110, 1981.

10. Lerat, A., Sidès, J.: "A New Finite Volume Method for the Euler Equations with Applications to Transonic Flows". Proc. IMA Conference on Numerical Methods in Aeronautical Fluid Dynamics, Reading, 1981, P.L. Roe (ed.), Academic Press, New York, 1982, pp. 245-288.

VIII A NOTE ON THE USE OF SUPERCOMPUTERS

8.1 Supercomputers as Driver of Computational Fluid Dynamics

Computational fluid dynamics has had a truly dramatic development process in the last decade. The first broad attempts to solve the Euler equations came at the begin of the Eighties, see for example Ref. 1, the proceedings of one of the first GAMM-Workshops in the field of Numerical Fluid Mechanics in 1979. Now even Navier-Stokes solutions are feasible for complex configurations.

Of course the pace of this development depended strongly on the pace of computer development. Especially the introduction of the vector computer was the major driving factor, which became many times stronger, when the first truly large in-core memories became available. The aerodynamic community in research and in industry reacted very positively on this development, although vector computers only slowly became available in industry. The big potential of numerical aerodynamics and the role of the supercomputer for it certainly was first recognized in the US, leading for instance to the establishment of the Numerical Aerodynamic Simulation (NAS) Program[2]. The experience with vector computers was disseminated fast in the community, see for instance Refs. 3 and 4, and today the vector computer can be considered as being accepted in the field, especially in view of the fact that minicomputers and even workstation already feature vector processors.

The crucial feature of vector computers is the vector compiler, or vectorizer. Of course, it is most important that an algorithm must have a basic structure, which is vectorizable at all. A lot can be found about this basic requirement in literature, for instance in Refs. 3 and 4. The vector compilers in use today have reached a high development level. However, experience[5] with many vector computers teaches, that the user must have a good knowledge about the hardware and compiler capabilities of a given vector computer[6]. A large hand-vectorization effort should then be spent, when the code is to be used for many computations, and net savings are to be expected. Otherwise the user may be content with a given auto-vectorizer result. An important observation, made already very early, is that a good vectorized program also runs considerably faster than before on a sequential computer.

8.2 Future Developments in Supercomputing: Parallel Processing

The real breakthrough of numerical methods in design aerodynamics will only come after a decrease of the specific computation costs by at least one to two orders of magnitude[5]. This decrease most probably will not result from advances in solution algorithms, but from further advances in computer performance. The mono-vector processor computer certainly still has a big potential in performance/cost improvements. However, it is to be expected that parallel processing finally will be the decisive answer to the cost problem.

It is not intended to review here the recent developments in parallel architectures. The interested reader is referred to Ref. 6. It seems, however, worthwhile, to recall the experiences with the first vector computers. Special program-language extensions obviously proved to be repulsive for potential users. Good autovectorisers, even if larger adaptations of a code were to be made, finally attracted them.

The same will happen with parallel computers. If the user is not bothered too much by the pecularities of such hardware, i.e., if the compiler does the distribution of the work load between the processors, they will easily be accepted. They will either allow to tackle computation problems much larger than today (see e.g. Sub-Chapter 12.5), or to reduce costs. It is telling that for instance the SUPRENUM company considers its computer as a universal computer. Massive parallel computers, at least for the time being, are seen as special machines for special purposes, which presently not necessarily include aerodynamic applications.

Modern vector computers come with up to eight or ten, or so, processors, working in general on a shared memory. It is claimed that the autoparallelizers for these machines have reached a level like that of modern autovectorizers. Considering the computer requirements given by Bailey, Ref. 2, the needs for non-linear inviscid codes are met today by such machines. Meanwhile, however, it is believed, that for instance the grid-point requirement he gives for this class of problems with 10^5, is too small.

It remains to be seen, how the development in computer hardware will affect the development and use of Euler solvers. As long as the cost of computing a given problem is not reduced drastically, Euler codes, although being a valuable aerodynamic design tool, will not become an everyday design tool and interdisciplinary topics (see Sub-Chapter 12.5) will be hampered in their development.

8.3 References

1. Rizzi, A., Viviand, H. (eds.): "Numerical Methods for the computation of Inviscid Transonic Flows with Shock Waves". Volume 3 of Notes on Numerical Fluid Mechanics, Vieweg, Braunschweig/Wiesbaden, 1981.

2. Bailey, F.R.: "NAS: Supercomputing Master Tool for Aeronautics". Aerospace America, Vol. 23, No. 1, 1985, pp. 118-121.

3. Gentzsch, W.: " Vectorization of Computer Programs with Applications to Computational Fluid Dynamics". Volume 8 of Notes on Numerical Fluid Mechanics, Vieweg, Braunschweig/ Wiesbaden, 1984.

4. Schönauer, W., Gentzsch, W. (eds.): "The Efficient Use of Vector Computers with Emphasis on Computational Fluid Dynamics". Volume 12 of Notes on Numerical Fluid Mechanics, Vieweg, Braunschweig/Wiesbaden, 1986.

5. Hirschel, E.H.: "Super Computers Today: Sufficient for Aircraft Design? – Experiences and Demands". In: H.W. Meuer (ed.): Supercomputer '88. Hanser, München/Wien, 1988, pp. 110-150.

6. Gentzsch, W., Neves, K.W.: "Computational Fluid Dynamics: Algorithms & Supercomputers". AGARDograph, AGARD-AG-311, 1988.

IX COUPLING OF EULER SOLUTIONS TO VISCOUS MODELS

Viscosity effects are present in every flow. If, however, the momentum forces acting in the flow are much larger than the viscous forces (large Reynolds number), the latter can be neglected. Flow with such properties - inviscid flows - can be described with the Euler equations, or even with the potential equation. The latter holds if the flow in addition is irrotational, and if compressibility effects are weak.

In reality viscosity effects, or more general "diffusive transport effects", are always present in the flow past or through a body, even at very large Reynolds numbers. They are, however, confined to a thin layer at the body surface, to the boundary layer. This boundary layer in practice can well be distinguished form the inviscid flow, also called the external flow. The inviscid flow thus is kept away from the body surface by the boundary layer.

If a boundary layer leaves the body surface - separation of a boundary layer - a wake forms. In such a case viscosity effects are present also in the flow away from the body surface. In general on an airplane several separations and hence wakes are present, forming in some cases a complex array of free vortex sheets and vortices.

Because diffusive transport effects are an ingredient of every fluid flow, they must be considered when deciding on a mathematical model for the description of the flow past or through a body. In the following it is discussed when and why a flow situation might demand the coupling of a pure inviscid flow model like the Euler equations to a viscous model. In addition a short overview is given over such models and the corresponding computation methods.

9.1 Diffusive Transport Effects in Fluid Flows

Mass, momentum, energy - and other properties - are transported by a flow[1]. The two basic mechanisms of transport are the convective transport and the diffusive transport*). The convective transport takes place along the streamlines of the flow and hence can be described by, for instance, the Euler equations alone. Diffusive transport is a molecular transport,

---

*) in literature instead of "diffusive transport" often only the word "transport" is used.

and in principle is not unidirectional. It follows gradients of scalar and vectorial properties of the flow. This means, the diffusive transport takes place also across streamlines, which is very important. In a boundary laryer the major diffusive transport effects are the transport of momentum (shear) and heat (heat conduction) in direction normal to the body surface and hence across the streamlines of the boundary-layer flow.

It must be added that in all flow cases of practical interest, especially in aerodynamics, the diffusive transport not only appears on a molecular level, but has an equivalent in the form of turbulent motion[1] (momentum transport), or connected with it (mass diffusion, heat conduction).

The transport phenomena are classified by similarity parameters[1]. The most important parameter in the present context is the, already mentioned, Reynolds number:

$$Re = \frac{\text{momentum force}}{\text{viscous force}} = \frac{\rho u L}{\mu} \,. \tag{9.1}$$

At $Re \to 0$ the flow is governed by viscous forces (Stokes flow), at $Re \to \infty$ it is governed by momentum forces (inviscid flow). In a boundary layer both types of forces are of equal magnitude: $Re = O(1)$ *). Boundary-layer theory is a widely and well developed field (see for example References 2 to 4).

In a boundary layer usually all other possible diffusive transport phenomena are also present if the appropriate flow situation exists. If the total enthalpy is high enough, heat transfer will occur in the boundary layer of compressible flow. The thickness of the temperature boundary layer is governed by the Prandtl number or by the Peyclet number. In air it roughly coincides with the viscous boundary layer. At even higher total enthalpies, when dissociation of molecules is present in the fluid[5], a diffusive mass transport might appear, and so on. There are countless other flow situations, combustion, two-phase flow, etc., where a boundary layer exists and where besides the viscous transport of momentum a diffusive transport of other entities occurs.

In the following the most important effects which a viscous boundary layer has on the flow past a body are reviewed.

---

*) This must not be confused with the fact that the boundary-layer equations are found in a limiting process[2] with $Re \to \infty$.

a) Attached boundary layer

In the attached boundary layer, which might be in a laminar or in a turbulent state, shear stresses act and a shear force is excerted on the body (viscous drag). In addition the boundary layer displaces the inviscid flow, which it sheathes from the surface, to a certain extend. The displacement thickness δ_1 is inversely proportional to some power of the Reynolds number:

$$\frac{\delta_1}{L} \sim \frac{1}{Re_L^{\,n}} \quad , \tag{9.2}$$

n = 0.5 for laminar flow,
n = 0.2 for turbulent flow.

Note that this relation holds for incompressible flat plate boundary-layer flow[2]. It can be used, however, for estimation purposes also in other flow situations.

If the displacement δ_1 of the inviscid flow is small compared to a characteristic body length L, a weak interaction is said to occur between inviscid flow field and boundary layer.

Finally in the boundary layer a diffusive transport of vorticity to and from the body surface takes place. The body surface acts with no-slip boundary conditions as a source or sink of vorticity[6].

b) Separated boundary layer

Due to an adverse pressure gradient the boundary layer may loose so much momentum that it separates from the body surface. Then a large displacement of the inviscid flow occurs, a strong interaction takes place between the inviscid flow field and the boundary layer, the distinction between them fails.

In such cases a small recirculation area might appear or the boundary layer might leave the surface alltogether. In the classical definition separation does not occur at a sharp trailing edge of a wing, if the boundary layers simply flow off the surface.

Here a more general view is taken[7]. Whether the boundary layer leaves a sharp edge (flow-off separation at the trailing edge of a wing, or at the leading edge of a delta wing at larger angle of attack), or gets separated at the body surface (free-surface or squeeze-off separation, the latter especially in three-dimensional flow situations), in each case the following effects appears:

- strong interaction of the separating boundary layer with the inviscid flow, which leads to a viscosity induced pressure drag,

- convective transport of vorticity away from the body surface in wakes (vortex sheets) and vortices (the induced drag of a finite-span lifting wing is related to this phenomenon).

If the flow past a body is to be computed, it follows from this discussion that a detailed investigation must show whether the ever present diffusive transport phenomena in the flow on parts of the body, or on the whole body and its vicinity -

a) -can all be neglected, then a solution of the Euler equations (or even of the potential equation) is appropriate to describe the flow,

b) -appear in the frame of a weak interaction of the viscous boundary layer with the inviscid flow, then a coupling of the Euler equations (or the potential equation) with the boundary layer equations is appropriate to describe the flow,

c) -appear in the frame of a strong interaction of the boundary layer with the inviscid flow, then the Euler equations (or the potential equation) even within a weak interaction coupling scheme are not appropriate to describe the flow,

d) -have led to vortex sheets and vortices behind or above the body (wing, delta wing, fuselage), then the Euler equations (or even the potential equation) are appropriate to a large degree to describe the flow with the embedded vortex sheets and vortices.

In the following Sub-Chapters 9.2 and 9.3 the approaches to compute weak (case b) and strong (case c) interacting flows are outlined. The whole chapter X is devoted to the simulation of vortex sheets and vortices by means of the Euler equations (case d), because of the importance, novelty, and the special problems of this approach.

9.2 Treatment of Weak Interaction Flow Problems

The problem of weak interaction of inviscid flow with a boundary layer can be treated completely in the frame of boundary layer theory[2-4]. There many computation methods are available, both in the form of integral methods (Pohlhausen type), and of, predominantly, finite-difference methods, both for two- and three-dimensional boundary layers.

The metrical treatment of general body shapes of all kind (wings, fuselages) is possible[8,9,10], thus allowing to compute boundary layers where ever boundary-layer theory is valid. Even second-order methods (see for instance Ref. 11) are available with only little extra computational effort over first-order methods.

Turbulence modelling with simple algebraic models in general appears to be of sufficient accuracy for attached boundary layers, for both two- and three-dimensional cases[12]. Strong three-dimensionality and also large surface curvature may change this picture. The situation is quite different concerning transition laminar-turbulent. Although linear stability theory has advanced considerably, even for three-dimensional boundary layers[13], criteria for transition are empirical, or at best semi-empirical (e^N-method)[14]. No non-empirical transition criterion for boundary layers exists.

Boundary-layer theory also allows to get an indication about the kind and the location of separation regions. Although strict boundary-layer theory can only give approximate results in this respect, it has proved to be quite useful (see e.g. Refs. 15 to 17).

Boundary-layer methods in general allow the computation of the following properties of laminar or turbulent boundary layers:

- skin friction,
- skin friction lines (three-dimensional cases),
- wall-heat transfer, wall temperature,
- surface curvature pressure corrections (second order theory),
- displacement thickness,
- equivalent inviscid source distribution,
- separation indication,
- etc.

Overviews over the possibilities and problems of boundary-layer theory and computation methods are given for example in References 18 and 19.

The weak coupling of an inviscid solution, found for instance with an Euler method, with a boundary-layer solution, in essence takes into account the displacement properties of the boundary layer. The body becomes virtually thicker by the displacement thickness, and in a certain sense extents to downstream infinity. The displacement thickness can be computed once the boundary-layer solution has been found and the mass- flux profiles are available. For three-dimensional flow a linear partial differential equation has to be solved[9]. The displacement thickness even might become negative in some regions on a body[17].

To change the body contour by the displacement thickness certainly is impractical. With the concept of the, to the displacement thickness, equivalent inviscid source distribution[20] the local surface normal mass flow[9] in general boundary-layer coordinates is found by

$$(\sqrt{a}\,\rho_o\,v_o^3)_{inv.} = (\sqrt{a}\,\rho_e\,v_e^{\alpha}\,\delta_1^{(\alpha)})_{,\alpha}\,, \qquad (9.3)$$

$$\alpha = 1.2\;,$$

where δ_1^{α} are the two local displacement thicknesses:

$$\delta_1^{\alpha} = \int_0^{\delta}(1-\frac{\rho\,v^{\alpha}}{\rho_e\,v_e^{(\alpha)}})dx^3. \qquad (9.4)$$

This surface-normal mass flow is introduced as surface boundary condition into the inviscid flow solution.

Usually first an Euler or potential flow solution is made, then the boundary-layer solution is performed, and the equivalent inviscid source distribution is computed. After this distribution is available as boundary condition, the inviscid computation is repeated. At high Reynolds numbers in principle only one repetition is necessary, more might be performed in a given case. In modern zonal solutions (see next sub-chaper), several boundary-layer solutions are performed within the convergence cycle of the inviscid solution (either Euler or potential methods). In this way one arrives at an equivalent inviscid flow[21] solution which consistently takes into account the weak interaction with the boundary layer without iteration of the converged inviscid solution.

Once this equivalent inviscid solution has bee performed, all boundary-layer information as listed above is available. It should be noted, that in case separation occurs on the body under consideration, the boundary-layer solution of course cannot be extended into that area (for inverse solutions see next sub-chapter). The equivalent inviscid source distribution, however, then can be extrapolated into that area, while phasing it out at the same time. In this way the whole computation process is not hampered, although the results must be considered with care. This is because the strong interaction region is not treated properly in this way and it will influence in general, execept for supersonic flow cases, globally also the weak interaction region. If this is the case, an appropriate other solution has to be applied (see the following sub-chapter).

9.3 Treatment of Strong Interaction Flow Problems

The most general model in the frame of continuum theory for the treatment of strongly interacting flows are the Navier-Stokes equations both for laminar and turbulent (Reynolds-averaged Navier-Stokes equations) flow. However, if the strongly interacting flow is a free-surface or squeeze-off separation[7], then the currently available turbulence models will not be sufficient. This holds in general even for transport-equation models, non the least because the empiricism incorporated there (correlations etc.) in essence comes from attached turbulent boundary layers. If it would be possible to fit coordinate surfaces to the skeleton surfaces of the separating wakes or vortex sheets, then even scalar turbulence models might be sufficient, except for the very vicinity of the region where the boundary layer actually leaves the body surface. The complexity of general three-dimensional separating flows will forbid this in the foreseeable future (a similar problem exists with the "fitting" of embedded shocks, where the current approach is to "capture" the shock, see Chapter III of this book). Also the large-eddy simulation of turbulence, or the direct simulation[22], which also would solve the transition problem, will not be available soon for practical problems.

In this context it should be noted that the knowledge of the topology of separated, especially three-dimensional viscous flows has advanced considerably over the last decade (see e.g. Refs. 23 to 26).

Solution methods for the Navier-Stokes equations meanwhile exist in considerable numbers. Since no reveiw is intended, the reader is referred to, for instance, Refs. 27 to 29.

Strong interaction flows, especially if no large squeeze-off separation areas appear, can also be treated by either interaction models, or by inverse solutions of the boundary-layer equations (see e.g. Refs. 30, 31). In two-dimensional and in some quasi-two-dimensional (infinite swept wing) cases such models are proving to be quite useful. In general three-dimensional cases with strong separation phenomena of complex topology they are not adequate.

Navier-Stokes solutions usually are obtained globally for a given configuration. The thin-layer approach[9], which takes of all the viscous terms into account only the boundary-layer terms leads to some savings in computational effort. In certain flow situations space marching versions, of the - imprecisely called - parabolized Navier-Stokes equations, can be used with large savings of computational effort.

A recently developed zonal approach to solve viscous flow problems is described in References 32 to 38. In regions of weak interaction the Euler and the second-order boundary-layer equations are solved in order to arrive at the equivalent inviscid flow (see the preceding sub-chapter). In regions of strong interaction a simultaneous local Navier-Stokes solution is embedded. The second-order boundary-layer solution is necessary in order to correct for pressure and metric variations in the boundary layer due to the body surface curvature[35]. In addition the rotational properties, which are not correctly described in the outer part of a boundary layer by first-order theory, are computed more accurately. All this is important at the coupling boundaries, at least for low to medium Reynolds numbers.

All involved governing equation solvers are closely coupled, and work on commonly structured nets. The resulting method is handled with no more complication than a global Navier-Stokes method. It is interesting to note that the Euler solution in the weak interaction region in many cases even acts as an accelerating pacemaker on the embedded Navier-Stokes solution.

This zonal approach leads for well suited flow problems to considerable savings in memory and computation time. For two-dimensional cases more than 50 per cent savings in computation time and 30 per cent in storage were found. In good conditioned three-dimensional cases much better numbers are obtained. An additional benefit is that the attached boundary layer can be much more refined than in a global Navier-Stokes solution. This makes the approach interesting for hypersonic flow problems, too. Here for instance also the entropy-layer swallowing on blunt bodies can be treated by an equivalent inviscid Euler/second-order boundary-layer solution in a large velocity/altitude intervall.

9.4 References

1. Bird, R.B., Stewart, W.E., Lightfoot, E.N.: "Transport Phenomena." J. Wiley & Sons, New York, London, Sydney, 1966.

2. Schlichting, H.: "Boundary-Layer Theory." McGraw-Hill Book Comp., New York, 1979.

3. Rosenhead, L. (ed.): "Laminar Boundary Layers." Oxford at the Clarendon Press, 1963.

4. White, F.M.: "Viscous Fluid Flow." McGraw - Hill, New York, 1974.

5. Vincenti, W.G., Kruger, C.H.: "Physical Gas Dynamics." J. Wiley & Sons, New York, London, Sydney, 1965.

6. Lighthill, M.J.: "Introduction Boundary-Layer Theory." In Ref.3, pp.46-113.

7. Hirschel, E.H.: "On the Creation of Vorticity and Entropy in the Solution of the Euler Equations for Lifting Wings." MBB-LKE122-AERO-MT-716, 1985.

8. Nash, J.F., Patel, V.C.: "Three-Dimensional Turbulent Boundary Layers." SBC Technical Books, Atlanta, 1972.

9. Hirschel, E.H., Kordulla, W,: "Shear Flow in Surface Oriented Coordinates." Vol.4 of Notes on Numerical Fluid Mechanics, Vieweg, Braunschweig/Wiesbaden, 1981.

10. Hirschel, E.H.: "Boundary-Layer Coordinates on General Wings and Fuselages." ZFW, Vol.6, No.3, 1982, pp.194-202.

11. Monnoyer, F.: "Calculation of Three-Dimensional Attached Viscous Flow on General Configurations with Second-Order Boundary-Layer Theory." ZFW, Vol. 14, No. 1/2, 1990, pp. 95-108.

12. Van den Berg, B., Humphreys, D.A., Krause, E., Lindhout, J.P.F.: " Three-Dimensional Boundary Layers - Calculations and Experiments." Vol.19 of Notes on Numerical Fluid Mechanics, Vieweg, Braunschweig/Wiesbaden, 1988.

13. Fischer, T., Ehrenstein, U., Meyer, F.: "Theoretical Investigation of Instability and Transition in the DFVLR-F5 Swept-Wing Flow." In: Zierep, J., Oertel, H. (eds.): Symposium Transsonicum III. Springer, Berlin-Heidelberg-New York, 1989, pp.243-252.

14. Bushnell, D.M., Malik, M.R., Harvey, W.D.: "Transition Prediction in External Flows via Linear Stability Theory." In: Zierep, J., Oertel, H. (eds.): Symposium Transsonicum III. Springer, Berlin-Heidelberg-New York, 1989. pp.225-242.

15. Hirschel, E.H.: "Computation of Three-Dimensional Boundary Layers on Fuselages." J. of Aircraft, Vol.21, No.1, 1984, pp.23-29.

16. Hirschel, E.H., Bretthauer, N., Röhe, H.: "Theoretical and Experimental Boundary Layer Studies on Car Bodies." J. Intern. Assoc. for Vehicle Design, Vol.5, No.5, 1984, pp.567-584.

17. Hirschel, E.H.: "Evaluation of Results of Boundary-Layer Calculations with Regard to Design Aerodynamics." In: Computation of Three-Dimensional Boundary Layers Including Separation. AGARD-R-741, 1986, pp.5-1 to 5-29.

18. Fernholz, H.H., Krause, E. (eds.): "Three-Dimensional Boundary Layers." Springer, Berlin-Heidelberg, 1982.

19. N.N.: "Computation of Three-Dimensional Boundary Layers Including Separation." AGARD-R-741, 1986.

20. Lighthill, M.J.: "On Displacement Thickness." J. Fluid. Mech., Vol.4, 1958, pp.383-392.

21. Lock, R.C., Firmin, M.C.P.: "Survey of Techniques for Estimating Viscous Effects in External Aerodynamics." In: Roe, P.L. (ed.): Numerical Methods in Aeronautical Fluid Dynamics. Academic Press, London, 1982, pp.337-430.

22. Schumann, U., Friedrich, R. (eds.): "Direct and Large-Eddy Simulation of Turbulence". Vol.15 of Notes on Numerical Fluid Mechanics, Vieweg, Braunschweig/ Wiesbaden, 1986.

23. Peake, D.J., Tobak, M.: "Three-Dimensional Interaction and Vortical Flows with Emphasis on High Speeds." AGARDograph No.252, 1980.

24. Dallmann, U.: "Topological Structures of Three-Dimensional Flow Separation." AIAA-Paper 83-1735, 1983.

25. Hornung, H., Perry, A.E.: "Some Aspects of Three-Dimensional Separation, Part 1: Stream-Surface Separation." ZFW, Vol.8, No.1, 1984, pp.77-87.

26. Wu, J.-Z., Gu, J.W., Wu, J.-M.: "Steady Three-Dimensional Fluid Particle Separation from Arbitrary Smooth Surface and Formation of Free Vortex Layers." ZFW, Vol.12, No.2, 1988, pp.89-98.

27. Shang, J.S.: "An Assessment of Numerical Solutions of the Compressible Navier-Stokes Equations." J. of Aircraft, Vol.22, No.5, 1985, pp.353-370.

28. Fujii, K., Obayashi, S.: "Navier-Stokes Simulations of Transonic Flows over a Wing-Fuselage Combination." AIAA-Journal, Vol.25, No.12, 1987, pp.1587-1596.

29. Kordulla, W. (ed.): "Numerical Simulation of the Transonic DFVLR-F5 Wing Experiment - Towards the Validation of Viscous Wing Flow Codes." Vol.22 of Notes on Numerical Fluid Mechanics, Vieweg, Braunschweig/Wiesbaden, 1988.

30. Melnik, R.E., Chow, R.: "Asymptotic Theory of Two-Dimensional Trailing-Edge Flows." NASA SP-347, 1975.

31. Dargel, G., Thiede, P.: "Viscous Transonic Airfoil Flow Simulation by an Efficient Viscous-Inviscid Interaction Method." AIAA-Paper No. 87-0412, 1987.

32. Hirschel, E.H., Schmatz, M.A.: "Zonal Solutions for Viscous Flow Problems." In: Hirschel, E.H. (ed.): Finite Approximations in Fluid Mechanics, DFG-Priority Research Programme, Results 1983-1985. Vol. 14 of Notes on Numerical Fluid Mechanics, Vieweg, Braunschweig/Wiesbaden, 1986, pp.99-112.

33. Schmatz, M.A., Hirschel, E.H.: "Zonal Solutions for Airfoils Using Euler, Boundary-Layer and Navier-Stokes Equations." In: Applications of Computational Fluid Dynamics in Aeronautics. AGARD CP 412, 1986, pp.20-1 to 20-13.

34. Schmatz, M.A.: "Simulation of Viscous Flows by Zonal Solutions of the Euler, Boundary-Layer and Navier-Stokes equations." ZFW, Vol.11, No. 4/5, 1987, pp.281-290.

35. Wanie, K.M., Schmatz, M.A., Monnoyer, F.: "A Close Coupling Procedure for Zonal Solutions of the Navier-Stokes, Euler and Boundary-Layer Equations." ZFW, Vol.11, No.6, 1987, pp.347-359.

36. Monnoyer, F., Wanie, K.M., Schmatz, M.A.: "Calculation of the Three-Dimensional Viscous Flow Past Ellipsoids at Incidence by Zonal Solutions." In: Deville, M. (ed.): Proc. 7th GAMM-Conf. on Num. Meth. in Fluid Mechanics. Vol. 20 of Notes on Numerical Fluid Mechanics, Vieweg, Braunschweig Wiesbaden, 1988, pp.229-238.

37. Schmatz, M.A., Monnoyer, F., Wanie, K.M., Hirschel, E.H.: "Zonal Solutions of Three-Dimensional Viscous Flow Problems." In: Zierep, J., Oertel, H. (eds.): Symposium Transsonicum III. Springer, Berlin-Heidelberg-New York, 1989, pp.65-74.

38. Wanie, K.M.: "Simulation dreidimensionaler viskoser Strömungen durch lokale Lösung der Navier-Stokesschen Gleichungen." Doctoral Thesis, München, 1988.

X MODELLING OF VORTEX FLOWS: VORTICITY IN EULER SOLUTIONS

In Chapter IX it was stated that wakes, respectively vortex sheets, and vortices can be handled in the frame of inviscid theory. This is done for a long time already in the frame of potential wing theory (see for instance Ref. 1). Panel methods, as discrete potential methods, can be seen as extension and generalization of - linear - potential wing theory to general and complicated aircraft geometries at subsonic and even supersonic speeds (see e.g. Ref. 2). With proper vortex sheet paneling, panel methods can also be employed on delta wings with leading-edge vortices, in this way allowing to compute non-linear lift problems with - linear - potential wing theory (see e.g. Ref. 3).

Nevertheless, it proved to be a surprise when in an Euler solution for a lifting wing, Ref. 4, most probably the first time, vortices were computed. According to theory with an irrotational free stream no vorticity should be created in an - inviscid - Euler flow model past a body, except for curved shock waves in supersonic flow. But the vorticity was there, was also found later in many other solutions. A good overview over the problem gives Ref. 5.

In the present chapter a consistent theory is given of this phenomenon. Potential wing theory regardless of its formulation has its base in the vortex laws[1], which were formulated a long time before the boundary layer was discovered by Prandtl in 1904. In this chapter it will be shown that the vortex singularities of potential wing theory can be understood as real boundary layers, wakes, respectively vortex layers, and vortices in the boundary-layer limit $Re_{ref} \to \infty$. Then, since no theory of the solution of the Euler equations for the present problem exists, analogies with potential wing theory are employed (Actually what is missing is a theory of discrete solutions of the Euler equations with regard to the present problem).

The explanation of the phenomenon, which is finally given, is seen to be consistent with an explanation on the base of weak solution theory[6]. Thus the appearance of vorticity in an Euler solution for lifting wings with finite span in steady flow can be confirmed as being correct in principle, although some unsolved problems still exist in this context.

10.1 Boundary Layers, Wakes and Vortices in their Inviscid Limit

Consider the generalized shear layer in Fig. 10.1. A curvilinear, orthogonal boundary-layer coordinate system (locally monoclinic surface-oriented coordinates[7]) has been placed with its origin $P(x^i=0)$ on the skeletal surface of it The x^α-coordinates ($\alpha=1,2$) lie in this surface. The x^3-coordinate is rectilinear and normal to the x^α-coordinates. The external inviscid flow lies above the upper edge of the shear layer ($\underline{v}_{e_u}$) at $x^3 \geq \delta_u$, and below the lower edge ($\underline{v}_{e_\ell}$) at $x^3 \leq \delta_\ell$.

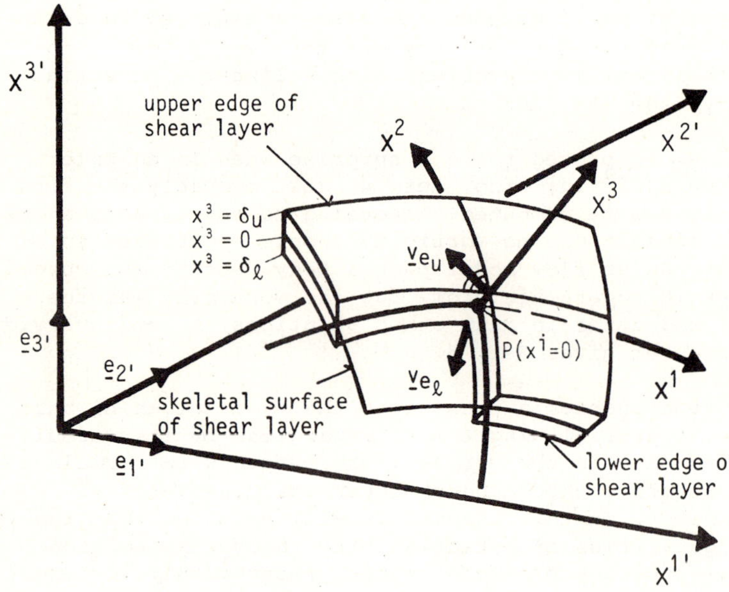

Fig. 10.1 Generalized shear layer in orthogonal locally monoclinic coordinates

The vorticity vector with contravariant components reads in this system[7] after the classical boundary-layer stretching

$$\tilde{x}^3 = x^3 \sqrt{Re_{ref}}, \quad \tilde{v}^3 = v^3 \sqrt{Re_{ref}} \tag{10.1}$$

has been introduced:

$$\begin{aligned} \underline{\omega} = [\omega^1; \omega^2; \omega^3] = \text{rot } \underline{v} = Re_{ref}^{1/2} (g)^{-1/2} \cdot \\ \cdot \left[Re_{ref}^{-1} \tilde{v}^3{}_{,2} - (g_{22} v^2)_{,\tilde{3}} ; (g_{11} v^1)_{,\tilde{3}} - \right. \\ - Re_{ref}^{-1} \tilde{v}^3{}_{,1} ; Re_{ref}^{-1/2} (g_{22} v^2)_{,1} - \\ \left. - Re_{ref}^{-1/2} (g_{11} v^1)_{,2} \right]. \end{aligned} \tag{10.2}$$

Here v^i are contravariant velocity components, and g is the determinant of the metric tensor. The Reynolds number Re_{ref} consists for instance of the freestream values u_∞, ρ_∞, μ_∞ and a characteristic body length L. Physical velocities and lengths are non-dimensionalized accordingly with u_∞ and L.

The components $g_{\alpha\beta}$ of the covariant metric tensor can be expressed in locally monoclinic coordinates completely[7] by the geometrical properties of the skeleton surface, i.e. by its metric tensor $a_{\alpha\beta}$, and its curvature properties $b_{\alpha\beta}$, b^γ_β. After introduction of the boundary-layer stretching they read

$$g_{\alpha\beta} = a_{\alpha\beta} - 2b_{\alpha\beta}\, Re_{ref}^{-1/2}\, \tilde{x}^3 + b_{\alpha\gamma}\, b^\gamma_\beta\, Re_{ref}^{-1} (\tilde{x}^3)^2, \quad \alpha = 1,2. \qquad (10.3)$$

Now the "local vorticity content of a shear layer"[8] is introduced as the integral of the vorticity across the shear layer in x^3-direction:

$$\underline{\Omega} = [\Omega^1;\ \Omega^2;\ \Omega^3] = \int_{x^3=\delta_\ell}^{x^3=\delta_u} \mathrm{rot}\ \underline{v}\ dx^3. \qquad (10.4)$$

After introducing Eqs. 10.1 and 10.3 into Eq. 10.4 and performing the limiting process $Re_{ref} \to \infty$, the local vorticity content reads:

$$\underline{\Omega}\big|_{Re_{ref}\to\infty} = (a)^{-1/2} [-a_{22}\, v^2;\ a_{11}\, v^1;\ 0]_{\tilde{\delta}_\ell}^{\tilde{\delta}_u}, \qquad (10.5)$$

where a is the determinant of the metric tensor of the skeletal surface. Exchanging finally the contravariant components F^α completely with physical components $F^{*\alpha}$ by employing the rule

$$F^\alpha = \frac{F^{*\alpha}}{\sqrt{a_{(\alpha\alpha)}}}, \qquad (10.6)$$

the local vorticity content reads in the large Reynolds-number limit

$$\underline{\Omega}\big|_{Re_{ref}\to\infty} = [\Omega^{*1};\ \Omega^{*2};\ \Omega^{*3}] = [-v^{*2};\ v^{*1};\ 0]_{\delta_\ell}^{\delta_u}, \qquad (10.7)$$

where the tilde on the upper and the lower bound was removed.

In the following for illustrative purposes relation (10.7) is discussed for several special shear-layer cases. Note that the form of the functions $v^{*\alpha}(x^3)$ inside a shear layer does not play a role as long as they are continuous, and continously blend into the external inviscid flow $v_e^{*\alpha}(\delta_{u,\ell})$.

— Two-dimensional wakes

In Fig. 10.2 two idealized wakes are sketched: a) the wake of an airfoil in steady subcritical motion, b) the wake of an airfoil in steady supercritical motion with a shock on the upper side. In both cases at the trailing edge the static pressure is the same on the upper and the lower side, regardless of whether the airfoil is lifting or not. This holds also in the limit $Re_{ref} \to \infty$ for both wakes, even if they are curved.

In case a) this leads to equal external inviscid velocities at δ_u and δ_ℓ: $v^{*1}_{e_u}(|=\underline{v}_{e_u}|) = v^{*1}_{e_\ell}(|=\underline{v}_{e_\ell}|)$. The local vorticity content of the wake then is, regardless of whether the airfoil is lifting or not:

$$\underline{\Omega}\Big|_{Re_{ref}\to\infty} = [0;\ 0;\ 0].$$

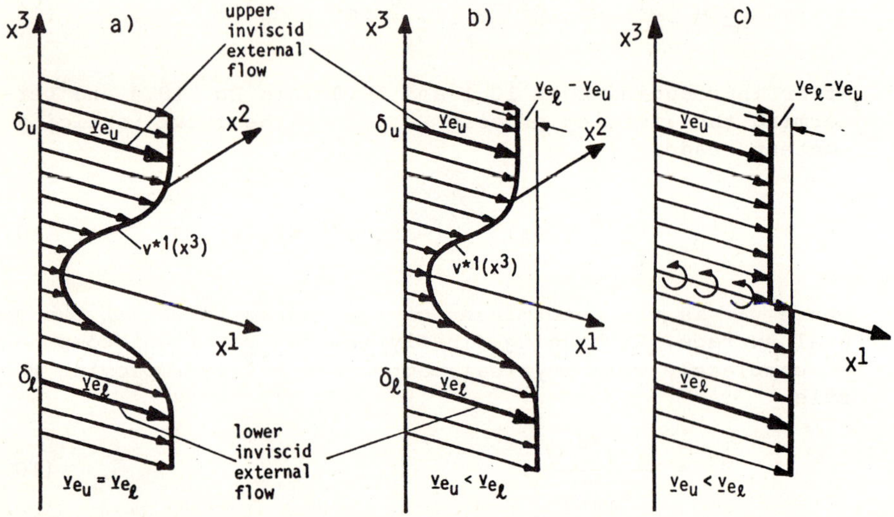

Fig. 10.2 Schematic of two-dimensional wakes:
a) wake of airfoil in steady subcritical motion;
b) wake of airfoil in steady supercritical motion;
c) wake b) in the limit $Re_{ref} \to \infty$

Although vorticity is present in the wake, it cancels out in the integral (10.4). Wake a) therefore is considered to be kinematically inactive. This is in accordance with theory[1], which says that once the steady motion of the airfoil has been established, the lifting vortex on the airfoil and the opposite starting vortex at infinity behind it do no more change the strength of their circulations. Note that the wake must by no means be symmetric, it is only the edge values which are symmetric.

In case b) the total pressure loss on the upper side of the airfoil leads to different velocities on the upper and the lower side of the wake because the static pressure is the same there: $v^{*1}_{e_u} (=|\underline{v}_{e_u}|) < v^{*1}_{e_\ell} (=|\underline{v}_{e_\ell}|)$. In this case a finite vorticity content is present:

$$\underline{\Omega}\big|_{Re_{ref}\to\infty} = [0;\ |\underline{v}_{e_u}| - |\underline{v}_{e_\ell}|;\ 0].$$

Of course also here no net vorticity can leave the airfoil. Therefore this kinematically active vorticity content must be cancelled elsewhere, although it is small in general compared to the vorticity content of the boundary layers on the airfoil[9].

Fig. 10.3 shows schematically the supercritical flow past an airfoil. It is evident that only the vorticity created by the

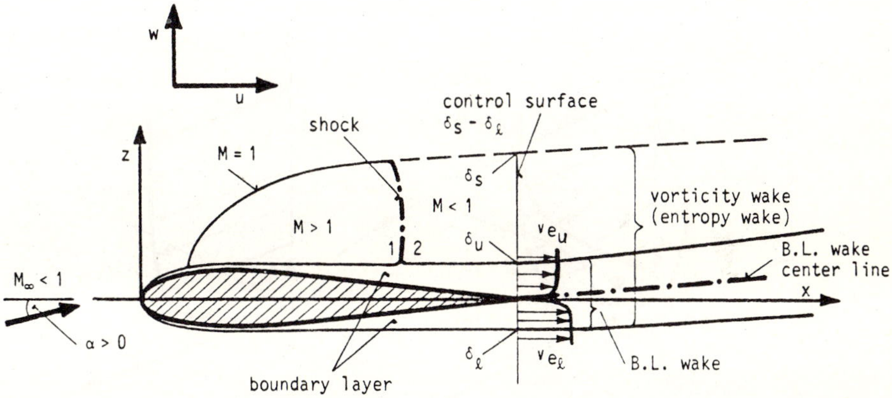

Fig. 10.3 Schematic of flow at the trailing edge of an airfoil at supercritical condition

shock can do the cancellation:

$$\underline{\Omega}_S = \int_{\delta_u}^{\delta_S} \text{rot}\ \underline{v}\ dx^3 = [0;\ -(|\underline{v}_{e_u}| - |\underline{v}_{e_\ell}|);\ 0]\ .$$

For kinematic reasons thus for lifting transonic wings the wake consideration must include the shock wake. The mechanism which brings about the necessary shock form and its change of strength to achieve the cancellation is not known yet.

Finally wake b) of Fig. 10.2 is related to potential theory in Fig. 10.2c. In the limiting process it collapses to a vortex discontinuity with the classical circulation amount per length element:

$$d\Gamma = (|\underline{v}_{e_u}| - |\underline{v}_{e_\ell}|)dx^1 = \Omega^{*2} dx^1 .$$

Although the vorticity at the discontinuity is infinity, the vortex discontinuity of potential theory has a finite vorticity content, which was proved with the present limiting process.

— Boundary layers

In Fig. 10.4 the profiles of a three-dimensional, Fig. 10.4a, and of a two-dimensional, Fig. 10.4b, boundary-layer are given. In both cases the external inviscid streamline is oriented with the x^1-coordinate of the local wake system of Fig. 10.1. At $x^3=\delta_u=\delta$ therefore $v^{*1}_{e_u} = |\underline{v}_e|$, the external inviscid speed, and $v^{*2}_u \equiv 0$ by definition. At the body surface, $x^3=0$, the no-slip condition holds with $v^{*1} = v^{*2} = 0$.

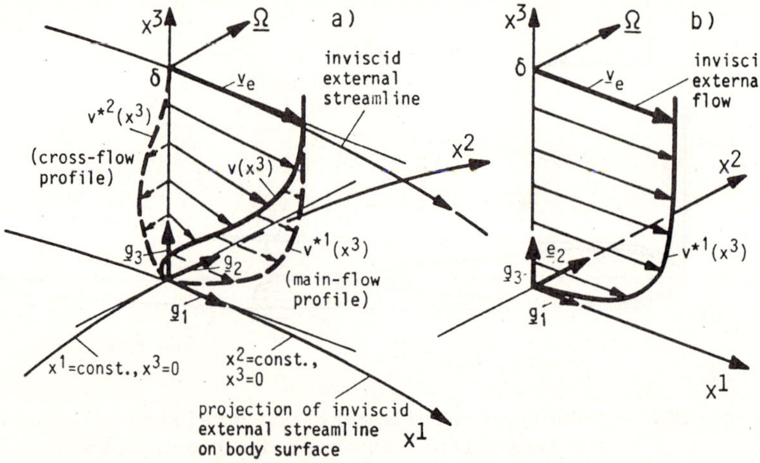

Fig. 10.4 Schematic of a) three-dimensional, b) two-dimensional boundary layer

In both cases therefore the local vorticity content vector lies in direction normal to the external inviscid streamline and has the magnitude $|\underline{v}_e| = v_e$:

$$\underline{\Omega}\big|_{Re_{ref} \to \infty} = [0;\ v_e;\ 0],$$

regardless of the shape of the profiles of the three-dimensional[10] or the two-dimensional boundary layer.

An application of this consideration to the boundary-layer flow on a lifting airfoil in steady subcritical motion, Fig. 10.5, yields by integration of the local vorticity content

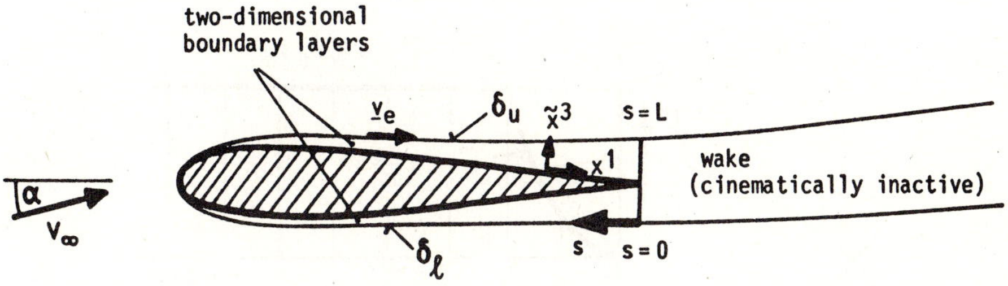

Fig. 10.5 Schematic of boundary layers on airfoil with lift (steady subcritical flow)

along the surface of the airfoil:

$$\int_0^L \Omega^{*2}\, ds \Big|_{Re_{ref}\to\infty} = \int_0^L v_e\, ds = \Gamma ,$$

because of $Re_{ref}\to\infty$: $\delta_u \to 0$, $\delta_\ell \to 0$.

This result says that the lifting vortex of potential theory with its circulation Γ actually consists of the vorticity in the upper and the lower boundary layer on the wing. This in effect is an application of the theorem of Stokes[11] in the limit $Re_{ref}\to\infty$. The wake in this case, as was shown above, carries not kinematically active vorticity.

— The Rankine vortex

In order to make the result for the airfoil in Fig. 10.5 more clear, the vorticity content of the core of the Rankine vortex, shown in Fig. 10.6 is computed. This seemingly trivial operation yields

$$\Gamma = 2\int_0^{r_o}\int_0^{2\pi} \underbrace{\left[\frac{1}{2}\,\frac{1}{r}\,\frac{d(rv)}{dr}\right]}_{\omega} r\,d\varphi\, dv = 2\pi r_o v_o .$$

This result says that with a given content of vorticity, which equates to a given circulation Γ, the rotational core, which drives the potential vortex part, can have any diameter, because $r_o v_o$=constant. To the potential vortex singularity thus belongs a finite vorticity content, and this singularity can be considered as the vortex core of the Rankine vortex in the limit $r_o \rightarrow 0$. If this is true, then of course the lifting potential vortex of the airfoil also somewhere must have a "vortex core", which, as was seen, are the boundary layers.

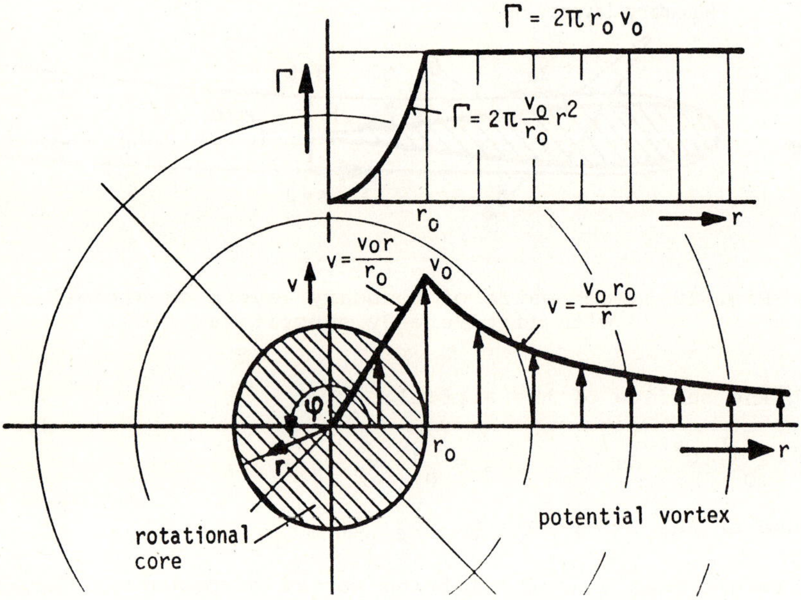

Fig. 10.6 Schematic of Rankine vortex and its velocity (v) and circulation (Γ) distribution

10.2 The Lifting Wing as Inviscid Computation Problem

Consider the lifting wing configurations in Fig. 10.7. Behind the wing with low leading-edge sweep, Fig. 10.7a, lies a wake. This wake, which lies behind every lifting wing, contracts to two discrete vortices further downstream. The wake, or vortex layer, "carries" the viscous drag and the viscosity induced pressure drag, and, as will be seen later, "carries" also the induced drag. It is formed by the boundary layers, which flow off the upper and the lower side of the wing at the trailing edge, where a strong interaction with the external inviscid flow occurs.

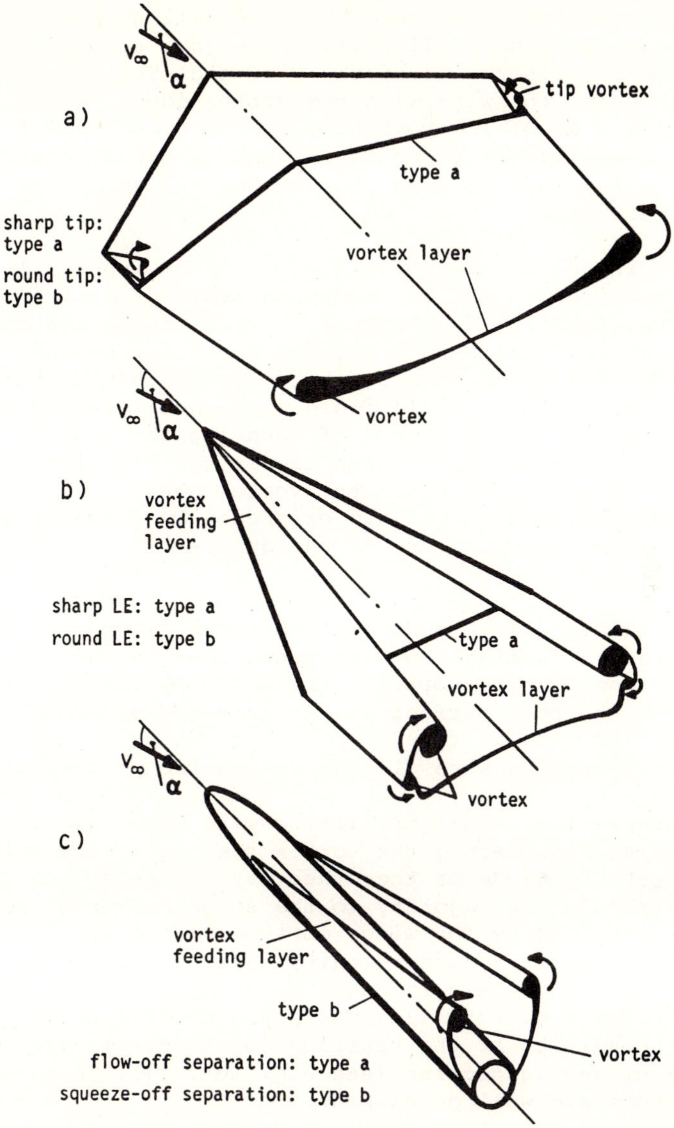

Fig. 10.7 Schematic of vortex layers and vortices at:
a) wing with small leading-edge sweep;
b) with large leading-edge sweep; c) fuselage

At the tip of the wing, Fig. 10.7a, an additional vortex phenomenon appears, the tip vortex. If the tip has a sharp edge, the vortex is formed via flow-off separation[8]. If the tip is well rounded, the boundary layer will go to a certain extend around it, and finally will undergo a squeeze-off separation[8] on the upper side of the wing tip. The tip vortex interacts behind the wing with the vortex sheet, which in any case carries most of its vorticity in the outer wing section part, and seems to get immersed in this sheet. It contributes somewhat to the induced drag, but is not the source of it, as is sometimes stated.

If the leading-edge sweep of a wing is large enough, leading-edge vortices appear at a certain angle of attack, Fig. 10.7b. These leading-edge vortices, which are much stronger than the tip vortices, are the cause for the non-linear or vortex lift. Their strength can be so large that on the leeside of a delta wing the flowfield becomes completely restructered[12], usually to the extend of changing the sign of vorticity in the vortex sheet which leaves the trailing edge of such wings, Fig. 10.7b. Again the vortex sheets which leave the leading edges may come into existence by flow-off separation at sharp leading edges, or by squeeze-off separation at round leading edges.

If the delta wing with round leading edges is made more and more slender, a fuselage body is approached in the limit, Fig. 10.7c. Now the leeside vortex pair is fed by vortex sheets which leave the body surface via squeeze-off separation. In this case, and also in case b) possible secondary and higher separation phenomena were disregarded in the discussion.

In any mathematical model of lifting wing flow, and hence also in any computation method the vortex sheets and the vortices must be present in one or the other way, at least the ones which contribute the majority to the sought-after effects. In potential wing theory for wings of type a) the wake is treated essentially as vortex discontinuity sheet[1]. This is also true for panel methods, where it might be discretized in one or the other way (see e.g. Ref. 13). There are details connected to the Kutta condition at the trailing (or leading) edge, which warrant a closer inspection (see e.g. Ref. 14). In general these methods are well developed.

At higher subsonic and at transonic Mach numbers - linearized - potential theory will no more suffice because of compressibility effects together with shocks. The full potential equation, also called gasdynamic equation, or derivates of it, can be used in this regime. However, field solutions in the complete space around the wing must be performed in each case.

A phenomenon like the shock decambering due to the total pressure loss in the shock on the suction side of the wing[9], cannot be treated in the frame of such models. Connected to the total pressure loss are the kinematic properties of the whole wake, as discussed in the preceeding sub-chapter. They also cannot be presented by potential theory. Therefore such models can only be used if only weak shocks are present, or if appropriate corrections can be made. In general three-dimensional cases these corrections become unpractical, as well as the treatment of the discontinuity surface behind the wing.

Can such lifting-wing flows be treated correctly by means of the Euler equations? Compressibility, shocks, decambering, vorticity etc. in principle pose no difficulties. Again there are problems: for instance in a discrete solution of the Euler equations the stagnation point at the trailing edge of an airfoil usually is not reached[9]. In reality it is not reached, too, but in an inviscid model it should. However, the larger question is, whether in a solution of the Euler equations forces on a body (the paradox of d'Alembert) can be computed (flows with shocks are disregarded here). In this regard, usually silently, the analogy to potential flow is accepted: if in the Euler solution for a wing a Kutta condition is introduced at the trailing edge, the lift can be computed, and then behind finite-span wings in addition also a discontinuity sheet appears, modelling the wake of reality.

This Kutta condition usually must be described explicitly in finite-difference solutions for lifting wings. However, in finite-volume methods it is implicitly present (which is not yet understood). This led to confusion with regard to the creation of lift in such solutions, because the existence of the accompanying discontinuity sheet in the case of finite-span wings (especially delta wings) became not clear, too. Before discussing the properties of this discontinuity sheet, which indeed appears in - discrete - Euler solutions, the properties of the wake, as it appears in reality, are investigated in the following sub-chapter.

10.3 The Structure of the Wake of a Lifting Wing

Consider the flow past the wing in Fig. 10.7a. Intuition tells that, because of the finite span of the wing, a pressure relaxation takes place around the wing tip between the high pressure on the lower side of the wing and the low pressure on the upper side of the wing. This in general leads to a flow direction towards the wing tip on the lower side of the wing, and to a flow direction towards the wing root on the upper side. Therefore locally the wake immediately behind the wing trailing edge has - on the right-hand side of the wing - the schematic structure shown in Fig. 10.8a.

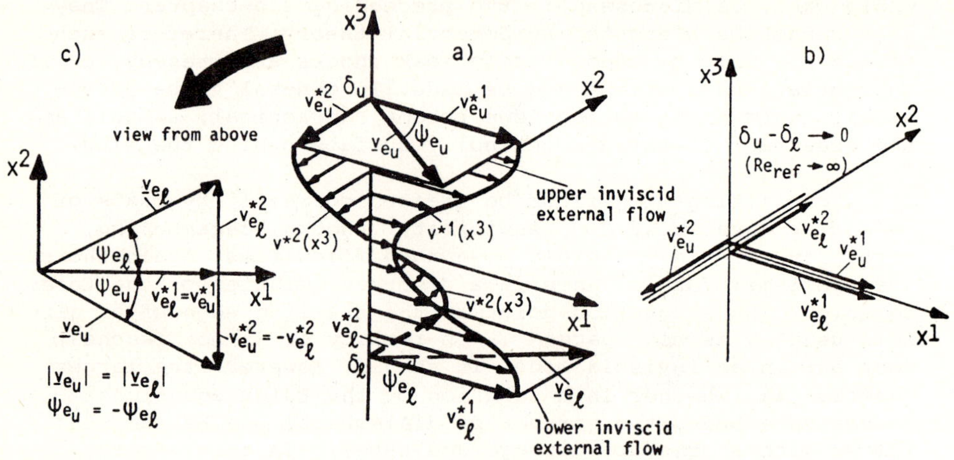

Fig. 10.8 Schematic of three-dimensional wake of a lifting wing in steady subcritical flow (right-hand side of wing): a) wake in reality; b) wake in the limit $Re_{ref} \to \infty$ as discontinuity layer, c) local wake coordinate system

It is assumed that the flow is subcritical (in any case shocks will contribute not much to the vorticity balance of a lifting wing[9]). Then in steady motion the inviscid flows at the upper and the lower side of the wake have the same magnitude (see also the two-dimensional cases discussed in Chapter 10.1). However, they are sheared against each other, because of the pressure relaxation discussed above.

The local wake coordinates in Fig. 10.8a are oriented with the bisector of the sheared upper and lower external inviscid flow. With this orientation the wake profile in x^1-direction has the shape of a two-dimensional wake, in contrast to the profile in x^2-direction, which resembles a vortex. Application of the local vorticity content relation Eq. (10.7) to this situation yields the result

$$\underline{\Omega}\big|_{Re_{ref} \to \infty} = [-(v^{*2}_{e_u} - v^{*2}_{e_\ell});\ 0;\ 0],$$

and with $v^{*2}_{e_u} = -v^{*2}_{e_\ell}$ because of the bisector orientation

$$\underline{\Omega}\big|_{Re_{ref} \to \infty} = [2v^{*2}_{e_\ell};\ 0;\ 0] = [-2v^{*2}_{e_u};\ 0;\ 0]. \qquad (10.8)$$

The profile in x^1-direction thus carries no kinematically active vorticity, whereas the profile in x^2-direction does. With regard to the wing drag thus the x^1-profile "carries" the viscous and the viscosity induced pressure drag, and the x^2-profile the induced drag, which is found in the flow field behind the wing. The x^1-axis in Fig. 10.8a actually is a vortex line with regard to the local vorticity content, because $\underline{v}\times\underline{\Omega}|_{Re_{ref}\to\infty} = 0$ holds there.

If the wake is collapsed with $\delta_u - \delta_\ell|_{Re_{ref}\to\infty} \to 0$, the familiar discontinuity sheet of potential wing theory appears, Fig. 10.8b.

Relation (10.8) can be rewritten in terms of the shear angles ψ_e in Fig. 10.8a:

$$\underline{\Omega}|_{Re_{ref}\to\infty} = [\,2|\underline{v}_{e_\ell}|\,\sin\psi_{e_\ell}\;;\;0\;;\;0] = [-2|\underline{v}_{e_u}|\,\sin\psi_{e_u}\;;\;0\;;\;0]. \qquad (10.9)$$

Wing potential theory postulates that the local change of the circulation distribution in span direction of a wing, $d\Gamma/dy$, is a measure of the vortex strength there of the wake behind the wing. Thus it holds that the kinematically active vorticity content of the wake, which leaves the trailing edge, at every span station equates negatively to $d\Gamma/dy$:

$$\frac{d\Gamma}{dy} = -\,\Omega^{*1} = -\,2|\underline{v}_{e_\ell}|\,\sin\psi_{e_\ell} = 2|\underline{v}_{e_u}|\,\sin\psi_{e_u}\,. \qquad (10.10)$$

Observation shows that along the trailing edge of a - lifting - wing the static pressure does not vary much in general. Therefore also the magnitude of the external inviscid velocities $|\underline{v}_{e_u}|$ and $|\underline{v}_{e_\ell}|$ will not change much in general. Because Γ on a wing goes to zero in span direction towards the wing tip with increasing $-d\Gamma/dy$, it follows from Eq. 10.10 that the local shear angle $|\psi_{e_u}| + |\psi_{e_\ell}|$, Fig.10.8a, must increase towards the tip.

This feature again is related to the pressure relaxation around the wing tip. However, already the basic flow past a wing at angle of attack, i.e. the hypothetical flow without a Kutta condition, and therefore without lift, has this feature, Fig. 10.9.

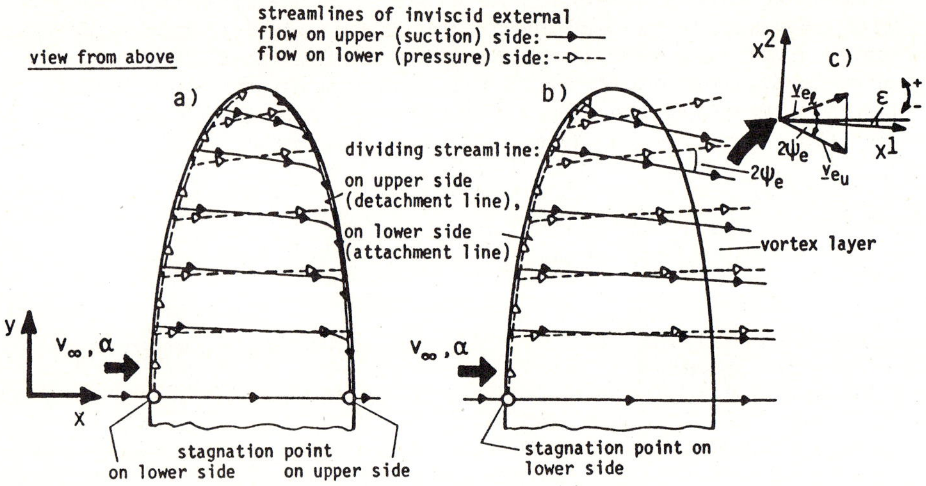

Fig. 10.9 Schematic of inviscid flowfield on wing (ellipsoid 3:1:0.125) at $\alpha=15^o$: a) ideal inviscid flow; b) expected real inviscid flow; c) local vortex-layer coordinate system (see Fig. 10.8) (after Ref. 17)

Fig. 10.9a shows the external inviscid streamlines on a very strongly flattened ellipsoid (3:1:0.125) at angle of attack $\alpha=15^o$. The streamlines were found with an analytical solution[15]. Note that a shearing is present between the flow on the upper and the lower side which increases with increasing span direction.

Experimental and theoretical/numerical results from flowfield investigations on many configurations suggest that a "locality principle" exists in so far as a change in body shape affects the flow only locally and downstream of that location. Of course the flow is also changed upstream - the elliptical properties of subsonic flowfields - but these changes in general are small. They can be significant if the wake of a body carries kinematically active vorticity. Then a global interaction occurs, which leads, for instance, to the induced drag of lifting wings.

If now the trailing edge of the ellipsoid is flattened sufficiently, or the flow is sufficiently real, i.e. viscous, it undoubtedly will break away from the surface at the trailing edge in form of a flow-off separation. Because of the locality principle the flow upstream of the trailing edge will retain the general shear between the upper and the lower side flow, which extends downstream into the wake, leading to a flow geometry like that shown in Fig. 10.9b. In Fig. 10.9c also the local vortex-layer coordinate system is shown. It may be inclined against the x-direction as indicated by the vortex-line angle ε, which in general is small, even to the extent that it can be set to zero in potential flow methods. It may also lie not in the x-y plane - the chord plane of the wing. Especially near the wing tip this must be taken into account.

Finally a quantitative proof in the frame of potential wing theory is given of relation (10.10). The HISSS-panel method[16] was used to compute[17] the inviscid flow past the Kolbe wing. The results concerning the present investigation are given here. The velocity field at the very trailing edge was evaluated with regard to Ω^{*1}, ψ_{e_ℓ} and ε. Fig. 10.10 shows that relation (10.10) is well fulfilled.

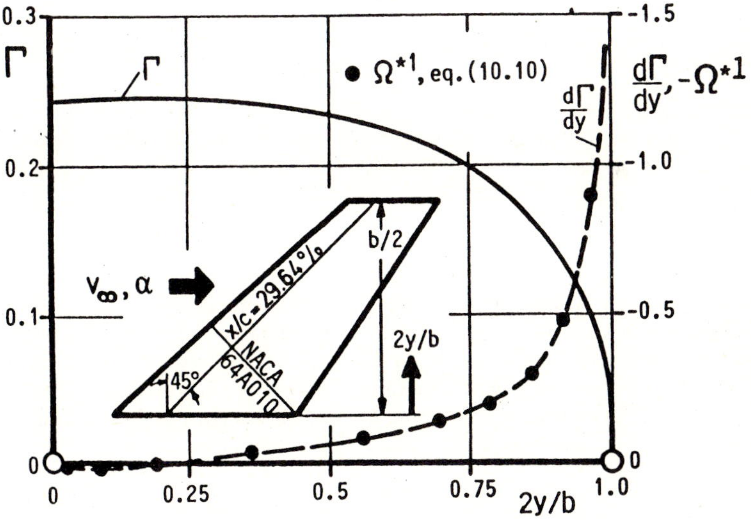

Fig. 10.10 Circulation distribution Γ, its derivative $d\Gamma/dy$, and local vorticity content Ω^{*1} of panel solution for the Kolbe wing at $\alpha=8.2^\circ$, $M_\infty=0.25$ (after Ref.17)

Fig. 10.11 shows how the shear angle ψ_{e_ℓ} increases with increasing span station. The vortex-line angle ε is positive, and indeed is small, and nearly constant over the span. Its sign is governed by the sweep of the trailing edge[17]: forward sweep leads to a negative value, backward sweep to a positive. Its magnitude is governed by the thickness of the wing, and in linear methods it is independent of the angle of attack.

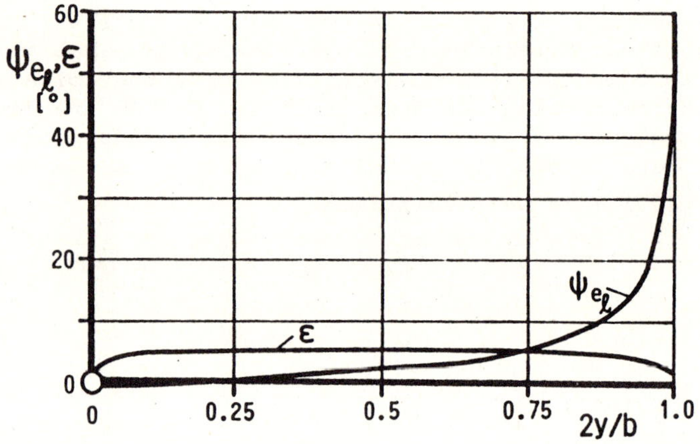

Fig. 10.11 Shear angle ψ_{e_ℓ}, and vortex-line angle ε of panel solution for the Kolbe wing at $\alpha=8.2^\circ$, $M_\infty=0.25$ (after Ref. 17)

These results show clearly that the solution contains the vortex sheet in the limit $Re_{ref} \rightarrow \infty$ as indicated in Fig. 10.8. It should be noted that some first-order panel methods give wrong velocity fields near the trailing edge of lifting wings[17] because of the vortex-lattice discretization of the vortex sheet. This does not affect the computed lift, induced drag etc. However, a boundary-layer computation will become erroneous as was demonstrated in Ref. 17.

10.4 Vorticity Creation and Entropy Rise in Euler Solutions for Lifting Wings

As it was discussed in Sub-Chapter 10.2 of course also the Euler solution for a lifting wing must exhibit the wake of reality in some form. The discussion in sub-chapter 10.3 has shown that this wake collapses to a discontinuity surface - a singularity vortex sheet - in potential wing theory. In this discontinuity surface the kinematically active vorticity of reality is hidden. In analogy to potential wing theory this discontinuity surface should appear in an ideal Euler solution too. That it is also possible as a weak solution of the Euler equations, is discussed in Ref. 6.

This discontinuity surface is smeared out in a discrete solution, like a captured shock is smeared out over some grid points, as is discussed elsewhere in this book. This situation is sketched in Fig. 10.12. The ideal discountinuity surface, Fig 10.12a, is widened up to a certain extend by an, in the strict sense "false"

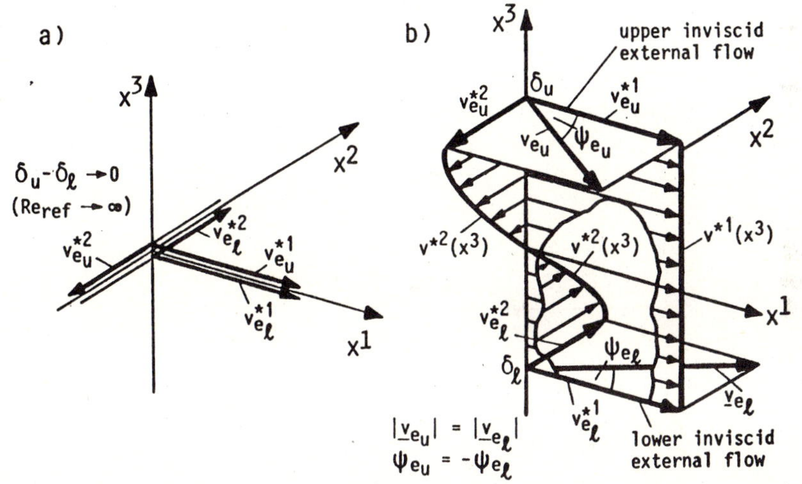

Fig. 10.12 Schematic of three-dimensional wake of a lifting wing in steady subcritical flow: a) vortex wake as discontinuity layer; b) discontinuity layer widened up by numerical diffusive transport of vorticity (Euler wake)

diffusive transport, due to the "capturing" of the vortex sheet by the Euler solution. It then appears as an "Euler wake", Fig. 10.12b, which, however, in principle has the right properties. The x^1-profile now is uniform, because in an inviscid model no viscous drag or viscosity induced pressure drag is present, and correspondingly, also no kinematically inactive vorticity.

The x^2-profile is the vortex-like profile familiar from Fig. 10.8. The precise form of this profile does not matter in this context. Important is the fact, that the kinematically active vorticity content, which was hidden in the discontinuity surface, now has reappeared. Because this vorticity content "carries" the induced drag, its appearance is compatibel with a discrete inviscid model of finite-span lifting wing flow.

The Euler wake, which appears in a discrete numerical solution for a finite lifting wing, provided a Kutta condition is present, thus is a necessary and sufffficient property of that solution (problems connected to it will be discussed in the next sub-chapter). The vorticity, which becomes visible, appears with the right amount, in principle. Because vorticity is present in the Euler wake, also an accompanying entropy rise or a respective total pressure loss is present.

In the following this is investigated with the help of Crocco's law[18], which for convenience is applied in its inviscid form. It reads for steady iso-enthalpic flow

$$\underline{v} \times \text{rot}\ \underline{v} = -T\ \text{grad}\ s\ . \tag{10.11}$$

Introduction of relation (10.2) for $Re \to \infty$ (note that $\tilde{v}^{*3} \to 0$ in that case, too) yields

$$\frac{1}{2}\left[(v^{*1})^2 + (v^{*2})^2\right]_{,\tilde{3}} = -T\ s_{,\tilde{3}}\ . \tag{10.12}$$

Introducing

$$T = T_t - \frac{(v^{*1})^2 + (v^{*2})^2}{2c_p}$$

for iso-enthalpic, thermally and calorically perfect flow into Eq. (10.12) and integrating it starting at the lower edge of the shear layer (Fig. 10.1) gives

$$s(\tilde{x}^3) - s(\delta_\ell)\Big|_{Re_{ref} \to \infty} = c_p\ \ell n\ \frac{-2c_pT_t + (v^{*1})^2 + (v^{*2})^2}{-2c_pT_t + (v^{*1}_{e_\ell})^2 + (v^{*2}_{e_\ell})^2}\ . \tag{10.13}$$

The external inviscid velocity at the lower edge can be written

$$|\underline{v}_{e_\ell}|^2 = (v^{*1}_{e_\ell})^2 + (v^{*2}_{e_\ell})^2\ ,$$

and the Mach number

$$M_{e_\ell} = \frac{|\underline{v}_{e_\ell}|}{(\gamma R\, T_{e_\ell})^{1/2}} \quad .$$

Then finally the entropy rise in the wake reads[8]

$$\left.\frac{s(\tilde{x}^3) - s(\delta_\ell)}{c_p}\right|_{Re_{ref}\to\infty} = \left.\frac{\Delta s(\tilde{x}^3)}{c_p}\right|_{Re_{ref}\to\infty} =$$

$$= \ell n \left\{\frac{\gamma-1}{2} M_{e_\ell}^2 \left[1 - \frac{(v^{*1}(\tilde{x}^3))^2 + (v^{*2}(\tilde{x}^3))^2}{|\underline{v}_{e_\ell}|^2}\right] + 1\right\}, \qquad (10.14)$$

and the equivalent total pressure loss

$$1 - \frac{p_t(\tilde{x}^3)}{p_{t_{e_\ell}}} = 1 - e^{-\frac{\gamma}{\gamma-1}\frac{\Delta s(\tilde{x}^3)}{c_p}} \quad . \qquad (10.15)$$

Entropy rise and total pressure loss thus are a function of the ratio of the square of the tangential velocity in the wake, and the edge velocity.

Eq. (10.14) reduces simply to two-dimensional cases, and yields, of course, also for the kinematically inactive wake in any case an entropy rise. It is zero for the thermodynamically singular case of zero Mach number.

For the Euler wake considered here (see Fig. 10.12):

$$v^{*1}(\tilde{x}^3) \equiv v^{*1}_{e_\ell} = \text{const.} \quad ,$$

Eq.(10.14) then reduces to

$$\left.\frac{\Delta s\ (\tilde{x}^3)}{c_p}\right|_{Re_{ref}\to\infty} = \ell n \left\{\frac{\gamma-1}{2} M_{e_\ell}^{\ 2}\left[1 - \left(\frac{v^{*1}_{e_\ell}}{|\underline{v}_{e_\ell}|}\right)^2 \left[1 + \left(\frac{v^{*2}(\tilde{x}^3)}{v^{*1}_{e_\ell}}\right)^2\right]\right] + 1\right\} \quad . \qquad (10.16)$$

This result shows that in an Euler wake with kinematically active vorticity, which appears due to the numerical viscous transport, also an entropy rise occurs. Without kinematically active vorticity, i.e. without shear in the wake

$$v^{*2}(\tilde{x}^3) = v^{*2}_{e_\ell} = v^{*2}_{e_u} \equiv 0 \;, \quad v^{*1}_{e_\ell} \equiv |v_{e_\ell}|, \text{ etc.},$$

no entropy rise would be observed (infinite swept wing case).

This sub-chapter closes with the discussion of the results of an Euler solution for a lifting wing[19]. It will be seen that the Euler wake indeed has the properties which the present theory demands.

In Fig. 10.13 the planform of the lifting wing is shown with the computed inviscid streamlines on the surface of the wing

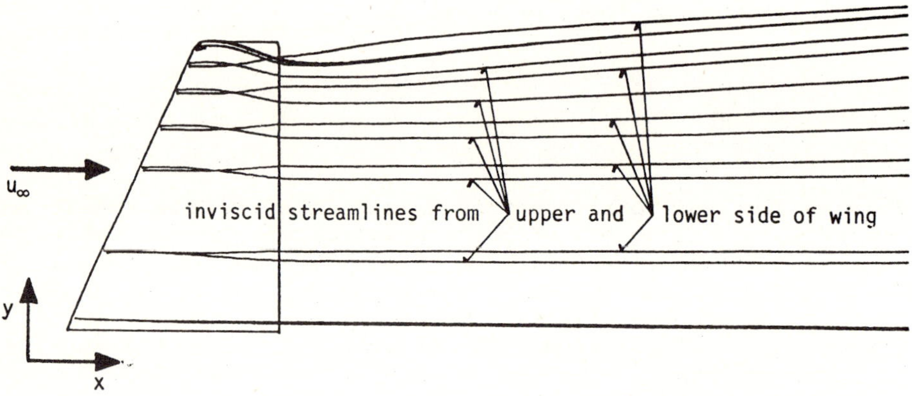

Fig. 10.13 Wing plan form (right-hand side), and inviscid streamlines seen from above, $\alpha=5^\circ$, $M_\infty= 0.3$ (Ref. 19)

at a free-stream Mach number $M_\infty=0.3$, and an angle of attack $\alpha=5^\circ$. On the wing the increase of the shear from the root to the tip is well discernable at the trailing edge. The vortex-line angle ε is very small because the trailing edge is un-swept. In Fig. 10.13 the streamlines lie in the skeletal plane of the wake, and therefore essentially in chord direction. This explains why no shear is discoverable there.

In Fig. 10.14 the circulation distribution in spanwise direction from the Euler solution is compared with the distribution found with a linear method. Because the linear method does not take into account the wing thickness, and was made for

M_∞= 0.01, the result is about 10 per cent lower, which is the right order of magnitude. However, in the mid-section of the wing, the Euler solution shows an irregularity, which seems to be due to the sharp apex of the wing, Fig. 10.13.

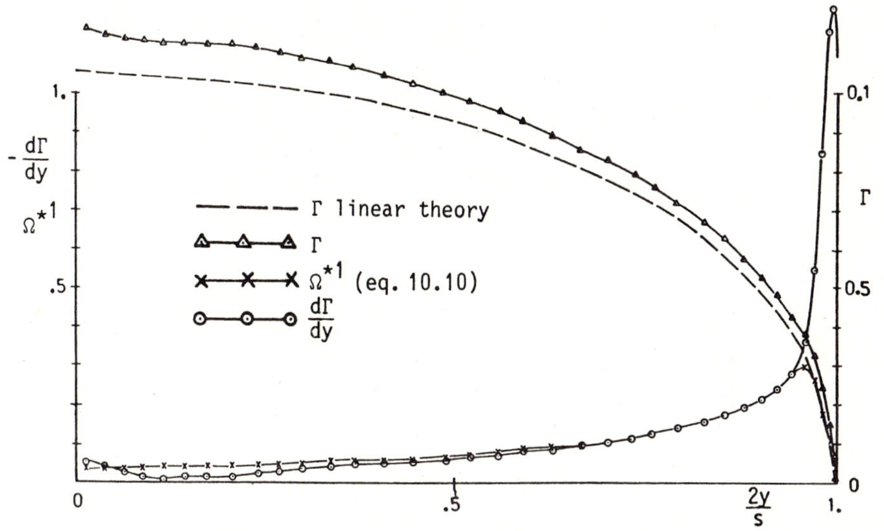

Fig. 10.14 Circulation distribution Γ, its derivative dΓ/dy, and local vorticity content of Euler solution for wing at $\alpha=5^\circ$, M_∞=0.3 (Ref. 19)

In Fig. 10.14 also relation (10.10) is evaluated. It is fulfilled to a high degree of accuracy from 30 per cent span up to 95 per cent span. Near the wing tip of course the wing tip flow disturbs the picture.

In Figs. 10.15 to 10.17 the Cartesian velocity components u, v, w are given at four spanwise stations each at the station of 5 per cent mid-wing chord behind the wing. Because the vortex-line angle ε is less than -1° at all span stations the components u and v are nearly the same as the components v^{*1} and v^{*2} in the local wake coordinate system, Fig. 10.8. Indicated in the figures is also the thickness of the Euler wake as seen in Fig. 10.18. It extends over roughly four cells. In Fig. 10.15 the kinematically inactive wake part indeed is nearly uniform. In Fig. 10.16 the shear of the flow between the upper and the lower side of the wake is seen. Obviously the resolution is not yet fine enough at the outermost span station. The asymmetry of the v-profiles at the inner stations is due to the small negative vortex-line angle ε. The w-component is stronger positive on the lower side of the wake than on the upper. At the location of the Euler wake it is nearly zero, Fig. 10.17, as was to be expected.

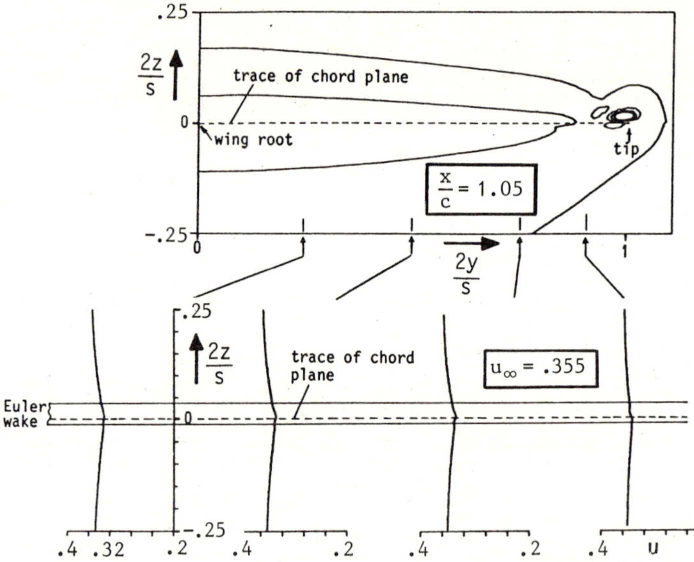

Fig. 10.15 Distribution of Cartesian velocity component $u(z)(\approx v^{*1}(x^3))$ at four span stations behind right-hand side of wing (x/c=1.05), $\alpha=5^\circ$, $M_\infty=0.3$ (Ref. 19)

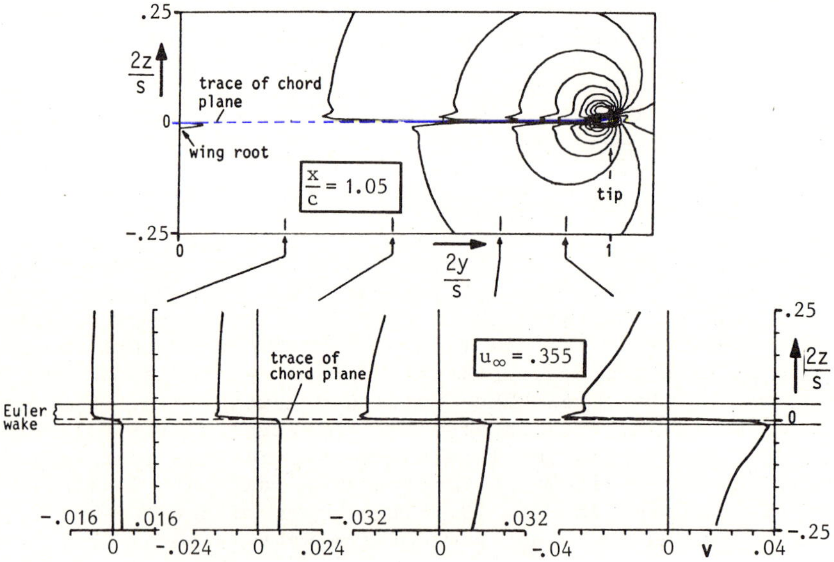

Fig. 10.16 Distribution of Cartesian velocity component $v(z)$ ($\approx v^{*2}(x^3)$) at four span stations behind right-hand side of wing (x/c=1.05), $\alpha=5^\circ$, $M_\infty=0.3$ (Ref. 19)

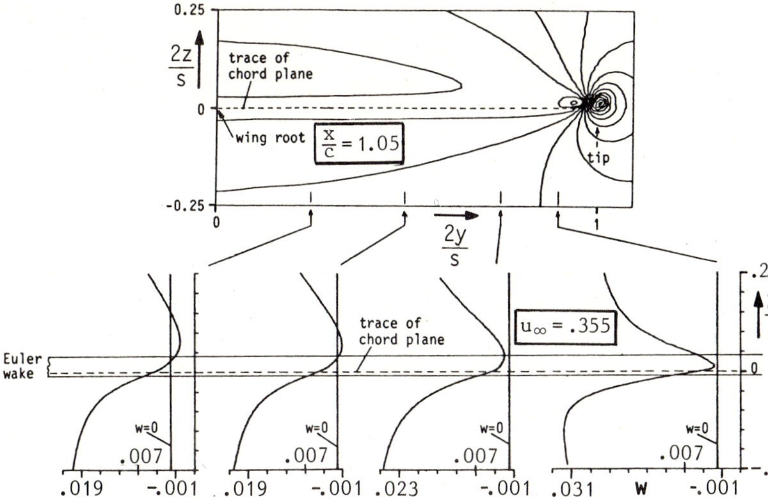

Fig. 10.17 Distribution of Cartesian velocity component w(z) at four span stations behind right-hand side of wing (x/c=1.05), $\alpha=5^\circ$, $M_\infty=0.3$ (Ref. 19)

In Figs 10.18 and 10.19 finally the distribution of the kinematically active vorticity and the total pressure loss are shown. The structure of the distributions is in accordance with the theory put forward here and compatible with the findings in Ref. 8. Numerical scatter, expecially at the outermost station, demands further work on the solution.

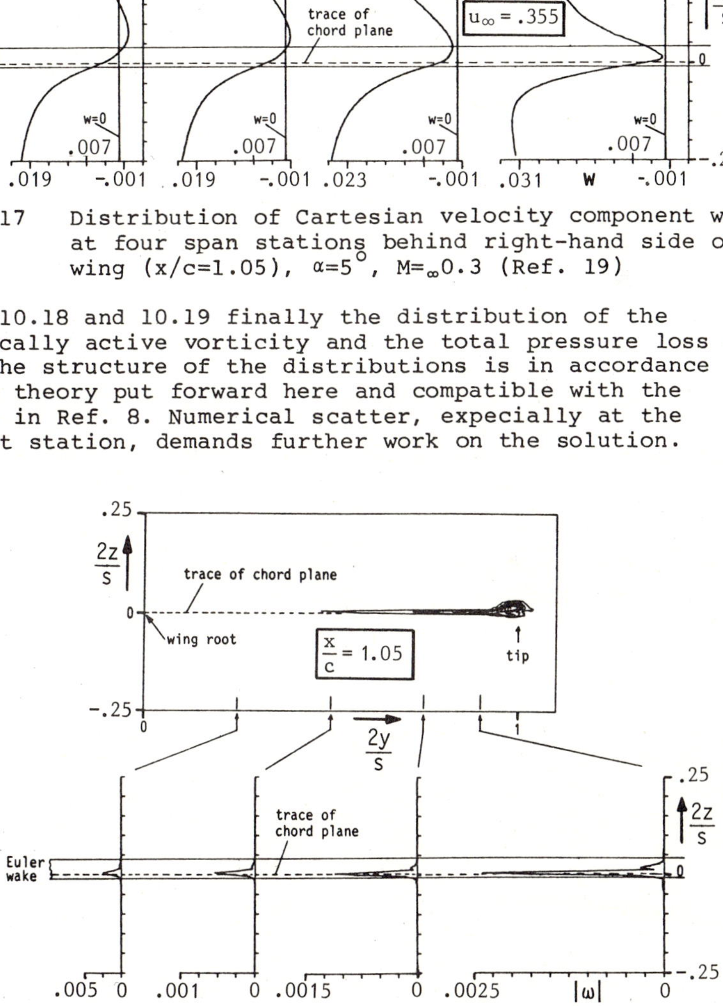

Fig. 10.18 Distribution of magnitude of vorticity $\omega(z)(=\omega_x(z) \approx \omega^{*1}(x^3))$ at four span stations behind right-hand side of wing (x/c=1.05), $\alpha=5^\circ$, $M_\infty=0.3$ (Ref. 19)

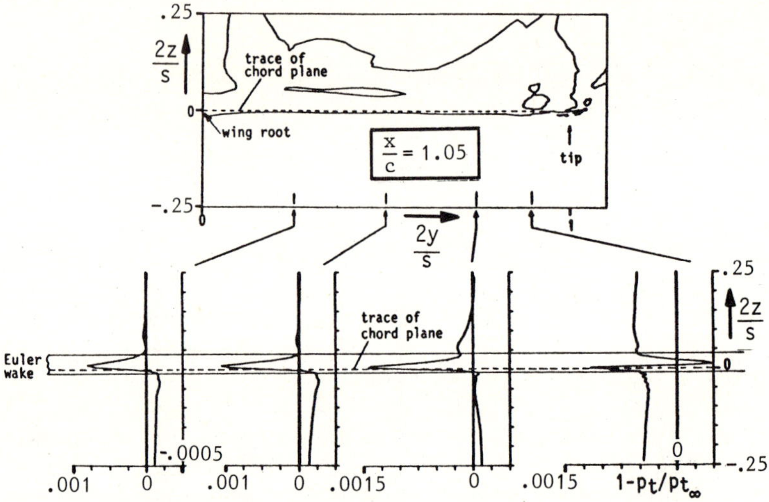

Fig. 10.19 Distribution of total pressure loss as function of z at four span stations behind right-hand side of wing (x/c=1.05), $\alpha=5^{\circ}$, $M_{\infty}=0.3$ (Ref. 19)

10.5 The State of the Art: a Critical Evaluation

In the preceeding sub-chapters it was argued that, in analogy to potential wing theory, an Euler solution for a lifting wing must exhibit a discontinuity surface, which is the wake of reality in the limit $Re_{ref} \to \infty$. True discontinous vortex sheets, as in potential wing theory, although exact, would hamper very much Euler solutions for general wing and aircraft configurations because they would have to be "fitted", i.e. presented by coordinate surfaces (see also the discussions in Chapter IX with regard to turbulence models). Therefore the appearance of an Euler wake of finite thickness, which is "captured" by the solution, is not only acceptable, but even to prefer, especially because the kinematically active vorticity content appearing then in principle is exact.

A careful study of present-day Euler solutions, however, shows at least two deficiencies connected to the phenomeneon Euler wake, as it is treated today:

a) The thickness of the Euler wake in the ideal case just should be that of the real wake in nature. It is immediately evident, that, if the Euler wake is smeared out over, say, four cells, as in Ref. 19, it depends on the size of the cells how thick it is. In the case of an isolated wing, Fig. 10.7a, the thickness of the Euler wake does not matter, as long as the right vorticity content is there. However, if vortex sheets or vortices interact with each other, or with a part of the configuration of an aircraft, or lie above the wing like in Fig. 10.7b, than, if the discretization in the Euler wake region is not fine enough, the solution will become erroneous.

b) The investigation of the integral properties of the Euler wake of Ref. 19 has revealed that it looses part of the kinematically active vorticity, and hence circulation behind the wing. Similar observations have been made elsewhere, too. Again in some applications this does not matter. In a panel solution for an isolated lifting wing, for instance, usually only a short length of the wake behind the wing is paneled, i.e. modeled with singularities, without undue loss of accuracy[20]. Therefore it can be expected, that lift, induced drag etc., can be computed to a sufficient degree of accuracy with an Euler solution for an isolated wing, as was proved for instance in Ref. 19. However, if, again, vortex sheets or vortices interact with each other, or with a part of the configuration of an aircraft, or lie above the wing like in Fig. 10.7b, the solution will become erroneous, if the kinematically active vorticity and hence the circulation is not conserved.

Deficiency b) certainly has to do with the diffusive properties of discrete Euler solutions which of course also influence deficiency a). In addition it must be apprehended that the far-field boundary conditions, which in most methods today are characteristic boundary conditions, may influence the loss of circulation. Even for lifting airfoils the circulation vanishes in many solutions near the outer boundary.

Applications of Euler methods to flow past both wings with low and high leading-edge sweep can be found in rather large quantity in literature.

The International Vortex Flow Symposium[5] was expecially concerned with Euler solutions for delta wings with lee-side vortices. The mechanism of vorticity creation, and hence of the lee-side vortices, as put forward here, cannot be so easily verified there because of the higher structural complexity of such flows compared, for instance, to that past the wing of Ref. 19 (large curvature of the feeding vortex layers, concentration to vortices in the vicinity of the wing surface).
In addition to the complexity, the two deficiencies discussed above, may influence the solutions, probably being responsible for the very large total pressure losses found in the vortex cores.

A much discussed observation is that Euler solutions for delta wings with round leading edges exhibit lee-side vortices strongly dependent on the discretization (systematic studies on this are reported for instance in Ref. 21). This is not observed to that extend on delta wings with sharp leading edges. In the latter case flow-off separation occurs in reality, which means that the location of separation is fixed. This is also the case in finit-volume methods with their implicit Kutta condition at sharp edges. Therefore sharp-edge solutions should be rather insensitive to discretization[22].

On a round leading edge squeeze-off separation occurs in reality, which depends on the flow field and on the Reynolds number. Figure 10.20 shows schematically the situation on a delta wing with lee-side vortices[8] (note that also the secondary vortices are sketched, which to our knowledge sofar have only been modelled in Euler solutions by W. Schwarz[23]). On the left-hand side laminar boundary-layer flow is indicated, and on the right-hand side turbulent flow. The laminar boundary layer in general will separate at a much higher velocity level than the turbulent boundary layer because it cannot negotiate so much

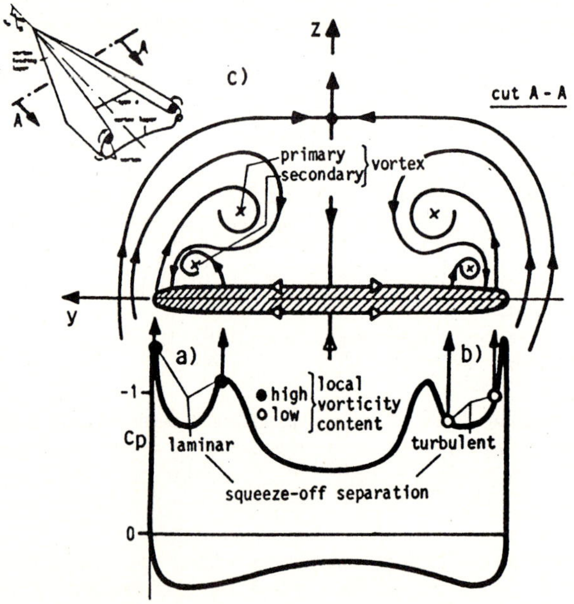

Fig. 10.20 Schematic of separation in cross-section of a slender delta wing with round leading edges. Pressure-coefficient distribution (without local interaction effects) and separation location of a) laminar flow, b) turbulent flow, c) location of primary and secondary separation and vortices (Ref. 8)

adverse pressure gradient as the latter. This holds both for the primary and the secondary separation shown in Fig. 10.20.

If the structure of separation, especially the shear between the upper and the lower inviscid flow at the leading edges, is not completely different in the two cases a) and b), then relation (10.9) shows: the vorticity content is larger in the laminar case because of the higher velocity level at separation found there, than in the turbulent case. Hence in the laminar case the vortices will be stronger in general than in the turbulent case.

From this discussion it must be concluded that at delta wings with round leading edges an explicit Kutta condition must be prescribed in any inviscid model solution. Otherwise the problem is not well posed. The location of the Kutta condition can be determined to probably sufficient accuracy by a boundary-layer solution. At low angles of attack no separation at all will occur, at higher angles of attack it depends on the flow situation[24]. Ideally an Euler solution should yield no vortices on a delta wing with rounded edges, if no explicit Kutta condition is prescribed. Why they appear in such cases is not yet clear.

Inviscid solutions, especially Euler solutions for lifting wing problems, can be of high value in practice. However, the following items must be considered before applying them (see Ref. 24 for very useful correlations of all kind of lee-side flow phenomena on delta wings):

1. Because an inviscid flow model is a simplified model of reality, the real flow situation must be known at least approximately before applying it: Do the primary separation and the primary vortices dominate the flow? Can they be modeled to a sufficient degree? Can secondary and higher-order separation phenomena be neglected, if they cannot be modeled realistically with an inviscid model? If flow-off separation is the source of the dominating vortex sheets, and vortices, does the Euler method to be used have the property of an implicit Kutta condition? Otherwise it must be described explicity as it must be in any case, if squeeze-off (free-surface) separation occurs (round leading edge)(see Sub-Chapter 11.2.4).

2. Are there vortex/geometry and/or vortex/vortex interactions where the deficiencies discussed above may play a role? Can they be taken care of?

3. Can vortex bursting be of importance? Can it be described with sufficient reliability and accuracy with the discrete Euler model? (For the state of the art see for instance Ref. 25).

10.6 A Note on the Solution of the Navier-Stokes Equations

Potential applicators in aircraft design sometimes express the view that if Euler solutions for lifting wings have the deficiencies discussed in Sub-Chapter 10.5 (both due to the model properties and due to the insufficient description of vortex sheets because of numerical effects), why not employ instead from the begin a Navier-Stokes solution.

Of course the Navier-Stokes equations are the most complete model of fluid flow. However, their employment in an ordinary design process is much too expensive on present-day computers[26]. On the other hand in a design process different levels of modelization are needed. After a proper assessment of the possibilities an Euler approach offers with regard to the flow situation given, it can very well fill the gap between the linear potential theory approach and a Navier-Stoke approach.

With regard to the deficiencies due to the numerical and possibly far-field boundary-condition effects encountered in Euler simulations of vortex sheets, most of the modern Navier-Stokes methods will encounter the same problems. This is because the convective part of the Navier-Stokes equations usually in effect is treated by an Euler solver, and the far-field boundary conditions are the same in both cases. Therefore also Navier-Stokes methods need an assessment with regard to the modelling capabilities of vortex sheets and vortices, and the conservation of vorticity and circulation[27]. The same improvements with regard to numerical diffusion and far-field boundary conditions are necessary in both Euler and Navier-Stokes methods.

10.7 References

1. Schlichting, H., Truckenbrodt, E.: "Aerodynamics of the Airplane". McGraw-Hill International Book Company, New York, 1979.

2. Fornasier, L.: "HISSS - a Higher-Order Panel Method for Subsonic and Supersonic Attached Flow About Arbitrary Configurations". In: Ref. 13, pp. 52-70.

3. Hoeijmakers, H.W.M., Vaatstra, W.: "A Higher-Order Panel Method for the Computation of the Flow About Slender Delta Wings with Leading-Edge Vortex Separation". In: Proc. Fourth GAMM-Conference on Numerical Methods in Fluid Mechanics 1981, (H. Viviand, ed.), Vol.5 of Notes on Numerical Fluid Mechanics, Vieweg, Braunschweig/Wiesbaden, 1982, pp. 137-149.

4. Eriksson, L.-E., Rizzi, A.: "Computation of Vortex Flow Around Wings Using the Euler Equations". In: Proc. Fourth GAMM-Conference on Numerical Methods in Fluid Mechanics, 1981 (H. Viviand, ed.), Vol. 5 of Notes on Numerical Fluid Mechanics, Vieweg, Braunschweig/Wiesbaden, 1982, pp. 87-105.

5. Elsenaar, A., Eriksson, L.-E. (eds.): "International Vortex- Flow Experiment on Euler Code Validation". FFA Bromma, 1987, ISBN 91-97 0914-0-5.

6. Powell, K.G., Murman, E.M., Perez, E.S., Baron, J.R.: "Total Pressure Loss in Vortical Solutions of the Conical Euler Equations". AIAA J., Vol. 25, No. 3, 1987, pp. 360-368.

7. Hirschel, E.H., Kordulla, W.: "Shear Flow in Surface Oriented Coordinates". Vol. 4 of Notes on Numerical Fluid Mechanics, Vieweg, Braunschweig/Wiesbaden, 1981.

8. Hirschel, E.H.: "On the Creation of Vorticity and Entropy in the Solution of the Euler Equations for Lifting Wings". MBB-LKE122-AERO-MT-716, 1985.

9. Hirschel, E.H., Lucchi, C.W.: "On the Kutta Condition for Transonic Airfoils". MBB-FE122-AERO-MT-651, 1983.

10. Hirschel, E.H.: "Evaluation of Results of Boundary-Layer Calculations with Regard to Design Aerodynamics" In: Computation of Three-Dimensional Boundary Layers Including Separation. AGARD-R-741, 1986, pp. 5-1 to 5-29.

11. Batchelor, G.K.: "Fluid Dynamics". Cambridge University Press, 1967.

12. Hummel, D.: "On the Vortex Formation Over a Slender Wing at Large Angles of Incidence". AGARD CP-247, 1979, pp. 15-1 to 15-17.

13. Ballmann, J., Eppler, R., Hackbusch, W. (eds.): "Panel Methods in Fluid Mechanics with Emphasis on Aerodynamics". Vol. 21 of Notes on Numerical Fluid Mechanics, Vieweg, Braunschweig/Wiesbaden, 1988.

14. Mangler, K.W., Smith, J.H.B.: "Behaviour of the Vortex Sheet at the Trailing Edge of a Lifting Wing". The Aeronautical Journal of the Royal Aeronautical Society, Vol. 74, 1970, pp. 906-908.

15. Zahm, A.F.: "Flow and Force Equations for a Body Revolving in a Fluid". NACA Report 323, 1930.

16. Fornasier, L.: "MBB Higher-Order Subsonic-Supersonic Singularity Method". AIAA-Paper No. 84-1646, 1984.

17. Hirschel, E.H., Fornasier, L.: "Flowfield and Vorticity Distribution Near Wing Trailing Edges". AIAA-Paper No. 84-0421, 1984.

18. Liepmann, H.W., Roshko, A.: "Elements of Gasdynamics". J. Wiley & Sons, New York/London/Sydney, 1966.

19. Hirschel, E.H., Rizzi, A.: "The Mechanism of Vorticity Creation in Euler Solutions for Lifting Wings". In: Ref. 5, pp. 127-162.

20. Fornasier, L., private communication, 1988.

21. Schwarz, W., Schmatz, M.A.: "Euler and Navier-Stokes Calculations for the Vortex-Flow Experiment - Interim Progress Report". MBB-FE122-AERO-MT-799, 1988.

22. Powell, K.G.: "Vortical Solutions of the Conical Euler Equations". NNFM 28, Vieweg, Braunschweig/Wiesbaden, 1990.

23. Schwarz, W.: "IEPG-TA15, Computational Methods in Aerodynamics, Status of Phase 3 Activities at MBB". MBB-FE122-Aero-MT-852, 1989.

24. Staudacher, W.: "Beeinflussung von Vorderkanten-Wirbelsystemen an schlanken Tragflügeln (Influencing Leading-Edge Vortex Systems of Slender Wings)". Doctoral Thesis, University Stuttgart, 1992.

25. Hitzel, S.M.: "Wing Vortex-Flows up Into Vortex-Break-Down". AIAA-Paper No. 88-2518, 1988.

26. Hirschel, E.H.: "Super Computers Today: Sufficient for Aircraft Design? Experiences and Demands". In: Meuer, H.W. (ed.): Super Computer '88. Carl Hanser, München-Wien, 1988, pp. 110-150.

27. Wanie, K.M., Hirschel, E.H., Schmatz, M.A.: "Analysis of Numerical Solutions for Three-Dimensional Lifting Wing Flows". To appear in ZFW, 1991.

XI METHODS IN PRACTICAL APPLICATIONS

In this chapter we present some results obtained from solutions computed to the Euler equations. They are a selection mainly from own work that covers a range of flow velocities, from very low to very high, and flow complexities, from simple academic cases to complicated industrial flow problems. These cases illustrate many different phenomena. Some are chosen to demonstrate the type and character of flow separation that is observed in Euler computations, other are chosen to demonstrate the practical utility of Euler solutions for industrial design.

11.1 Near-Incompressible Flow

We turn our attention first to flow at speeds slow enough so that compressibility effects can be ignored. We therefore include strictly incompressible flow as well as compressible flow whose Mach number is very small, $M_\infty \ll 1$. Such flows are known to possess severe problems for convergence to steady state. The reason for this is due to the very high velocities of the sound waves. In general there are two proven methods to circumvent this problem. One is the use of an implicit time integration scheme which does not represent the sound waves accurately[1]. The other is the artificial compressibility method in which the sound waves are removed from the governing equations even before computation begins (see Ref.2). In effect both methods ignore the sound waves.

11.1.1 Transverse Circular Cylinder

Incompressible flow past a transverse circular cylinder is two dimensional and is described by the analytical solution to the Laplace equation. It is also a bona fide solution to the Euler equations. Whether other solutions exist, e.g. the so-called Batchelor solution, with vortex-sheet separation, does not concern us here. The question we wish to pose is, can a numerical method solving the Euler equations produce a solution that approaches the irrotational solution to the Laplace equation?

In addressing this question Rizzi[3] used the artificial compressibility method to obtain the numerical solution to the incompressible Euler equations. That method of removing the sound waves is not described here, but the details are given in Ref. 2. The crucial point for us is whether the solution is irrotational or vortical. Since the body is completely smooth, the only source of vortical flow features must be numerical truncation errors.

Figure 11.1a demonstrates how, when these errors are large due to a poor mesh distribution and an inappropriate second-order artificial viscosity, the flow, as it begins to recompress on the lee-side of the cylinder, looses total pressure and cannot reach the rearmost stagnation point[3]. Instead it stagnates somewhat before that point, separates, and then because of continuity, fills in behind forming a recirculation region. Completely bogus, this is simply a feature of poor numerical accuracy. When the mesh is improved and a more suitable fourth-order artificial viscosity is used, a solution is obtained (Fig. 11.1b) closely approximating the exact analytical one. It seems that, unlike an airfoil, the circular geometry does not excite any transient vortex sheets, which then lead to a steady vortical flow.

11.1.2 Some Numerical Experiments on Transverse Cylinders

The study Ref. 4 on the other hand was devoted to the question how to simulate separation in two-dimensional, near-incompressible flow by means of solutions of the Euler equations. An early EUFLEX code[5] was employed. In the following four examples from the study are discussed.

1) Basic flow past a circular cylinder

First this flow was studied in order to make sure that the analytical solution can be recovered with the code (see also 11.1.1). The solution at $M_\infty=0.1$ with $160\times90=9600$ cells is given in Figure 11.2. In Fig. 11.2a rather flat cells near the cylinder surface lead to symmetric steady recirculation areas at the leeward side. When the cells are quadratic, especially at the surface, a very good approximation to the separation - free Laplace solution is found, Fig. 11.2b.

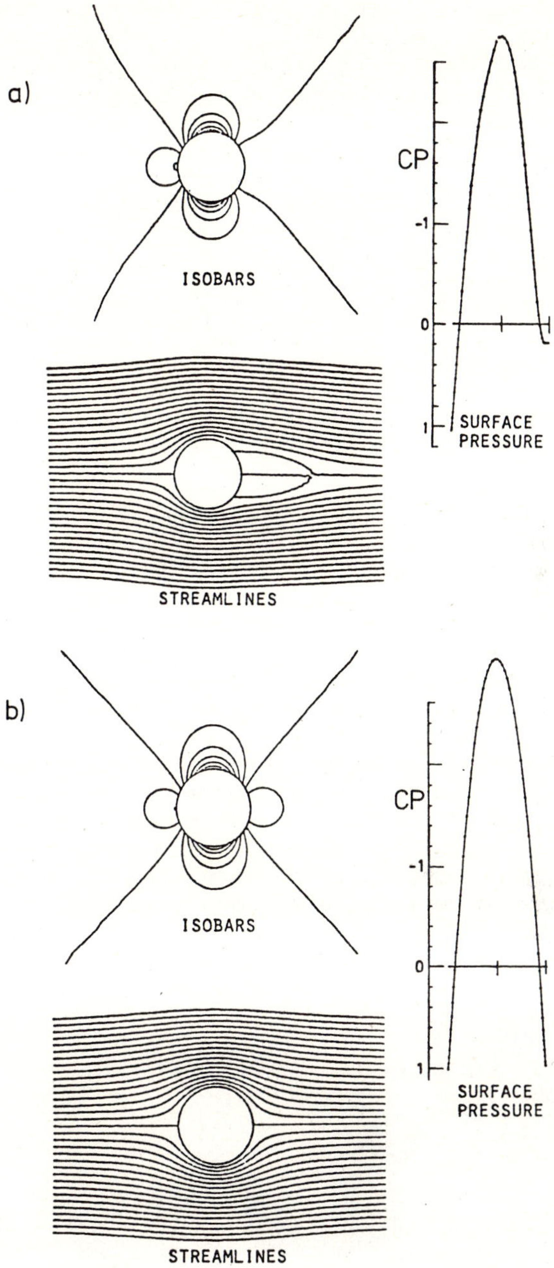

Fig 11.1 Comparison of accurate inviscid isobars, streamlines, and pressure distribution with ones having large total pressure loss and artificial separation for incompressible flow past a cylinder[3]. a) large truncation errors, b) accurate inviscid solution

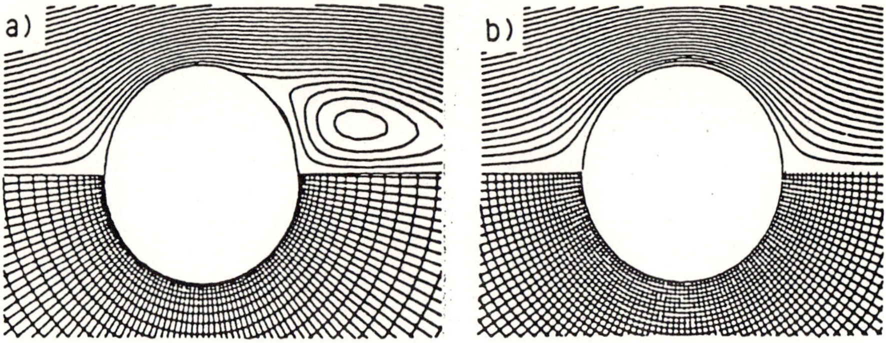

Fig. 11.2 Streamlines (upper side) and discretization (lower side) from Euler solutions[4] on circular cylinders (M_∞=0.1), a) standard discretization, b) discretization with quadratic cells

2) Flow with explicit Kutta conditions

Explicit Kutta conditions in form of fences, corners, stagnation conditions, and blowing cells were studied. Figure 11.3 gives the stream-lines past a circular cylinder with two corners, each with a hight of 0.053 of the cylinder radius (the corner bases extend over two surface cells) at ϕ_K=110°, at M_∞=0.1. The interesting result is that for fine nets (here 19602 cells) no steady solutions were found, Fig. 11.3a. The instantaneous stream-line pattern in Fig. 11.3.a resembles very much that of a von Kármán vortex street as given for instance in Ref. 6, for a Reynolds number Re=140. Reduction of the number of cells to 1000, which amounts to larger cells and hence to larger numerical viscosity, led to a steady solution, Fig. 11.3.b. The separation pattern resembles that for Re=26 given in Ref. 6.

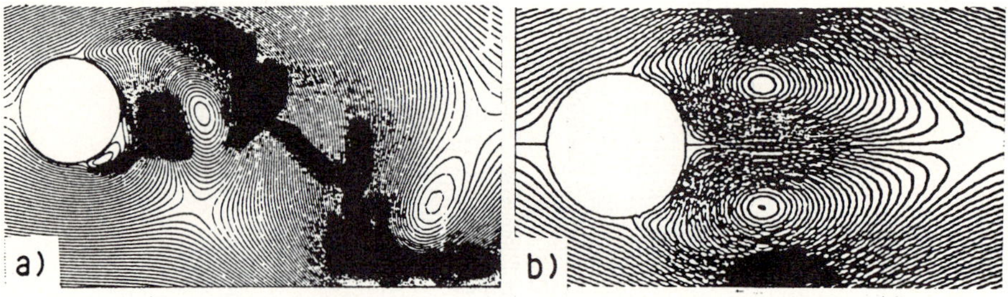

Fig. 11.3 Streamline pattern of flow calculated[4] with two explicit Kuttta conditions (K) at ϕ_K=110°, a) fine net with 19602 cells, b) coarse net with 1000 cells (M_∞=0.1)

3) Flow with only one explicit Kutta condition

Surprisingly the omission of the lower Kutta corner led to a flow pattern like that of a potential vortex superimposed a circular-cylinder flow. Fig. 11.4 shows the evolution of the solution. The Kutta corner lies at $\phi_K=150^\circ$ at the leeside.

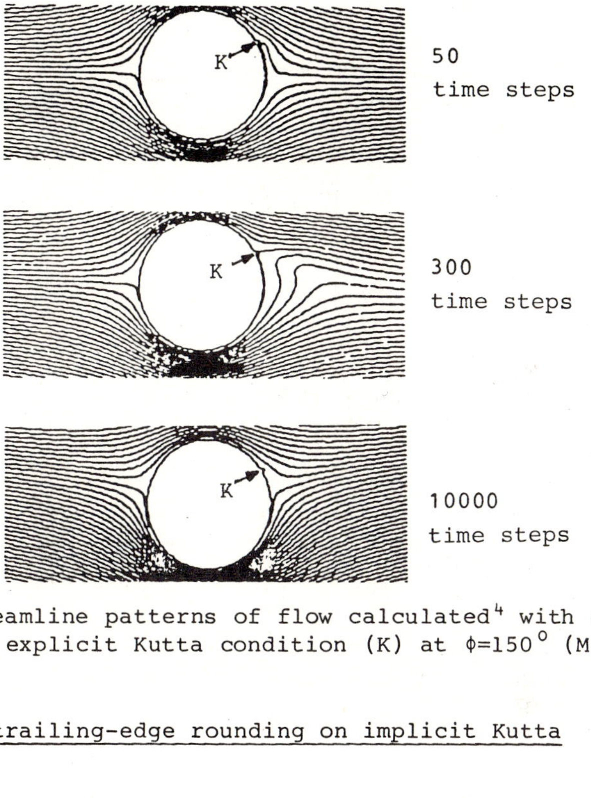

Fig. 11.4 Streamline patterns of flow calculated[4] with only one explicit Kutta condition (K) at $\phi=150^\circ$ ($M_\infty=0.1$)

4) Effect of trailing-edge rounding on implicit Kutta condition

The EUFLEX code as a finite-volume code produces at sharp trailing edges an implicit Kutta condition. It was studied whether a rounding radius of the trailing edge exists, which could be considered as a critical one in the sense that for larger radii no implicit Kutta condition occurs, while for smaller ones a fully effective exists. Fig. 11.5 shows the result for the NACA 0012 airfoil at $M_\infty=0.1$ and an angle of attack $\alpha=5^\circ$ with different locations x_s for the rounding r. Fig. 11.5.6 shows that obviously no critical ratio $r/\Delta x$ exists. There is a smooth transition from nearly no lift for $x_s=0.7$ to the fully developed lift case ($x_s=1$).

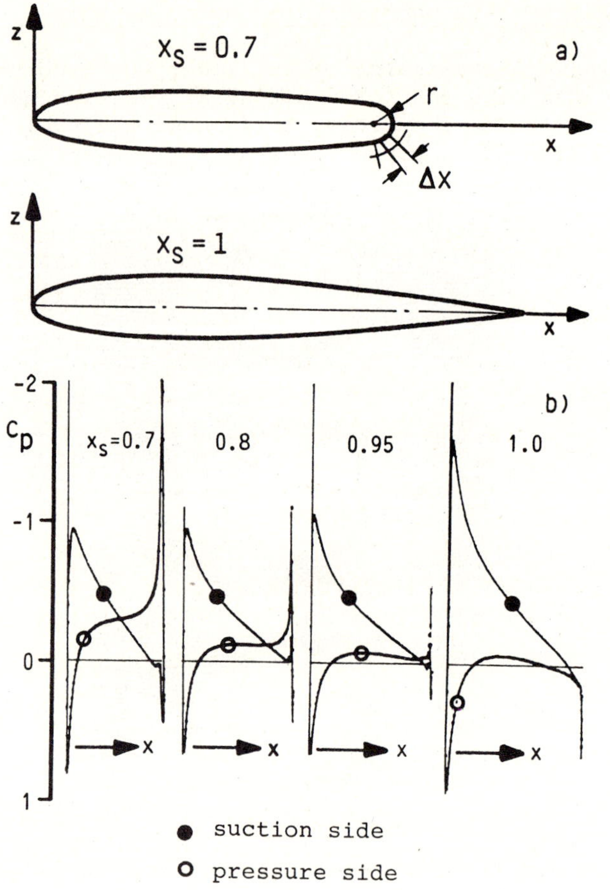

Fig. 11.5 Effect of trailing-edge rounding on pressure distribution[4], NACA 0012, M_∞=0.1, α=5°,
a) schematical of trailing-edge rounding,
b) pressure distributions for 4 cases x_s

These studies finally led to the development of an effective explicit Kutta condition, which presently is employed in many Euler similations of separated flow (see for instance also Sub-Chapter 11.2.4).

Krukow has extended his work, Ref. 4, and included a better representation of geometrical properties into the EUPLEX code (see also Chapter VI). A result taken from Ref. 7 is given in Fig. 11.6. It is demonstrated with Fig. 11.6 that modern Euler solvers indeed can handle extreme flow situations.

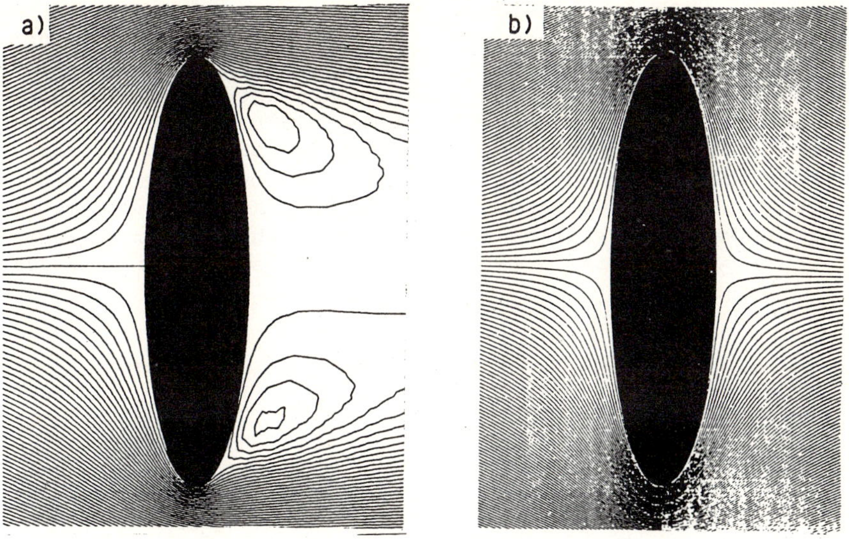

Fig. 11.6 Flow past an ellipse[7], M_∞=0.1, EUFLEX a) without, b) with quadratic interpolation of fluxes to equidistancy

11.1.3 Airfoil with Lift

The flow is incompressible and the geometry is the NACA 0012 airfoil at 5° incidence. Measured by the decay of the average and maximum time difference of pressure in the entire field and by the evolution of lift and drag, the convergence of the solution computed upon a mesh with 128 cells around the airfoil and 28 outward is given in Fig. 11.7. Although the comparison of computed c_p on the airfoil with that of an accurate boundary-integral ('singularity', or so-called 'panel') method (Eriksson[8]) is generally good (Fig. 11.8) there are small discrepancies at the leading edge suction peak and larger ones at the trailing edge, the latter due undoubtedly in part to the mesh being unable to resolve completely the flow singularity there.

But perhaps the best gauge of accuracy is the degree to which the Bernoulli relation along a streamline in steady flow, $p+0.5\ V^2=p_t$(constant), is satisfied. In this flow all streamlines originate from a constant free stream, so the total pressure p_t takes the same value on every streamline. The total-pressure coefficient $(p_t-p_\infty)/(p_{t\infty}-p_\infty)$ in Fig. 11.9 confirms that there are errors in the vicinity of the leading and trailing edges, but they are small in magnitude and confined locally to these two regions. This loss, however, should not be interpreted as a physical quantity because it is not transported along streamlines, but rather as an error perturbation upon the solution where the flow gradients are especially large, which may be a possible source for the injection of vorticity. In regions where the flow gradients are not severe, an accurate value for the Bernoulli constant is produced. We conclude that an almost-irrotational solution is obtained with very nearly the correct circulation (as given by the panel method). Furthermore, even without our invoking a Kutta condition the flow leaves the trailing edge smoothly, which is an essential feature of potential flow showing a realistic amount of lift.

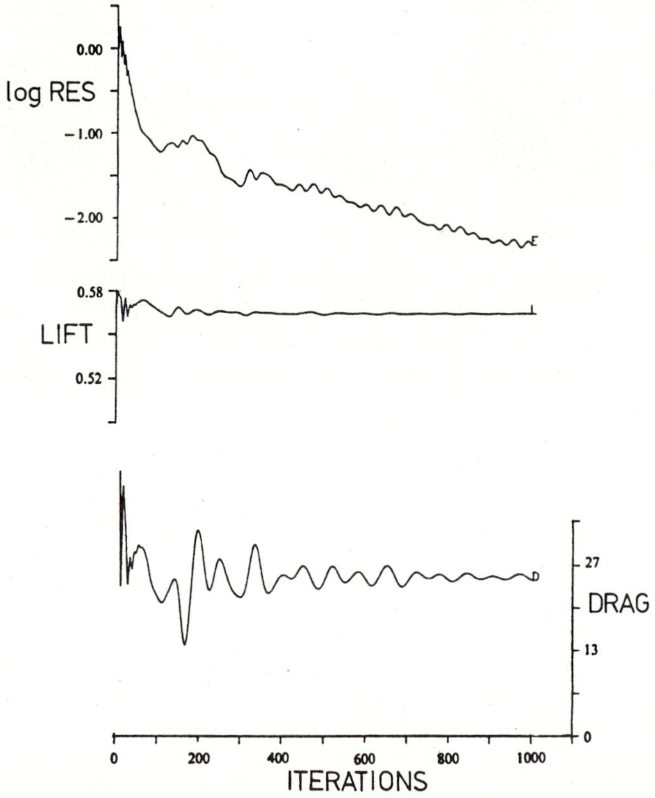

Fig. 11.7 Convergence of the solution of incompressible flow past the NACA 0012 airfoil indicated by the decay of the average time difference of pressure and the evolution of lift and drag[2], $M_\infty=0$, $\alpha=5^\circ$

The central question before us is therefore 'how does the Euler-equation method arrive at the correct circulatory flow?' (i.e. flow separation at the trailing edge without applying a Kutta condition. We called this "implicit" Kutta condition in the preceding sub-chapter).

It is generally believed now that if the fluid were to flow around the sharp trailing edge, the velocity and pressure gradients would be so large that the numerical truncation error, including artificial viscosity, would dissipate much of the energy in the fluid in a way that adjusts the flow to separate at the edge. The mechanism here is not physical, but the end result is the desired one. It appears to produce the realistic solution provided that the edge is sharp, or at least with a relatively small radius of curvature (see also preceeding sub-chapter).

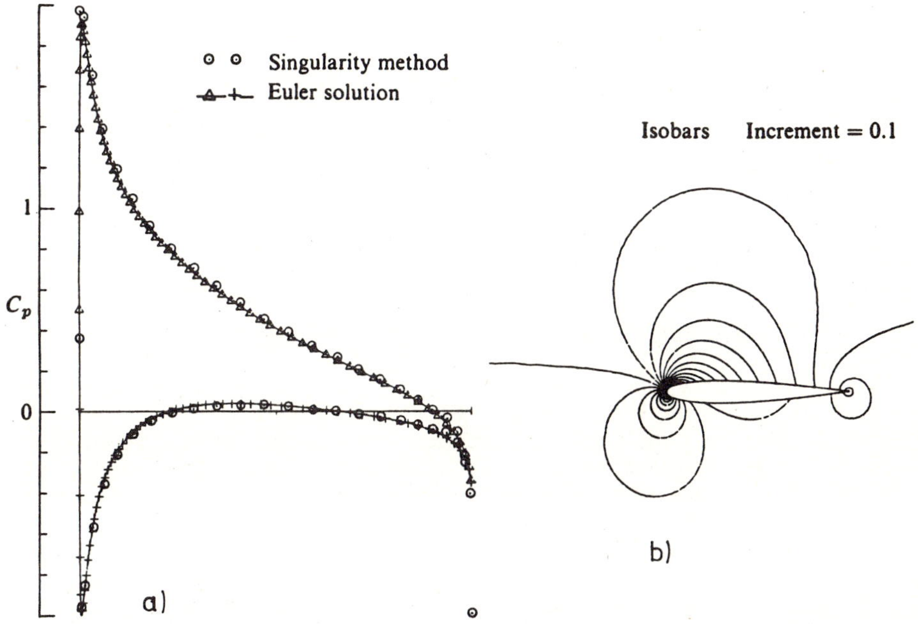

Fig. 11.8 The pressure field of the computed Euler-equation solution (Ref. 2). a) The surface distribution is compared with the highly accurate results from a potential singularity method. NACA 0012 airfoil, $M_\infty=0$,
b) the isobars of the Euler solution

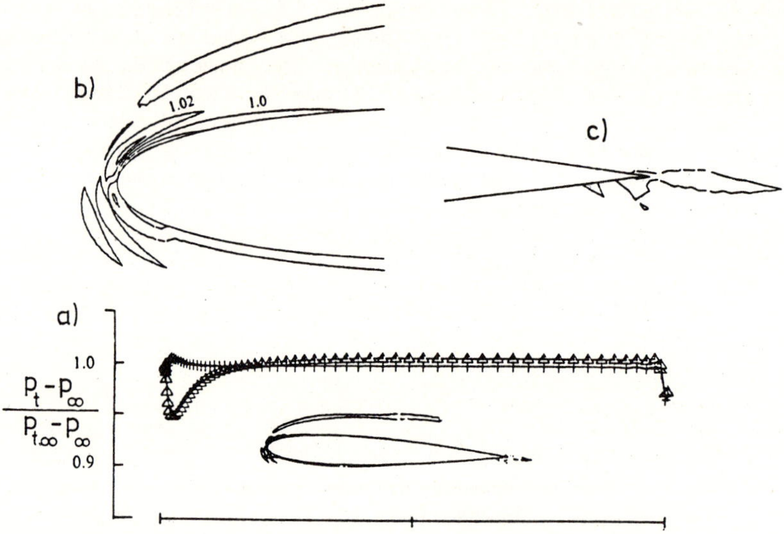

Fig. 11.9 Accuracy of the computed Euler-equation solution indicated by total-pressure coefficient $(p_t-p_\infty)/(p_{t\infty}-p_\infty)$. NACA 0012 airfoil, M=0, $\alpha=5^\circ$.
a) Total pressure loss on the surface,
b) and c) isolines of total pressure coefficient

11.1.4 Vortex Flow Over Sharp-Edged Delta Wing

This subsection presents the computed solution of incompressible flow around a 70° swept delta wing of zero thickness and unit length at 20° angle of attack. The steady flow separates from the leading edge in a vortex sheet, which then, under the influence of its own vorticity, rolls up to form a vortex over the wing. Owing to the lack of experimentally measured data for this case, a comparison with measurements is not carried out. But Rizzi and Eriksson[2] do compare results with those from Hoeijmakers's potential boundary-integral (panel) method, which inserts a vortex sheet, adjusts it to the surrounding irrotational flow field, and allows it to roll up under its own influence for several turns, and then models the remaining core by an isolated line vortex (see Hoeijmakers and Vaatstra[9]; Hoeijmakers and Rizzi[10]). The position and strength of the vortex sheet and isolated vortex are determined as part of the solution, sometimes termed "fitting" the rotational-flow features. They are true discontinuities, infinitesimally thin, and for this reason a very good choice for comparison, because the sheet and vortex in our solution are not infinitesimally thin but smeared or "captured "over a number of computational mesh cells. That Euler solutions in principle can describe the flow past lifting wings was discussed in detail in Chapter X of this book. The comparison therefore offers a good control on the position of our computed vortex and the diffusion of the sheet. Furthermore such panel-method results have been found to agree reasonably well with measurements (see Hoeijmakers et al.[13]).

The thickness of the rotational flow features captured in the solution to the Euler equations varies directly with the size of the mesh cells. The simplest way, therefore, to minimize the diffusion of vorticity is to use as dense a mesh as possible. The computation uses an O-O type constructed by Eriksson's[11] interpolation method that places a polar singular line at the apex and a parabolic singular line at the tip of the trailing edge and has 80 cells around the half-span, 40 each, on the upper and lower chord, and 24 outwards, for a total of 76 800 cells (Fig. 11.10). This particular grid topology focuses cells along the leading and trailing edges, as well as the apex where the flow changes most rapidly. It requires, however, a slight rounding of the wingtip.

Rizzi and Eriksson[2] produced a solution for incompressible flow using this mesh and reported the results that included a leading-edge vortex over the wing as well as the trailing-edge vortex, both of which interact with each other downstream of the wing.

Comparison of these results with those of the 3D panel method which fits the vortex sheet to the surrounding potential flowfield was quite favorable, but the Euler solution indicated a peculiar structure within the vortex core just ahead of the trailing edge that was not seen in the potential results.

Isograms of the Euler results viewed axially through the core indicate the approximately conical nature of the flow starting at the apex. At about the 80 per cent chord position, however, the leading-edge vortex lifts up slightly, and an abrupt change takes place which might be interpreted as an additional, and unexpected vortex phenomenon. The exact cause of this feature is still not known. It may be a precursory effect of the development of the trailing-edge vortex. If the conjecture of such a flow process is correct, then our rounding of the tip just exacerbates it.

There are grounds to believe that such a curious feature on an otherwise conical vorticity field may in fact be physically realistic. It may well be the often-observed phenomenon of vortex bursting where the core of the vortex suddenly bellows out, and the circumferential velocity decreases dramatically. At the trailing edge the flow experiences a substantial upwash which lifts the vortex core and may cause it to burst.

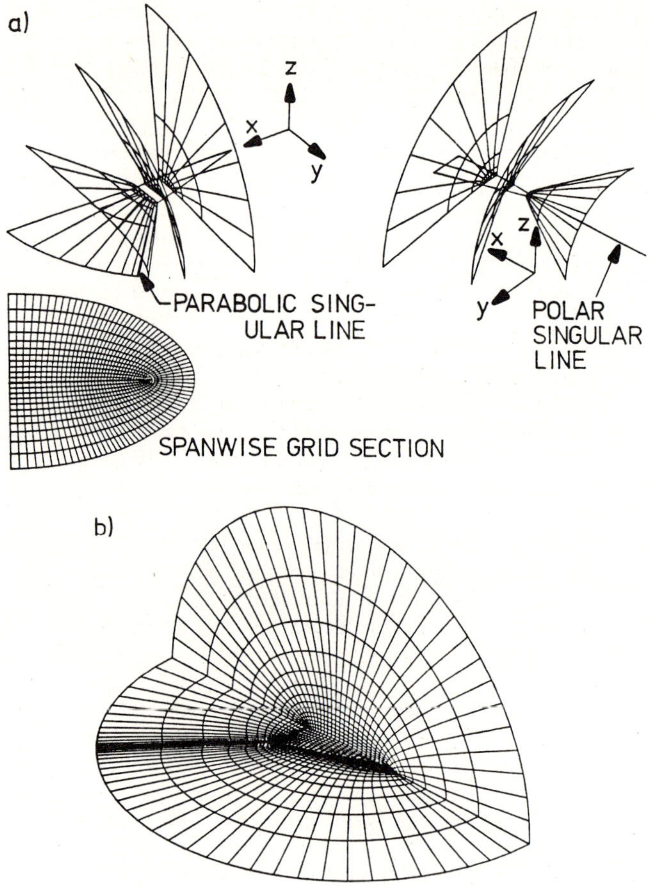

Fig. 11.10 a) Grid generated around a delta-shaped small aspect ratio wing has an 0-0 topology. The polar singular line produces a dense and nearly conical distribution of points at the apex which is needed to resolve the rapidly varying flow there. This mesh is well-suited for computing the flow around wings of combat aircraft (Ref. 3).
b) Three-dimensional view of the delta wing mesh

The upwash gives curvature and therefore torsion[12] to the vortex core. This uplifting phenomena can be clearly seen in the vortex-element solution of Hoeijmakers[13] (Fig. 11.11), and it suggests a second possible mechanism at work to explain the curious structure in the Euler solution ahead of the trailing edge. The torsional force induced by the upwash may set in motion helical waves throughout the region of the vortex core. Under certain conditions such motion may become unstable and lead to the so-called multiple-vortex phenomenon[14].

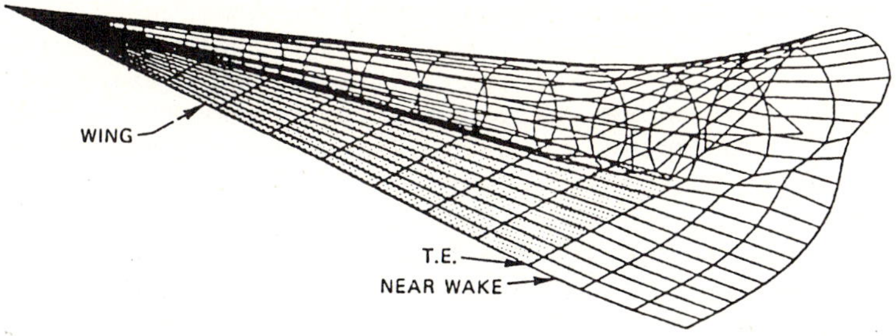

Fig. 11.11 The vortex sheet fitted as a discontinuity to the potential solution shows the curving of the vortex structure due to the upwash at the trailing edge[10]

Rizzi has investigated this question by mesh refinement. He halved the mesh spacing in each direction. The mesh then is $160 \times 48 \times 80$ cells. The previous mesh will now be referred to as the medium mesh, and this mesh as the fine mesh.

The computations were carried out on a CYBER 205 vector computer in 32-bit precision at the rate of 6 μs per cell per iteration which translates to over 125 mflops. The results for the fine mesh have been integrated for 2500 time steps where the solution is steady. The working data set was nearly 14 M 32-bit words in size and resided entirely in real memory. His experience indicates that it is very effective to run large-scale computations like these in a machine with ample real memory but under virtual-memory management. This is because at the start of the computations there are some initialization tasks that require additional scratch arrays, but which can be discarded after the iteration cycle has begun. At start-up practically 18 M words are needed, but the demand reduces to 14 M once the main cycle has begun. Virtual memory is one way to handle this initial overflow of 4 M words.

The overall features of the flows in the medium and the fine meshes are compared in Fig. 11.12 by isograms, drawn in plane projection, of the computed solution in three non-planar mesh surfaces $x/c=0.3$, 0.6, and 0.9 over the wing, one surface in the wake at $x/c=1.15$ and one cutting axially through the core of the vortex. The two solutions do agree and reveal qualitatively the leading-edge vortex over the wing, as well as the trailing-edge vortex that develops from the trailing-edge sheet. The comparison shows that the broad features of the flow are represented in both grids and that they do not change substantially under mesh refinement. The isograms viewed along the axis of the core indicate the approximately conical nature of the flow starting at the apex. In both solutions at about

80 per cent chord position, however, the leading-edge vortex lifts up slightly because of the rising pressure gradient beyond the trailing edge where the flow must eventually return to freestream pressure. This uplifting breaks the conical symmetry of the vortex and coincides with the asymmetric structure seen in the vorticity and total pressure fields ahead of the trailing edge (Figs. 11.12 and 11.13). Is this feature vortex bursting? Experiments with this wing have shown that bursting occurs over the wing at a higher angle of attack, about 30^0. Also the characteristic trait of bursting is a sudden thickening or bellowing out of the vortex core. The results here do not seem to suggest vortex bursting.

Instead Rizzi[15] believes more in another possible physical explanation. Since the pressure gradient beyond the trailing edge is increasing, it forces the vortex core to begin to lift up from the wing even before the edge. It gives the core curvature, Betchov's analysis then suggests that the core will spiral, and may become unstable. The most striking feature of these results is the abrupt change in the contour patterns that takes place ahead of the trailing edge between the two sections x/c=0.6 and 0.9. It might be interpreted as a multicelled vortex phenomenon. Although the cause of this feature may still be numerical, we should expect the core to undergo a helical disturbance phycially that could excite an instability like the one discussed by Snow[14] which results in multiple vortices. At least it is a plausible explanation of the numerical results. In any event more of the local details are brought out in the fine mesh, but the streamwise position and overall dimension of the phenomenon do not change with mesh size. Mesh convergence for the broad features has been obtained.

No loss occurs in the total pressure in the tracked vortex-sheet solution, but substantial losses do appear in the solutions where the sheet is captured over a number of mesh cells. The losses occur in two distinct locations and for different reasons. Near the leading edge of the wing the loss is attributed to the numerical effect of capturing the vortex sheet that is shed from the edge. Theoretically the loss should be zero on each side of the sheet, even though the velocity is in shear. But the numerical solution has to support this shear with a continuous profile over several mesh cells through the sheet, and any sort of reasonable profile (say a linear one) connecting the velocity vector on one side with the one on the other side immediately implies a total pressure loss for the profile even if the velocities at both sides are correct, Chapter X. This is the explanation also given by Powell et al., Ref.16. The loss can be seen as an order-one error made in capturing the sheet. It is unavoidable and is just like the error that occurs when capturing a shock wave. What is important is to obtain the magnitude of the shear across the sheet reasonably correctly, just as one must calculate the jump in pressure across a shock correctly. And the comparison here with the tracked-sheet solution seems to suggest that this is the case.

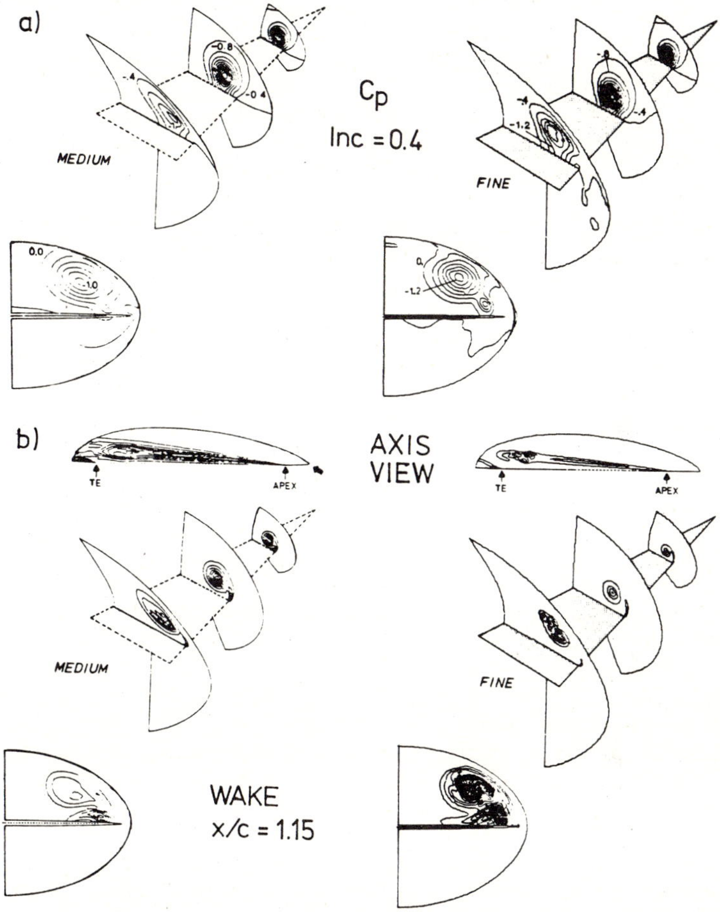

Fig. 11.12 Comparison of contour maps of the medium-mesh (80×24×40) and fine-mesh (160×48×80) solutions to the Euler equations for flow past a 70° swept flat plate delta wing. They are drawn in four non-planar mesh surfaces at the x/c=0.3, 0.6, 0.9,and 1.15 stations and in one mesh surface which passes approximately through the axial core of the vortex

a) Isobars of pressure coefficient c_p.

b) Vorticity magnitude contours (not the same increment for medium and fine contours),

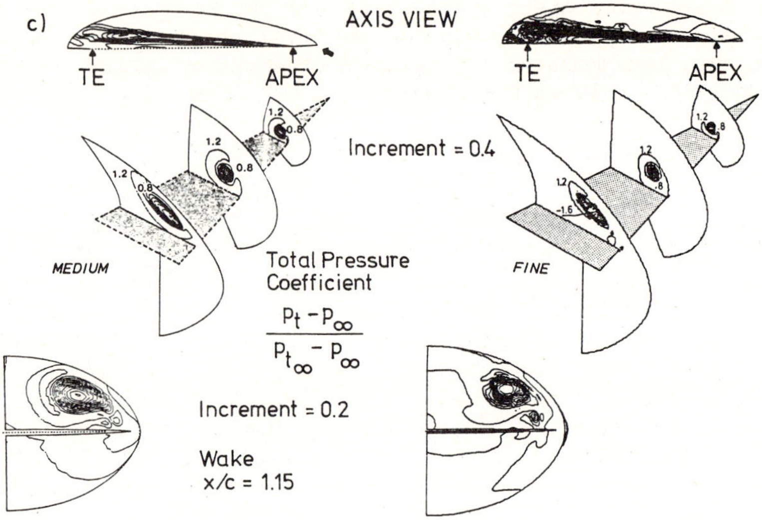

Fig. 11.12 c) Contours of total pressure
(contd.) coefficient. Increment = 0.4

The vortex core itself evolves from the sheet as it spirals inward tighter and tighter. This is a diffusive process, and since the method here is only artificially diffusive, the details cannot be correct. Moreover, with the kind of grid resolution offered here, even in the fine-mesh solution, barely one turn of the spiral is resolved before all trace of the jump in velocity shear is diffused into a region of smooth vorticity distribution. The vorticity reaches a maximum at the center of this region where the circumferental motion is transformed into axial motion. This region is the computed core, of which the local structure must be considered as fictitious. The essential role here of the artificial viscosity, or the truncation error, is to ensure that the circumferential or swirl velocity vanishes in the core. But the level of loss in total pressure found in the core is not random, it is set by the amount of shear produced at the leading edge which we believe is obtained with some accuracy through capture of the sheet. The amount of loss then is related directly to the strength of the sheet. The fine mesh supports more of the spiral, but the sheet must eventually disappear in this mesh too, in the same way, producing about the same loss at the center of a now smaller diameter core. It is the diameter therefore of the contour rings, but not their number, that varies with mesh spacing. In this way the method always produces a viscous model of the core, which of course is not physically precise in its detail, but it does have the correct behaviour of reducing the swirl velocity to zero at the center of the core.

Figure 11.13 presents the shape of the tracked vortex sheet (dashed lines) from the potential method superimposed upon the vorticity magnitude contours of the Euler-equation solution in three cross-flow planes. We see, when looking at the mesh in Fig. 11.10, that the vorticity captured in the field is diffused over 5 or 6 cells in both the medium and fine mesh solutions, and that, in general, the vortical flow region occupies a larger volume than that enclosed by the vortex sheet fitted to the potential solution. But the positions of the vortex cores in the comparisons and even the curvature of the sheets

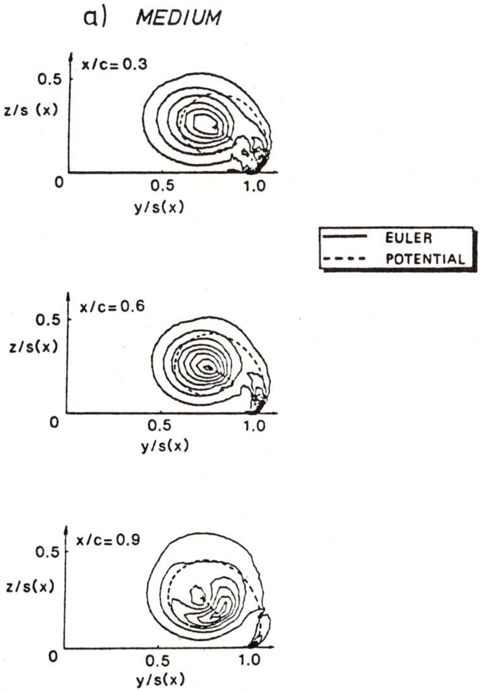

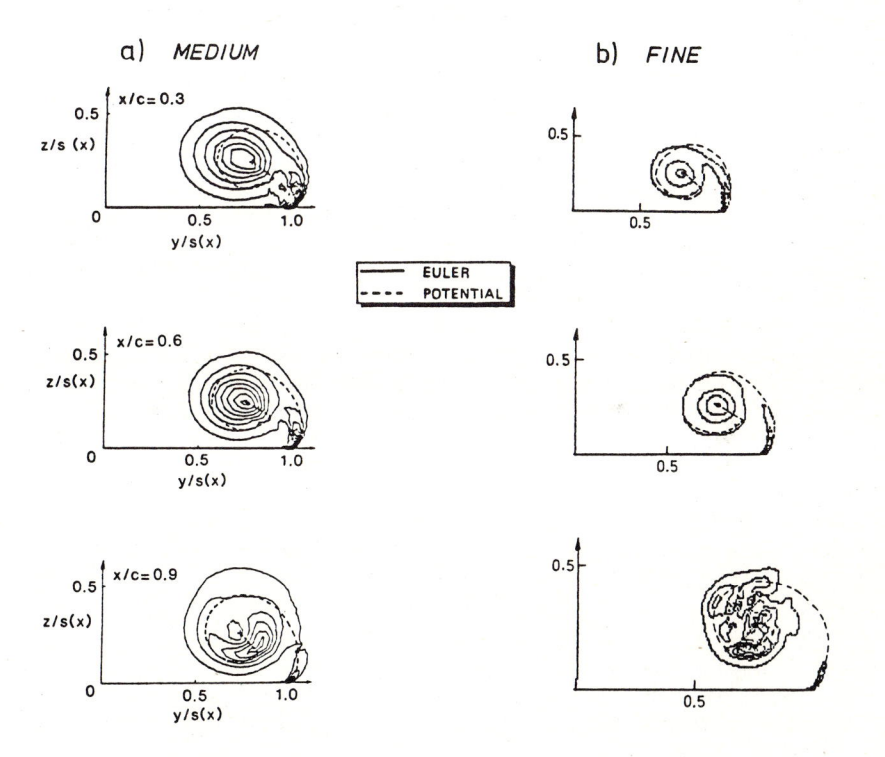

Fig. 11.13 Comparison of the vorticity fields indicated by vorticity magnitude contours (solid lines) computed with the Euler equations, using the medium a) and b) fine mesh, and the shed vortex (dashed lines) that is fitted as a discontinuity to the surrounding potential solution obtained by the 3D panel method (Ref. 15)

near the leading edge agree remarkably well. The vorticity in the fitted sheet is largest near the leading edge where the curvature of the sheet is singular, and the Euler-equation solutions indicate the same trend. The sheet appears to depart tangentially from the lower surface of the leading edge. This comparison with mesh refinement confirms that a stable vortex sheet separating from a swept leading edge can be captured in the vorticity field of the Euler-equation solution with a reasonable degree of realism. The curious distortion of the contours in the x/c=0.9 station and the associated cellular pattern of vorticity are a better representation of the phenomenon in Fig. 11.12 which Rizzi[15] called a standing torsional wave on the vortex core that gives rise to subsidiary vortices. Notice that the region of vorticity in this section does not bellow out appreciably, as one would expect if the vortex had actually burst. And the position and size of the vortical region does not change very much with mesh refinement. This converging (but still not yet converged in the core) sequence of three computations strongly suggests that this feature ahead of the trailing edge belongs to the true solution of the equations.

Figures 11.14a,b present isograms on the wing surface together with the more quantitative graphs of spanwise distributions at three x/c=constant stations and compare them with the potential values. In the sets of computed isobars (Fig. 11.14a) the pressure trough under the leading-edge vortex has about the same shape, position and width, and the three agree rather well. The peak level of the suction along the entire trough on the upper surface is somewhat lower in the medium-mesh Euler results, and shifted slightly inboard at x/c=0.3, but the fine-mesh results show a trend toward the potential solution. The fine-mesh results portray a pronounced waviness that may be a reflection of the character of the vortex as it approaches the trailing edge. A vortex core in helical motion might well produce such a pattern on the upper surface. This waviness, rather surprisingly, is not present in the circumferential $(v^2+w^2)^{1/2}/V_\infty$ velocity components on the upper surface (Fig. 11.14b). The contours of the circumferential velocity do reach a slightly higher value in the fine mesh than in the medium mesh, as one would expect, because the core is better resolved, the diffusion is less, and the swirl velocity is somewhat greater in the outer core region.

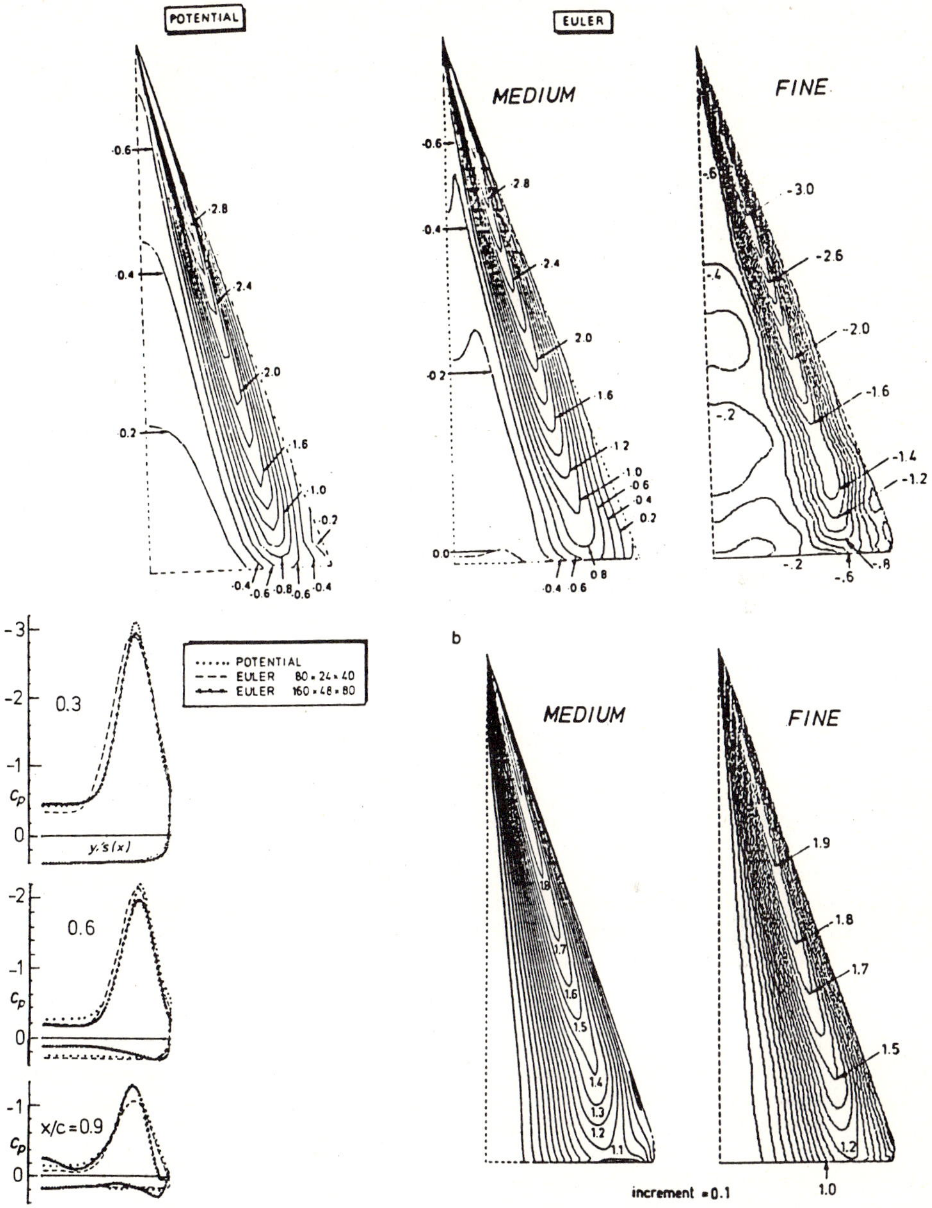

Fig. 11.14 Isograms of the computed medium-mesh and fine-mesh solutions on the upper surface of the wing[15].

a) Isolines of pressure coefficient c_p compared with the potential solution together with three corresponding graphs versus local semispan at x/c=0.3, 0.6, and 0.9. Increment = 0.2,

b) Circumferential velocity $(v^2+w^2)^{1/2}/V_\infty$ contours

11.1.5 Flow Through a Francis Water Turbine

Solution of the incompressible Euler equations has also found application to industrial rotating machines[17]. We illustrate this with the prediction of the flow in a Francis water turbine. The simulation is performed by the IMHEF group on the EPFL CRAY 1/S supercomputer. Of particular interest is the ability to predict the details of the flow as it passes the rotating turbine runner. An axial view of a Francis turbine is shown in Fig. 11.15.

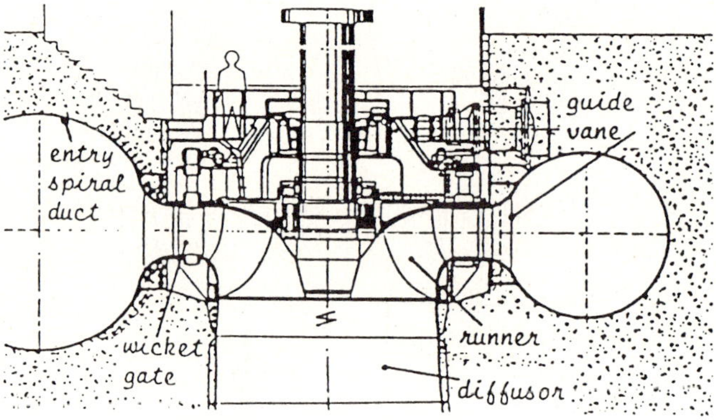

Fig. 11.15 Axial view of a Francis turbine[17]

The runner geometry and the surface mesh upon it are shown in Fig. 11.16. The flow domain is discretized by an H-H type mesh generated by transfinite interpolation. The computational grid generated to calculate the flow passing between two adjacent blades, i.e. the interblade fluid volume, which represents the computational domain considered, is shown in Figs. 11.17 and 11.18. Here an area marked "inlet surface" represents an imaginary surface upstream of the turbine runner over which a relative velocity distribution, as seen by an observer fixed with the rotating runner, is prescribed. The flow over the areas marked "periodic surface" is periodical in nature, since any one interblade volume is adjacent to another one, with identical flow behaviour. The flow over the exit surface of the computational domain cannot be prescribed a priori. Over this surface only the pressure distribution can be given. On solid surfaces represented by the blades, the crown and the band, the velocity vector must be tangential.

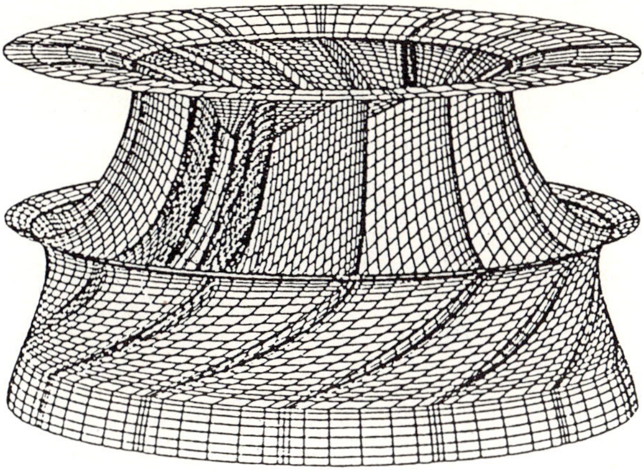

Fig. 11.16 Francis runner[17]

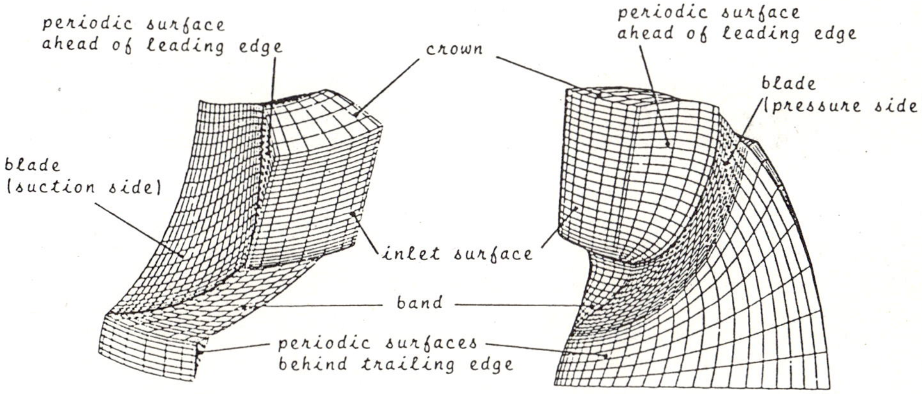

Fig. 11.17 Francis runner: interblade control volume[17]

The resulting topology given by this type of mesh is indicted in Fig. 11.18.

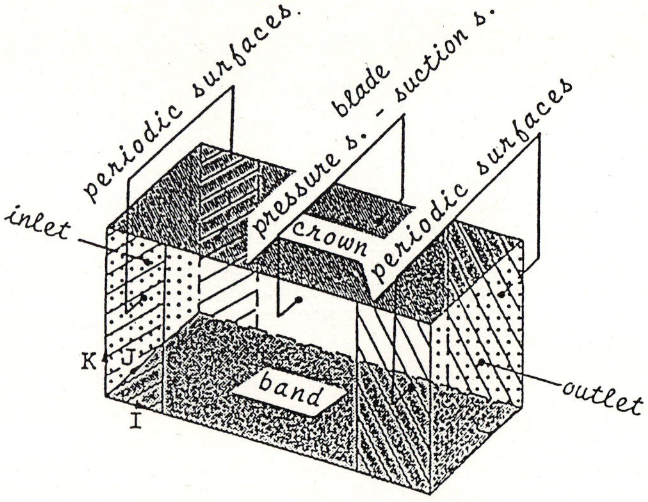

Fig. 11.18 Computational space resulting from the H-H mapping of a Francis runner[17]

The computer program developed[17] has been fully vectorized to suit the particular architecture of the CRAY computer. This means that one flow calculation case involving around 6000 grid points can be performed in less than 10 minutes. The graphical representation of the solution takes an additional minute to obtain, and can be given in different forms to suit any particular aspect in a performance analysis of the turbine. One example of such a graphical representation is shown in Fig. 11.19, illustrating the pressure distributions and the relative velocity distribution over the suction and the pressure side, respectively, of the blade surface. Here the arrows indicate the direction of the flow, and the length of the arrows are proportional to the velocity. From the calculated pressure distribution the torque exerted on the runner can be evaluated and is found to be within 1 per cent of the measured value.

Such simulations can be extremely useful for industrial design. Worldwide there exist today an estimated 300 supercomputers, of which there are about 50 in Europe. However, the enormous potential of these machines is not yet fully appreciated by the European water turbine industry, even though their impact is already felt in the aerospace and car industry. Professor Ryhming, among others, advocates combining adequate and convenient access to a supercomputer with a high tech flow simulation tool for industrial usage, and he believes it can make the difference in a situation where serious competition is felt.

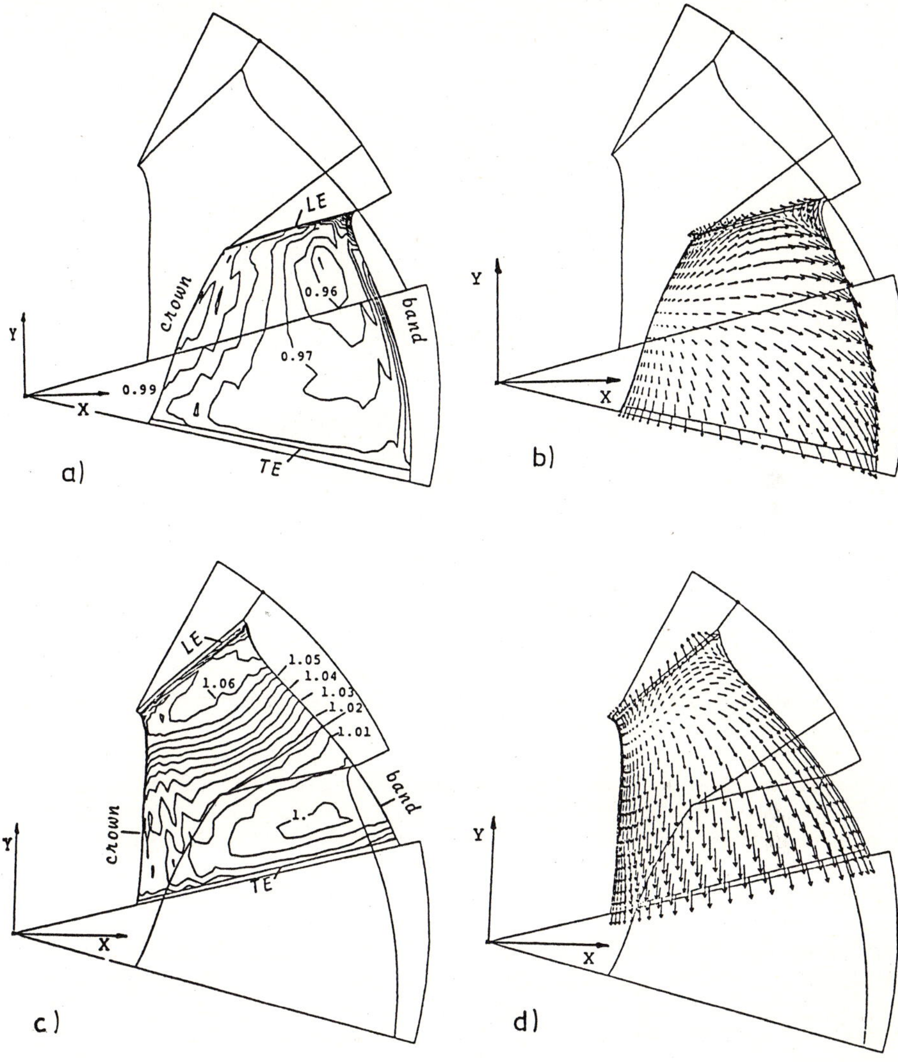

Fig. 11.19 Pressure field a) and relative velocity vectors b) on the suction side, c) and d) on the pressure side[17]

11.1.6 Flow Past an Automobile

As a final example for Euler computations of near incompressible flow results of Eberle and Schäfer, Ref.18, are given. They used the EUFLEX code to compute the flow past an automobile at $M_\infty=0.18$. The grid used for the symmetric half of the car was generated from three bi-harmonic equations for x, y, z, and contained 10^5 cells, Fig. 11.20.

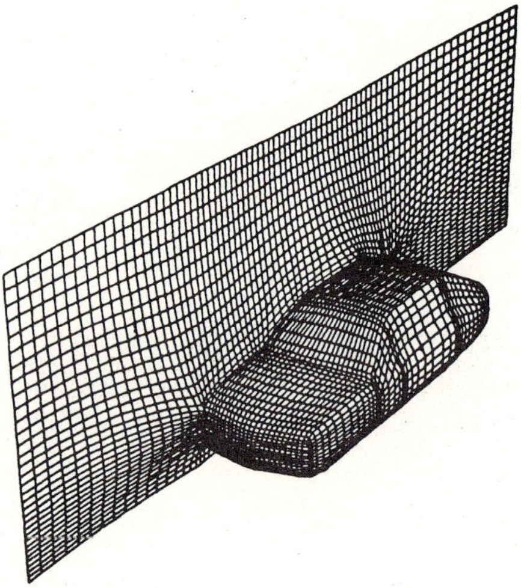

Fig. 11.20 Schematic of grid for the Euler solution[18] for the flow past an automobile

Especially it was investigated how second-order and third-order characteristic flux averaging, see also Ref. 19, influences the solution. Figure 11.21 shows some of the results. Flow regions, which are dominated by convection (front and mid part of the car) are well predicted especially by the third-order scheme. The second-order scheme underpredicts somewhat the pressure in maxima and minima compared to the measured data, Fig. 11.21a. On the lower side of the car a front spoiler in the experiment was not modelled in the computations, so that here larger discrepancies exist.

At the rear end of the car the corners and edges acted as Kutta conditions, so that the Euler results come quite close to the experimental results. The pressure compares rather well, vortex structures become visible, Fig.11.21b. The latter should not be overemphasized, because the situation is not quite as clear as with lifting wings. Interesting in this respect is also the big discrepancy in the calculated total pressures behind the car, Fig. 11.21c. The second-order result appears to be very diffusive compared to the third-order re

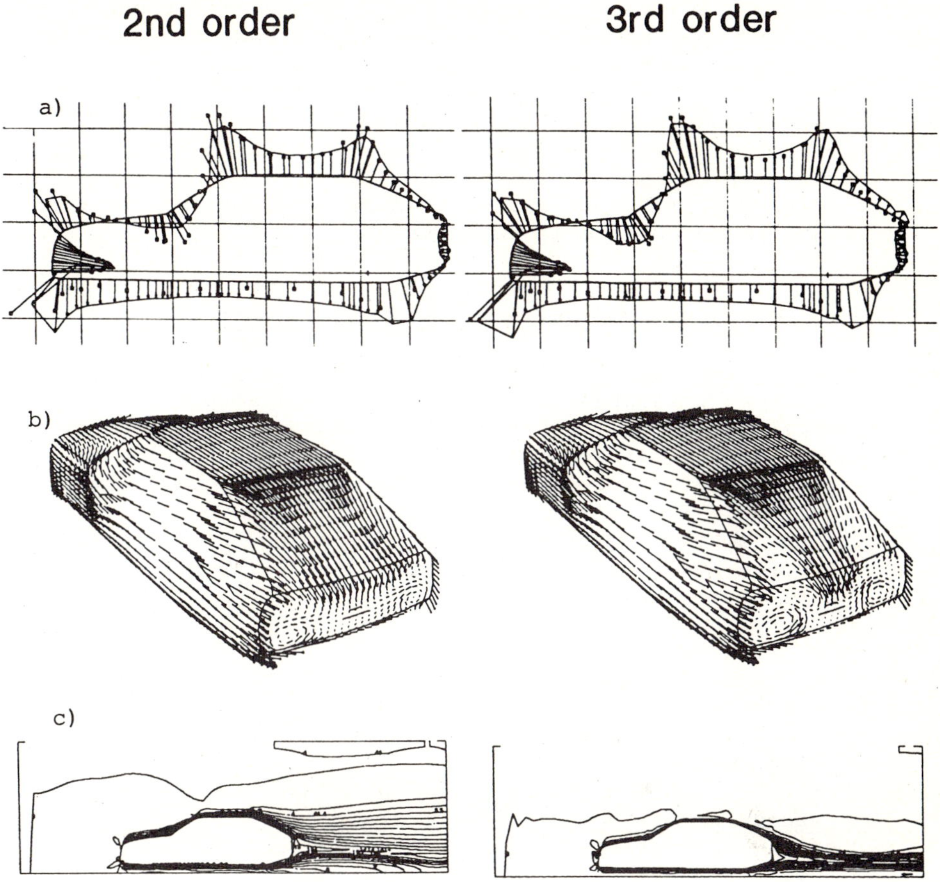

Fig. 11.21 Comparison of second- and third-order solution[18]:

a) pressure distribution in symmetry plane,
b) surface velocity vector field,
c) total pressure distribution in symmetry plane

sult. All together, the results, especially with the third-order approach, appear very interesting with regard to engineering applications. However, further investigations are necessary in order to make Euler solutions a reliable tool for the prediction of rear-part flows at cars.

11.2 Subsonic/Transonic Flow

The subsonic to transonic speed range is the traditional one for applications of Euler solutions. The speed of the sound waves is not substantially faster than the flow speed, and convergence to steady state is markedly faster than in low subsonic flows. Here we look at applications to flow past airfoils and wings. The report[20] of the AGARD Working Group 7 gives a good overall indication of what the methods can do on this class of problems. First, however, a comparison of different Riemann solvers is conducted.

11.2.1 Comparison of Different Riemann Solvers

There are many ways to incorporate a Riemann solver, chosen out of those presented in Sub-Chapters 6.13 to 6.36, into a numerical scheme solving the Euler equations. Here we only consider finite-volume schemes. A crude classification of possibilities to apply a Riemann solver for the calculation of the cell-face fluxes was already given in Sub-Chapter 6.4, where the conversion of the Courant/Isaacson/Reese scheme to a conserative numerical scheme is considered. The main possibilities are:

- flux-vector splitting,
- flux-difference splitting,
- mean-flux plus matrix viscosity,
- direct Riemann solution followed by the conventional flux calculation.

Flux-vector splitting is based on the calculation of two individual fluxes on either side of the cell face, the sum of which is the final flux. Flux-difference splitting assigns parts of the conventional flux differences across a cell face to either the left or the right cell contribution to the total flux difference there. The mean-flux plus matrix viscosity method resembles very much the artificial viscosity methods applied for centered schemes. In contrast to the latter, however, the coefficients of viscosity are obtained from the plus/minus split Jacobian matrix. Furthermore in this case each line of the Euler equations gets a numerical viscosity formed from differences of all flow variables, i.e. not only of the flow variable to be updated by the particular line under consideration. As an example let us take the continuity equation. This is the equation for the density update. The viscosity added consists not only of a higher difference of the density but also of all the remaining flow variables each multiplied by its own coefficient of viscosity. The direct Riemann solution followed by the conventional calculation of the flux is by far the best out of the possibilities considered here, since the entropy error is always the lowest as has been found out by Eberle in many numerical experiments. So only this method is considered in the comparisons to follow. However, still some problems with the definition of the left and right state remain to be considered, if a Riemann solver is chosen, the entries of which are non-conservative flow variables. A variety of those is described in the Sub-Chapters 6.13 through 6.21.

Here the question arises on which quantities the high-order base point interpolation should be applied. Several concepts are at our disposal:

a) Interpolate the conservative flow variables — convert the left and right state to those non-conservative variables needed for the Riemann solver chosen.

b) Convert the conservative flow variables to those non-conservative flow variables needed for the Riemann solver chosen — apply high-order interpolation for obtaining the left and right state being entry to the Riemann solver.

c) Convert the conservative flow variables to characteristic variables — apply high-order interpolation for obtaining the left and right state — convert the left and right state to the variables needed as entries to the Riemann solver.

These three approaches (a,b,c) are only a little choice out of a tremendous amount of possibilities showing a considerable non-uniqueness in solving a higher-order Riemann problem. They all suffer from not exploiting the homogeneous property of the Euler equations for improved accuracy.

To circumvent this difficulty it is highly recommended to use the simple linear conservative Riemann solver of Sub-Chapter 6.22 in conjunction with the high-order interpolation applied to the conservative flow variables. Whether the coefficients are taken from Roe's average, or the arithmetic mean, or the near-exact Riemann solver of Sub-Chapter 6.14, is of less importance for the accuracy, provided the coefficients are bounded inside the interval given by the left and right state.

Roe's average does not always fullfill this requirement, and cannot be recommended as the ultimate solution to that problem.

A good choice is to run the Riemann solver of Sub-Chapter 6.14 in parallel with the conservative Riemann solver using the high-order interpolated left and right state, and to use the solution of the non-conservative solver (6.14) as coefficients for the conservative solver.

There follow now some results, the first of which is taken from an airfoil code with the Osher-type solver (6.16) being incorporated. The airfoil is a typical supercritical wing section with 12 percent thickness. In Fig. 11.22 we see a cut out of the H-type grid used identically for all calculations to follow.

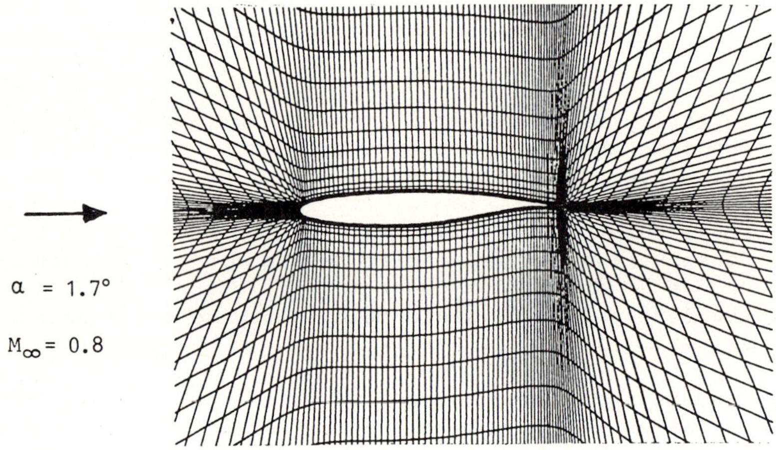

Fig. 11.22 H-type grid for a supercritical airfoil

The pressure distribution is displayed in Fig. 11.23. It shows a crisp shock representation.

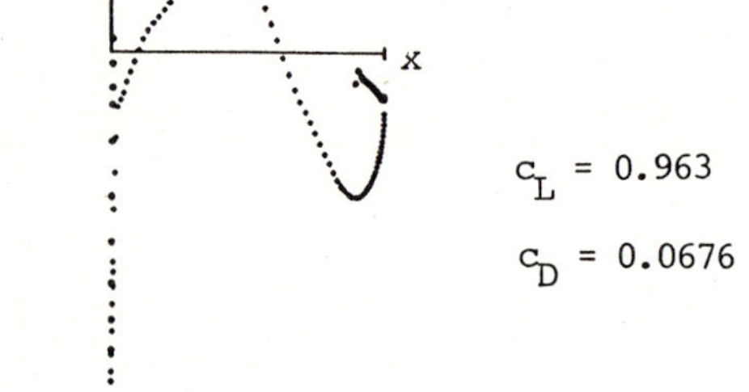

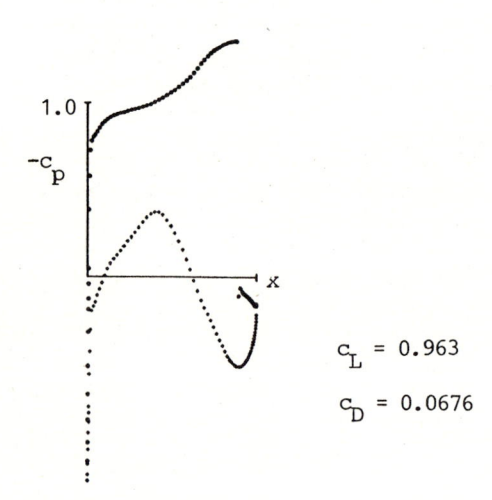

Fig. 11.23 Pressure distribution past supercritical airfoil

To obtain a feeling how accurate the calculated solution is, the diagrams of the total-pressure error and the total-enthalpy error are given, Fig. 11.24.

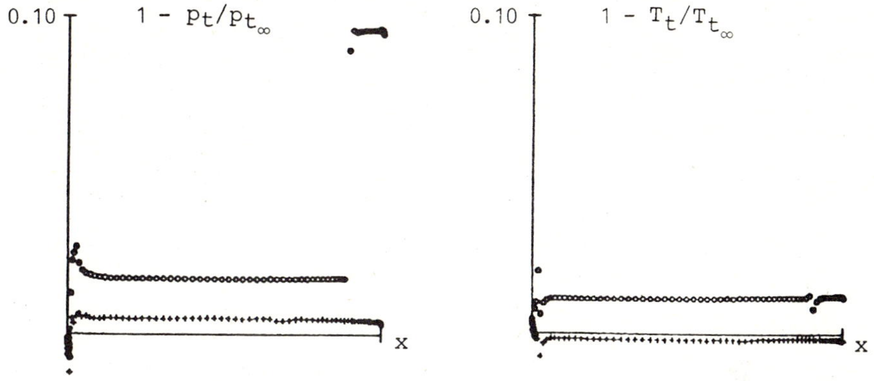

Fig. 11.24 Total pressure/enthalpy error of a finite volume scheme using the Osher-type Riemann solver of Sub-Chapter 6.16

While the total pressure is about one per cent off its correct value, the total enthalpy error is also too large.

The situation remains the same if the integration paths of the original Osher solver are interchanged, leading to the Riemann solver described in Sub-Chapter 6.18, Figure 11.25.

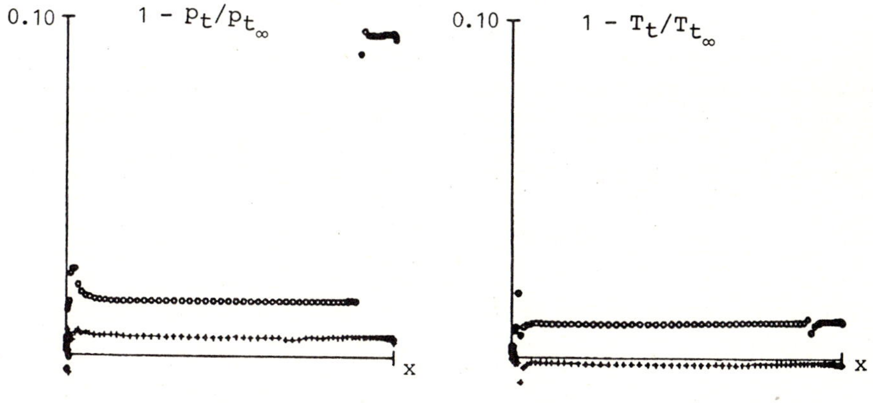

Fig 11.25 Total pressure/enthalpy error using the Riemann solver of Sub-Chapter 6.18

The differences to Fig. 11.24 are not visible at this scale. Both results were calculated with the high-order interpolation of the non-conservative variables followed by the solution of the Riemann problem.

Now we turn our attention again to the two Riemann solvers investigated, however with the high-order interpolation applied to the conservative variables prior to their conversion to the non-conservative variables defining the left and right state. The result for the original Osher solver is displayed in Fig. 11.26.

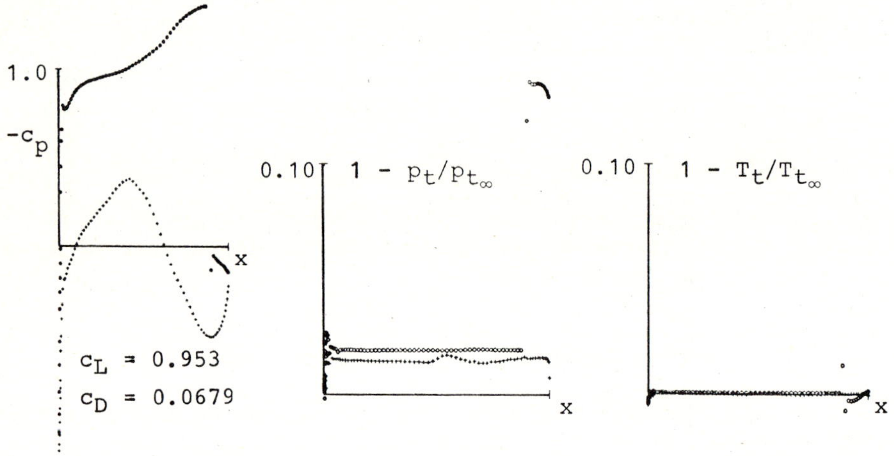

Fig. 11.26 Osher's Riemann solver with interpolation of the conservative variables

As can be seen, the total-enthalpy error is now acceptable, but the total-pressure error probibits the use of this approach, also the scheme tends to not converge to the steady state. The CFL number had to be chosen very low in the explicit range, although the update was performed by the implicit point Gauss-Seidel procedure of Sub-Chapter 6.59. After this experience the same exercise using the alternative Osher-type Riemann solver (6.18) has been dropped.

Although the linear Newton-type Riemann solver of Sub-Chapter 6.20 is based on sound theory, its performance shows that not all Riemann solvers do work well, Fig. 11.27.

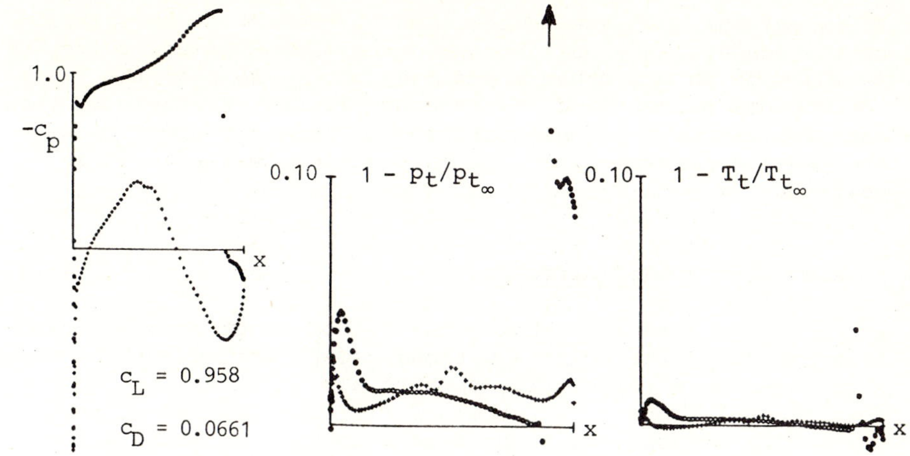

Fig. 11.27 Result obtained with the Newton-type Riemann solver of Sub-Chapter 6.20

It is evident that the scheme is no more capable to reach a steady state. As pointed out by Pandolfi in a lecture, Ref.21, in the piecewise parabolic method (PPM) by Colella and Woodward, Ref. 22, which uses a similar Riemann solver as the present, the fluxes have to be augmented by an extra artificial viscosity. Thus, we believe all the advantages of a characteristics based method are lost.

Finally we look at the result obtained with the fully homogeneous approach making use of the linear conservative Riemann solver described in Sub-Chapter 6.22, Fig. 11.28.

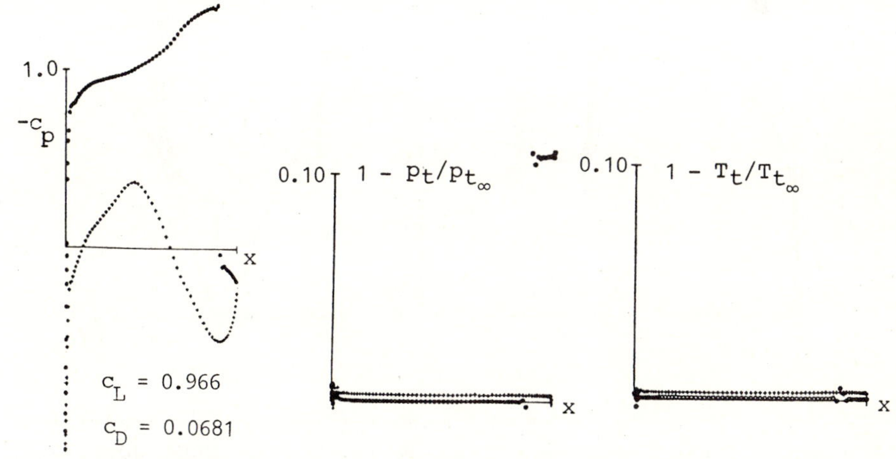

Fig. 11.28 The result obtained with the homogeneous Riemann solver of Sub-Chapter 6.22

It is evident that the homogeneous approach — interpolation of the conservative variables, conservative Riemann solver — is superior to all other approaches not exploiting the homogeneous property of the Euler equations. Furthermore this approach allows the implementation of implicit updating with the least arithmetic work for calculation of the Jacobian split matrix.

11.2.2 Flow Around Airfoils

Aware of the need for gauging the methods most commonly used today to solve the Euler equations, AGARD set up Working Group 7 to organize a competition open to all to obtain the best possible solutions to a number of different 2D and 3D test cases. The purpose was to produce a body of definitive solutions against which other methods could be measured. The goal was the same as that of the GAMM Workshop[23] held in Stockholm in 1979, some four years earlier, except that now the competition was restricted to the Euler equations only. Many of the test problems are identical, and the two competitions indicate the rate of progress in the development of the methods.

Eleven solutions were submitted for the case of flow $M_\infty=0.85$, $\alpha=1^\circ$ past the NACA 0012 airfoil. Four were eliminated as being too inaccurate, and the remaining seven are plotted against each other in Fig. 11.29, the major difference occurring in shock position.

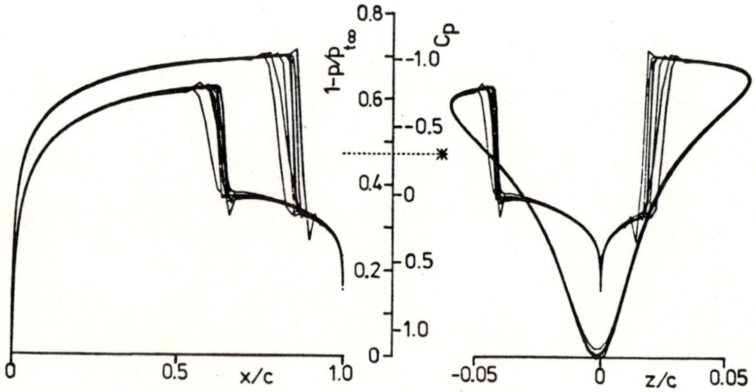

Fig. 11.29 Comparison of the computations contributed to the AGARD Test Case 02 (Ref. 20). NACA 0012, $M_\infty=0.85$, $\alpha=1^\circ$ shows the seven most accurate: a) pressure,

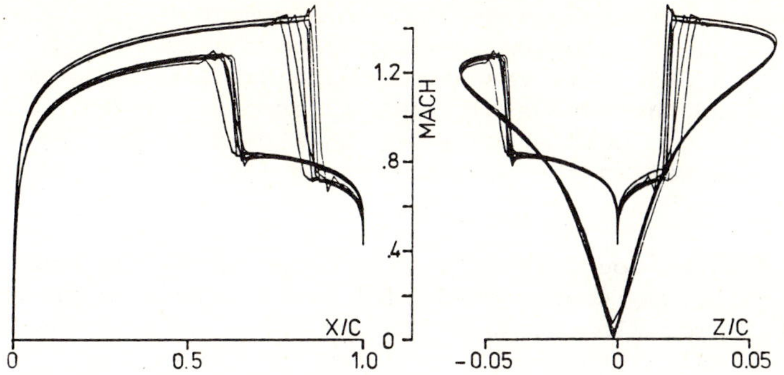

Fig. 11.29 (contd.) b) Mach number distributions on the airfoil

With a 128×28 O-type mesh Rizzi[24] produced the solution shown in Fig. 11.30, displaying the distribution of the properties over the surface and contours of the isobars in the field. As a comparison the isobars of the Dornier/Jameson results obtained with a mesh of four times the cells is also presented.

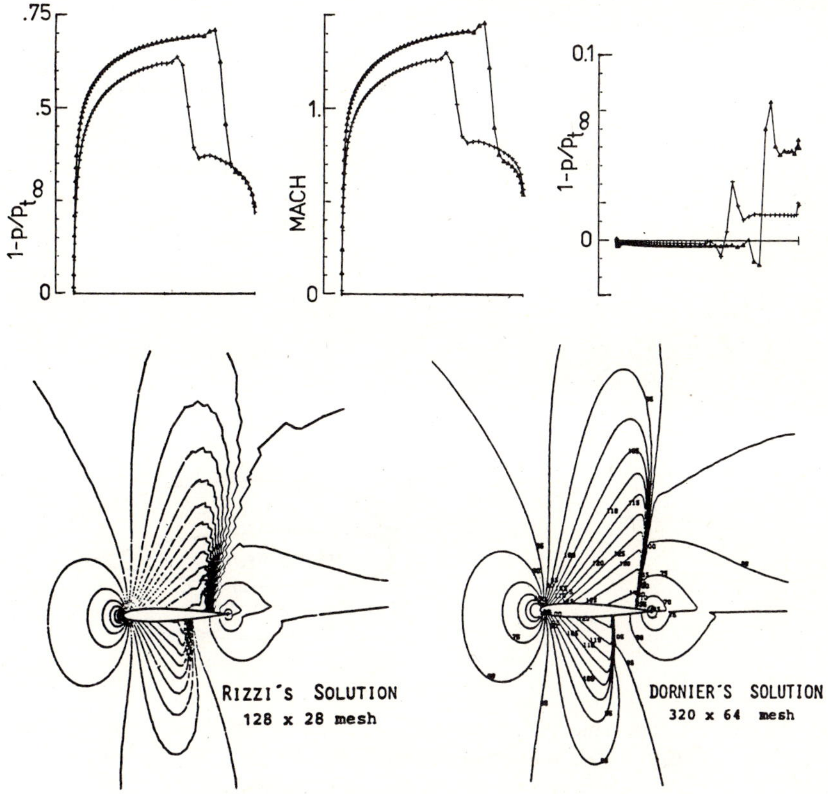

Fig. 11.30 Flow past the NACA 0012 airfoil $M_\infty=0.85$, $\alpha=1^\circ$ (Ref. 24). Normalized pressure, Mach number, and total pressure coefficient on surface, and isobar contours compared with the fine-mesh solution of Dornier

A supercritical shock-free flow is a very good test case because the exact solution is known from the hodograph theory. It's also a difficult case to compute because it is an isolated solution. Because small entropy errors tend to build up and form into a weak shock wave, this case is particularly good for revealing the spurious entropy production of the numerical method.

With the outer boundary at five chords, an O-type mesh of 160×32 cells was constructed, and the solution is presented in Fig. 11.31 for flow at $M_\infty=0.721$, $\alpha=-0.194^0$ around the NLR 7301 supercritical airfoil.

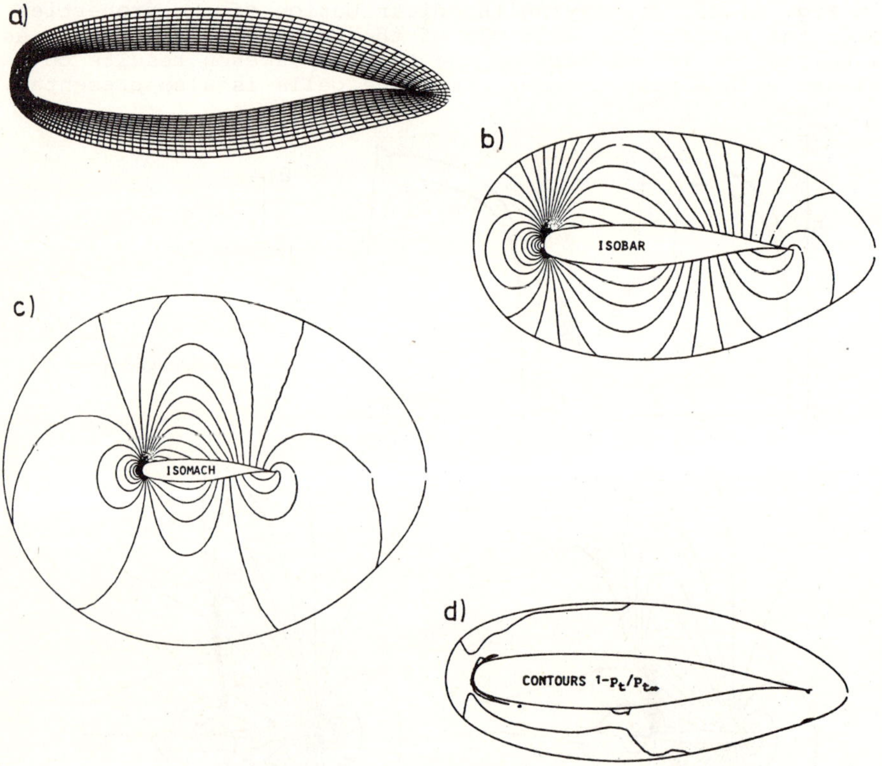

Fig. 11.31 Isograms of shock-free flow computed around the NLR 7301 airfoil (Ref. 24), $M_\infty=0.721$, $\alpha=-0.194^0$: a) near-surface grid, b) isobars, c) iso Mach contours, d) total-pressure contours

The computed lift is 0.597 compared to the exact value of 0.5939. The computed drag is just two counts. Figure 11.32 shows the comparison of computed and exact surface pressure and sonic line location. This demonstrates that highly accurate solutions to the Euler equations can be reached by the centered scheme with artificial viscosity.

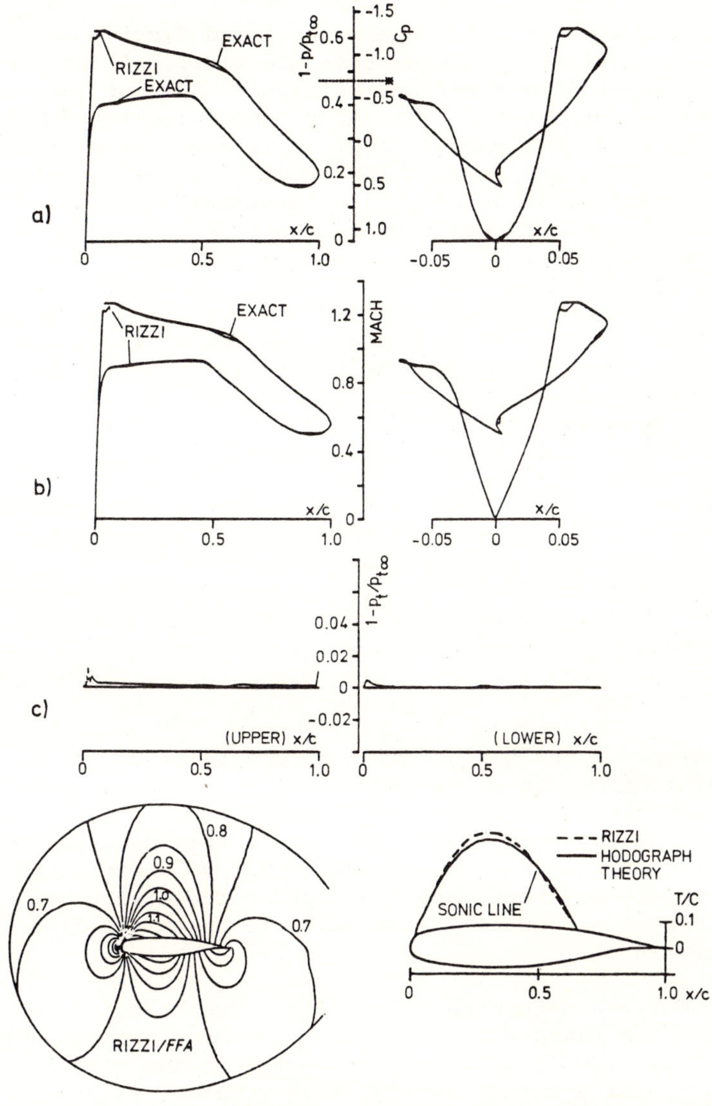

Fig. 11.32 Comparison of the exact hodograph results and the results computed on the surface of the NLR 7301 corresponding to the field contours in Fig. 11.31, $M_\infty=0.721$, $\alpha=-0.194^\circ$: a) pressure, b) Mach number, c) total pressure

11.2.3 Vortex Flow Over Sharp-Edged Delta Wings

Here we consider transonic flow at $M_\infty=0.7$, $\alpha=10^0$ around two delta wings both with the same plan form, 70^0 sweep angle, but different cross sections. The first wing is flat, it has zero thickness and a sharp leading edge. Its geometry is conical.

It is the wing on which Hoeijmakers and Rizzi[10] carried out their comparison for incompressible flow, and further results were discussed by Rizzi and Eriksson[2]. Some of these are presented in Sub-Chapter 11.1. Rizzi and Purcell[25] studied the wing in transonic vortex flow. It is also the wing for which Murman, Rizzi and Powell[26] obtained very good agreement at supersonic speeds between two-dimensional conical and fully three-dimensional computations, both using very high-resolution meshes.

The second wing has a maximum thickness which is 6 per cent of the local chord and has a round leading edge but a sharp trailing edge. We compare here results computed for both wings with a very dense grid in order to study the effect of a round versus a sharp leading edge. For the problem of round leading edges see Chapter X, and Sub-Chapter 11.2.4.

The mesh is one of 0-0 topology. Figure 11.33 presents chord and span sections, as well as a three-dimensional projection of the mesh. The mesh is drawn for dimensions 80×24×40 cells, but we calculate here with 192×56×96 for just over one million cells. The outer extremity of the mesh is a hemisphere (assuming centerline symmetry) with a radius of about 3 root chords c_R of the wing. The same topology and number of grid cells is used for the round leading-edge wing.

The computations begin with the freestream values as initial conditions on the so-called coarse grid of dimensions 48×14×24 obtained by using only every fourth point in the original mesh. One thousand iterations are carried out, the results are interpolated to the medium mesh 96×28×48 where another 1.000 iterations are performed, and this successive global mesh refinement is repeated once again to reach the fine mesh of 192×56×96 cells.

Let us first determine if the fine-mesh solution for flow around the sharp-edge delta is converged in the sense of mesh spacing. Side-by-side surveys of the medium-mesh and fine-mesh flow fields are suitable for this purpose. Figure 11.34 shows the streamlines that are integrated from the steady velocity field by selecting a number of positions distributed along a line near the leading edge as starting points and then tracing their path downstream. Alternatively, a line x=constant along the span can also be chosen. In either case the differences

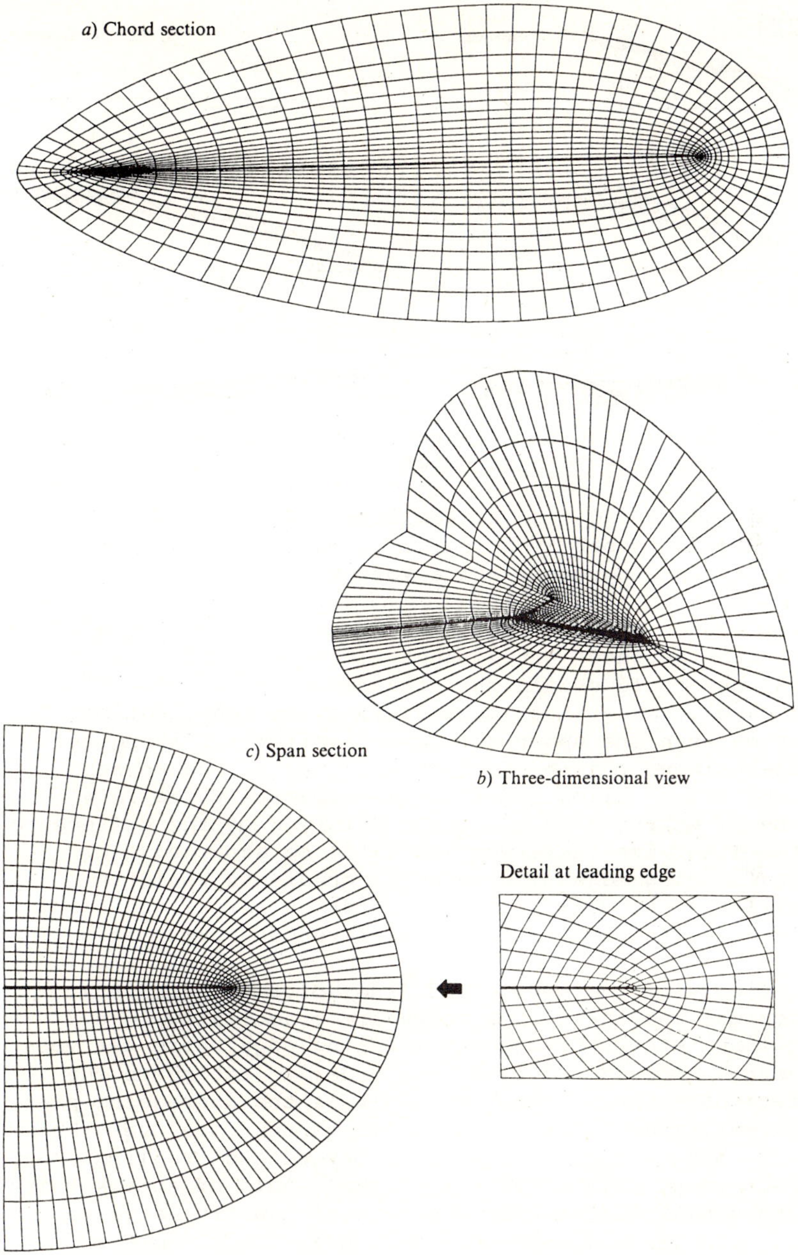

Fig. 11.33 Grid generated around the sharp edged delta wing has an 0-0 topology that concentrates points around the edges

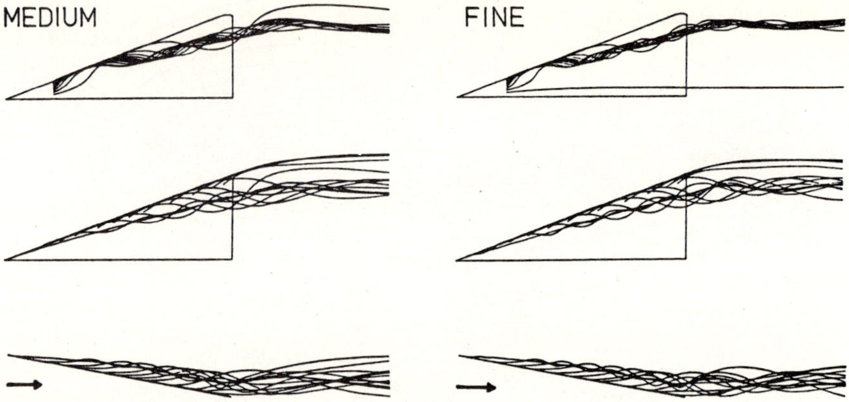

Fig. 11.34 Comparison of integrated streamlines from medium- and fine-mesh solutions

between the medium and fine streamlines are very small. In both flow fields, separation begins at the apex and continues all along the leading edge, and a vortex evolves.

There is no doubt that the numerical method is selecting the solution with the shed vortex sheet. It would be futile to think that one could obtain the attached potential solution for this case because that would require the cancellation of the shear between the upper and lower surface all along the geometrical singular line of the leading edge. The method prefers instead to allow the shear to separate there in a vortex sheet, which is an allowable solution of the difference equations. Otherwise there would have to be attached flow around the leading edge. The velocity gradients would be extremely large, as would be the truncation error, and hence the numerical dissipation, of the difference approximation.

The flow fields of the medium and fine solutions are very nearly conical and the vortex cores lie practically in a straight line, except beyond the trailing edge where the vortex is turned inboard and upward in order to return to the free-stream direction further back. It is the increasing pressure gradient, from a low value in the vortex core over the wing to the higher free-stream value beyond the trailing edge, that brings about this effect. At higher angle of attack this three-dimensional disturbance can be sufficiently large to produce a spiralling of the core and, eventually, vortex breakdown.

The isocontours of the medium and fine solutions in Fig. 11.35 confirm the conclusion that the flow is smooth and nearly conical and that the differences between the two of them are only small. The isobars (lines of constant ratio of local static to free-stream total pressure, $1-p/p_{t_\infty}$) in Fig. 11.35a indicate a low-pressure region in the core that changes little from the medium to the fine solution. The detail in the wake structure is sharper in the fine solution, and the axial view of the core might suggest that the fine core lies a little closer to the wing than the medium core. The feature just past the trailing edge in this view is a function of the curved mesh surface, in which the contours are drawn, cutting across the vortex. It should not be mistaken for a spiralling motion. The survey of the vorticity magnitude $|\Omega|$ in Fig. 11.35b

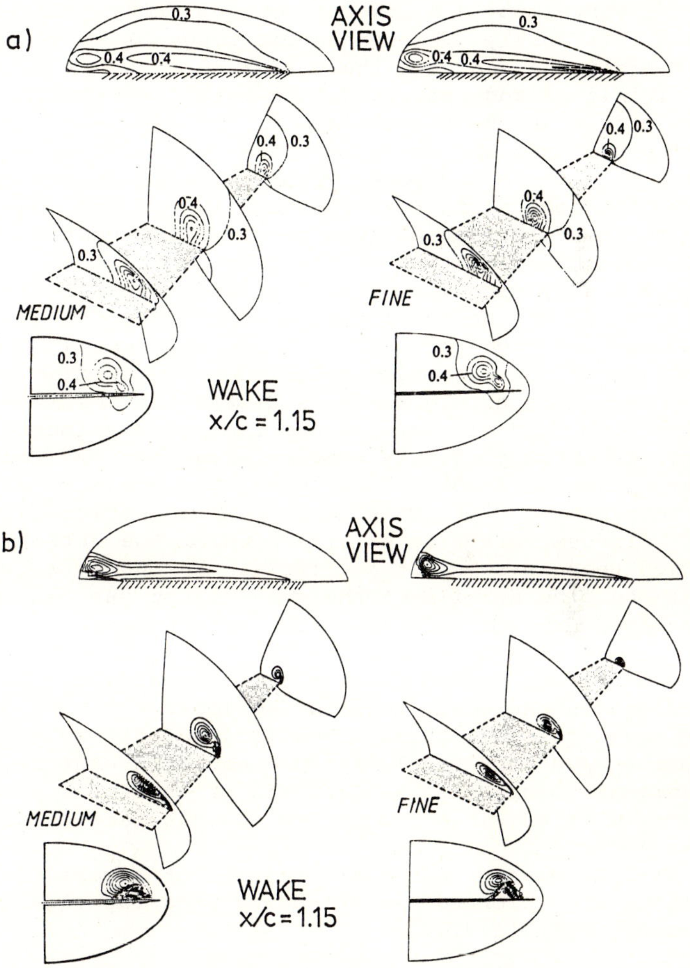

Fig. 11.35 Isograms comparing the medium (96×28×48) mesh and fine (192×56×96) mesh solutions drawn in a number of mesh surfaces around the sharp edge delta wing: $M_\infty=0.7$, $\alpha=10^\circ$, increment=0.05

a) Pressure $1-p/p_{t_\infty}$, b) vorticity magnitude $|\Omega|$

reveals a sharper resolution of the vortex sheet separating from the leading edge, and the interaction of the leading-edge and trailing-edge vortices in the wake. Aside from the crisper resolution of the sheet, the agreement in the overall features here implies that the vortex dynamics can be represented in numerical solutions upon a mesh to a good degree of accuracy. Both solutions indicate that the vorticity in the sheet falls off rapidly over a short distance from the leading edge. This would suggest that even before completing the first turn of its spiral the strength of the sheet is rather small. It might not be very important, therefore, to resolve the coils of the spiral distinctly. The concentric rings of increasing vorticity magnitude indicate the vortex core, where swirl velocity is diminished and converted into axial velocity. One can see this effect in the streamlines in Fig. 11.34. This is a dissipative process, and presumably the result of either truncation error or numerical viscosity, or both. It is dependent on mesh size; the core shrinks as the mesh is refined, but it produces the realistic effect of zero swirl at the centre of the core. Apparently the location of the core centre does not change significantly with mesh size (see also Fig. 11.34). The shear in velocity at the leading edge determines the swirl that is swept up into the core. How it then diffuses in the core is a function of the mesh size there. One can surmise that on the fine mesh more of the sheet is distinctly resolved as it spirals up, and the core, where the swirl velocity goes to zero, is correspondingly smaller. Even in the fine mesh, however, it seems doubtful that more than one turn of the spiral is represented distinctly. Mesh-computed solutions must produce diffuse vortex cores within which the details have to be looked upon as somewhat artificial. Furthermore, we presume that the fine details of the core do not substantially alter the broad dynamics of the vortex structures.

In summary we conclude that for these cases of low helicity (defined as the angle between the local velocity and vorticity vectors) and small three-dimensional departure from conical flow we have very nearly reached mesh convergence with the numerical solutions. The question of the round leading edge seems to be of minor importance in this particular case.

In Ref. 27 measurements, and results of several Euler calculations are reported for the delta wing/fuselage configuration shown in Fig, 11.36.

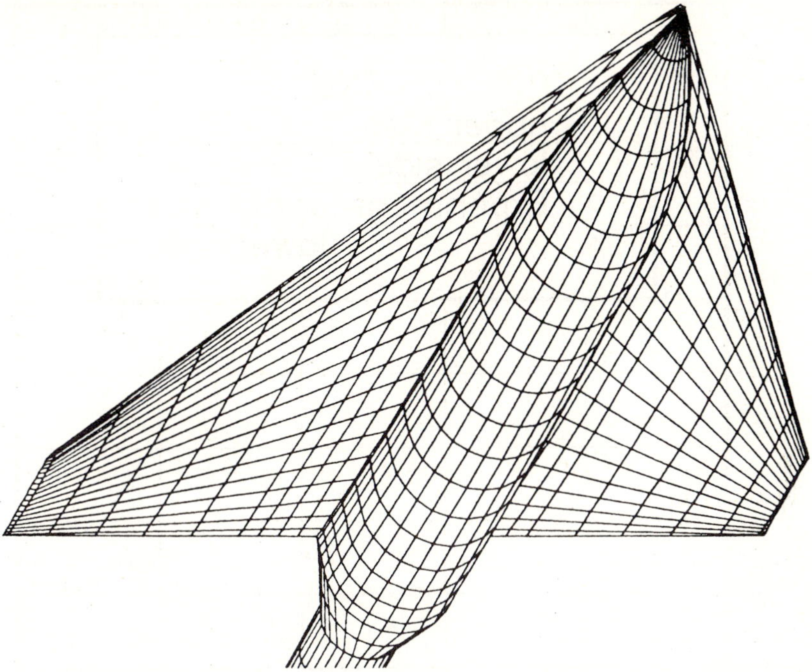

Fig. 11.36 Vortex-flow model seen from below, Ref. 27

Calculations were made by W. Schwarz[28,29] with the EUFLEX code for this configuration with sharp and round leading edges, with and without the fuselage. He studied the influence of mesh topology (C-H, H-H) and also of mesh density. In closing this sub-chapter a few comparisons of computed with measured[30] surface pressure data are presented. The data were obtained for the wing-body combination with sharp leading edge at $M_\infty=0.85$ and an angle of attack $\alpha=10^\circ$. Schwarz employed for the calculations an H-H grid with $135\times45\times58$ grid points.

Figures 11.37 to 11.39 show at three cross-sections calculated and measured pressure distributions on the upper side and on the lower side of the wing.

At the lower side of the wing a very good agreement was reached with the measured data. Obviously viscous effects play only a minor role here in the experiment (Reynolds number based on wing-root chord: $Re = 9\cdot10^6$).

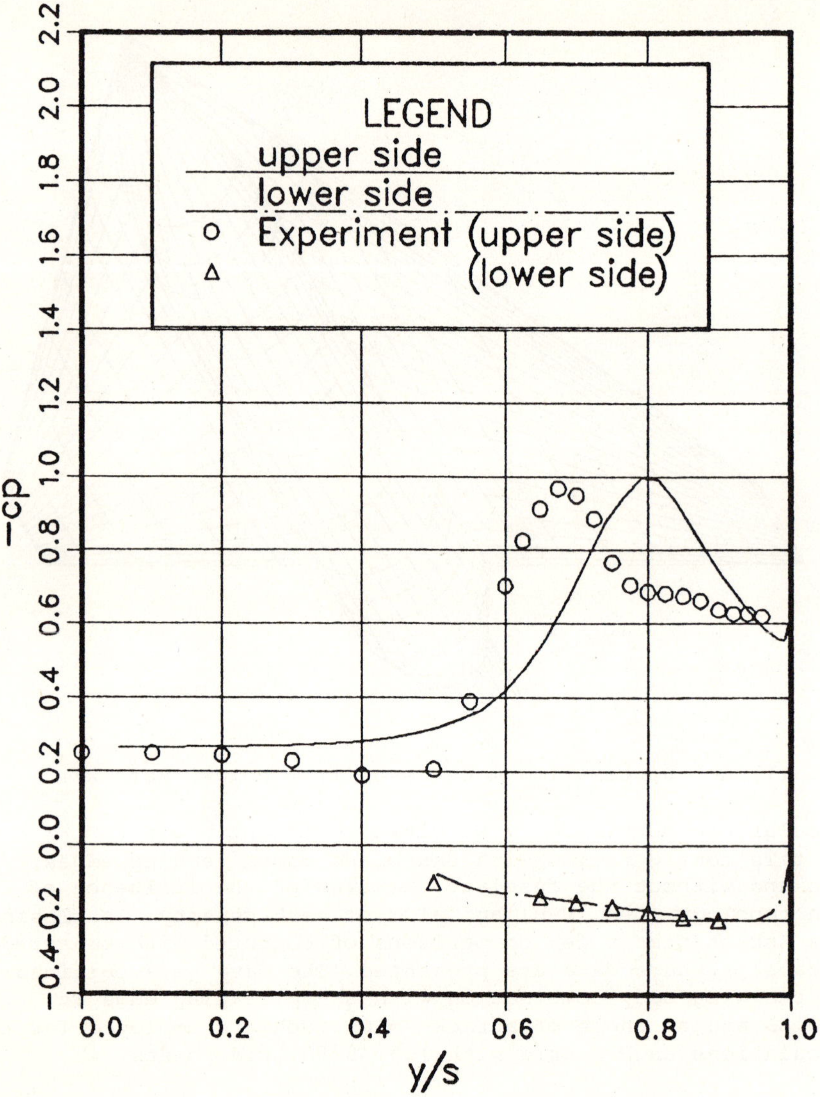

Fig. 11.37 Measured and calculated pressure distribution on vortex-flow model[28,29], M_∞=0.85, α=10°, x/c_R=0.3 (wing/body combination, sharp leading edges)

On the upper side significant and consistent differences occur, which can be attributed to secondary and maybe higher-order separation, because only the primary separation occuring at the sharp leading edges (implicit Kutta condition) is modelled in the calculation (see also Chapters IX and X). The

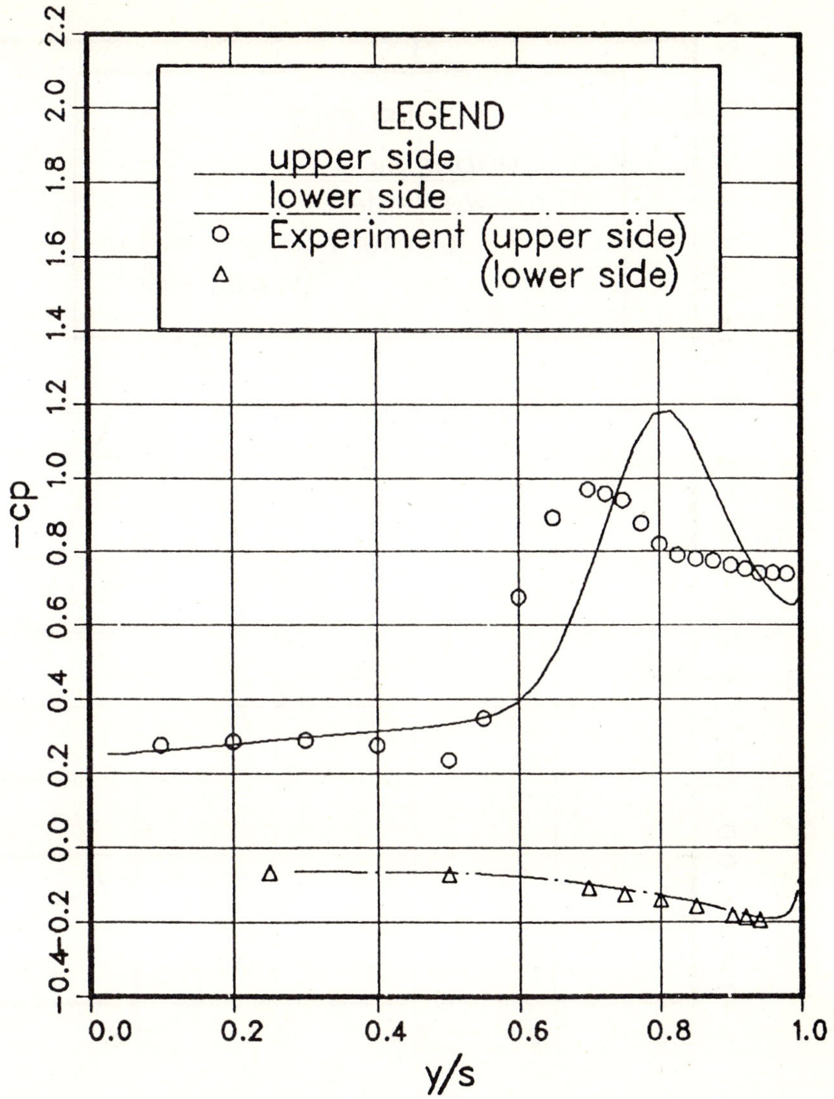

Fig. 11.38 Measured and calculated pressure distribution on vortex-flow model[28,29], M_∞=0.85, α=10°, x/c_R=0.6 (wing/body combination, sharp leading edges)

computed suction peak lies at y/s ≈ 0.8 compared to y/s ≈ 0.7 in the experiment (note the almost conical behaviour of the flow).

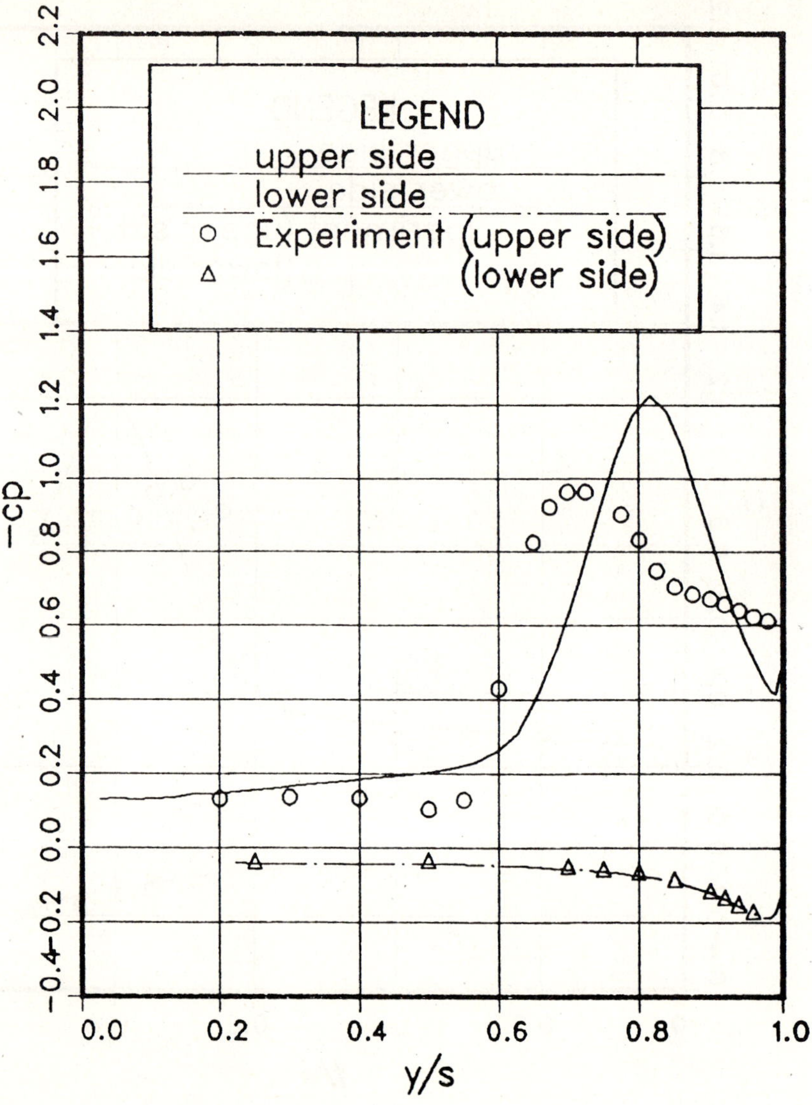

Fig. 11.39 Measured and calculated pressure distribution on vortex-flow model[28,29], M_∞=0.85, α=10°, x/c_R=0.8 (wing/body combination, sharp leading edges)

In any case also the computed suction peak is higher than the measured one. At the foremost shown station x/c_R=0.3, however, the influence of the secondary separation, which is still weak there, is partly compensated by the influence of the fuselage.

The solution can be considered as relatively free from grid and numerical influences, and certainly represents the best one can get with an Euler model for such flow situations. It remains to be seen how far secondary separation can be modelled in the frame of Euler models. Compared to potential solutions with vortex modelling such simulations are a progress, because transonic cases can be handled now in design aerodynamics, althought with certain, but consistent, errors.

11.2.4 Vortex Flow Over Round-Edged Delta Wings

Next we consider the delta wing[3] with round leading edges. A corresponding calculation has been carried out with the $192 \times 56 \times 96$ mesh for the same conditions $M_\infty = 0.7$ $\alpha = 10^0$. Fig. 11.40 compares the results computed for the round-edge wing with those for the sharp-edge wing (see Sub-Chapter 11.2.3).

The concentric isobar patterns over the wing are typical of shed vortex flow. The vortex begins in a flow singularity at the apex of the wing and remains highly conical over the centire wing. The effects of flow compressibility do not appear to be significant here. But the results in Fig. 11.40 demonstrate the dramatic effect of changing the wing cross-section to one of 6 per cent thickness and a round leading edge. Much reduced in size the vortex pattern now is non-conical. It begins somewhere past the 50 per cent chord, not at the apex, and the actions of compressibility are larger. Spanwise views of the v-w velocity vectors indicate that the flow is indeed separating in a vortex sheet from a geometrically smooth wing surface. A non-potential feature, it strongly suggests that, if the body shape has a bounded but sufficiently high curvature, a vortex sheet forms in the flow, presumably by the action of numerical dissipation.

Of course, at a round edge there exists in a pure inviscid model in principle the ambiguity in the location, where actually the primary vortex sheet forms. This was discussed in detail in Chapter X. Depending on the geometry, the vortex sheet may leave the surface at a velocity level, and hence with a strength (Sub-Chapter 10.5) which is substantially wrong.

In principle on round edges, where separation has to be modelled, an explicit Kutta condition must be prescribed.

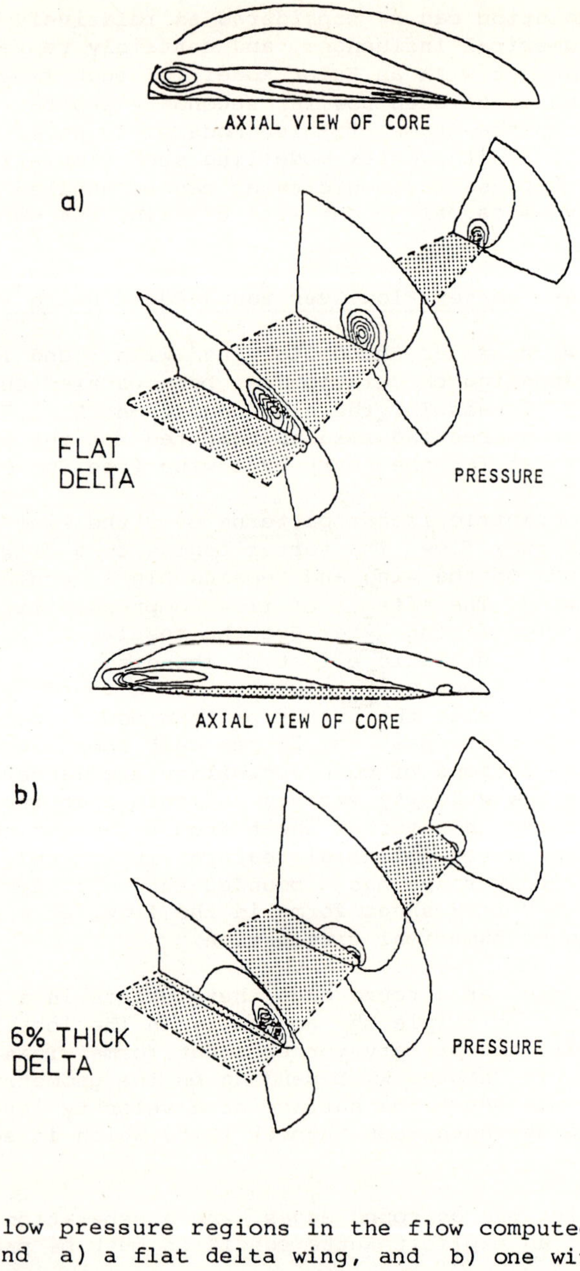

Fig. 11.40 The low pressure regions in the flow computed around a) a flat delta wing, and b) one with thickness and rounded leading edges, show that a vortex develops in both cases[3] (M_∞=0.7, α=10°)

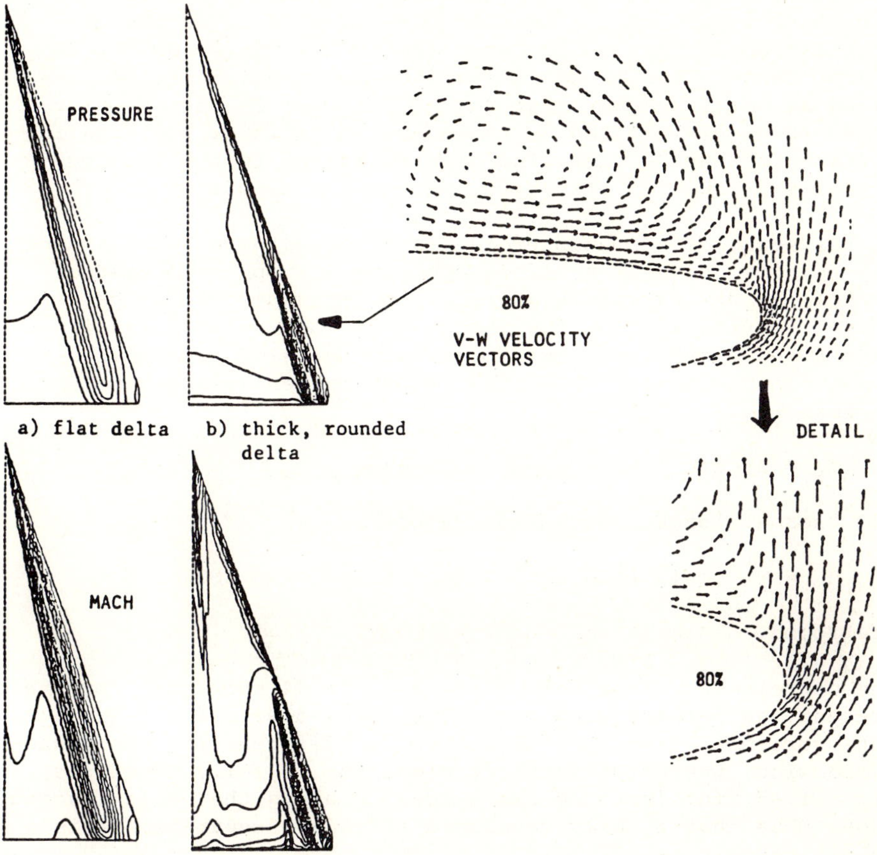

Fig. 11.40 (contd.)

Computed surface pressure and Mach-number contours indicate differences in start and position of vortex. Velocity vectors reveal presence of vortex sheet separation from round leading edge (Ref. 3)

In Ref. 29 tests are reported for the round-edged wing of the vortex-flow model (see the preceeding sub-chapter), where the primary separation line was modelled by an explicit Kutta condition. At the location visualized in the experiment[31] the surface-near cells of the C-H-grid were treated as solid cells.

Without the explicit Kutta condition a chance separation occurs on a part of the leading edge, depending on the discretisation in the leading-edge area, Figs. 11.41 and 11.42 (Refs. 28 and 29). Of course, without an explicit Kutte conditon no vortices should appear at all, see Chapter X. Fig. 11.43 from Refs. 28 and 29 shows the inviscid wall streamline pattern for such a case. The discretization in the leading-edge area, however, cannot be considered as adequate.

In the sense of Sub-Chapter 10.5 one should have an idea about the physical situation before applying a computation method. In Ref. 32 new extensive correlations of experimental data are given. For the wing of the vortex-flow model with a leading-edge sweep $\phi_o=65^\circ$, an angle of attack of $\alpha=10^\circ$, and a free-stream Mach number $M_\infty=0.85$ the two characteristic parameters "normal angle of attack"

$$\alpha_N = \arctan (\tan\alpha/\cos\phi_o) = 22.64^\circ,$$

and "normal leading-edge Mach number"

$$M_N = M_\infty \cos\phi_o \, (1 + \sin^2 \alpha \tan^2 \phi_o)^{0.5} = 0.3833$$

are found.

The correlations from Ref. 32 in Fig. 11.44 show that for the present sharp-edged wing obviously fully developed leading-edge vortices exist. The situation is different for the round-edged wing, where the vortices might not be fully developed, Fig. 11.45, for instance not extending along the whole leading edge. Note that a fully developed secondary separation exists in all cases.

Presently work is underway to improve the capabilities to prescribe explicit Kutta conditions in Euler solutions for both primary and secondary separation lines ($Euler^+$ solution). However, while the location of the primary separation line at a round leading edge may be found with sufficient accuracy with a boundary-layer computation - experience is favourable in this regard, Ref. 33 - the location of the secondary separation line must come from other sources, because of strong interaction phenomena there (vortex/boundary layer/shock(?) interaction). Probably correlations also from Ref. 32 will be helpful in solving this problem.

a)

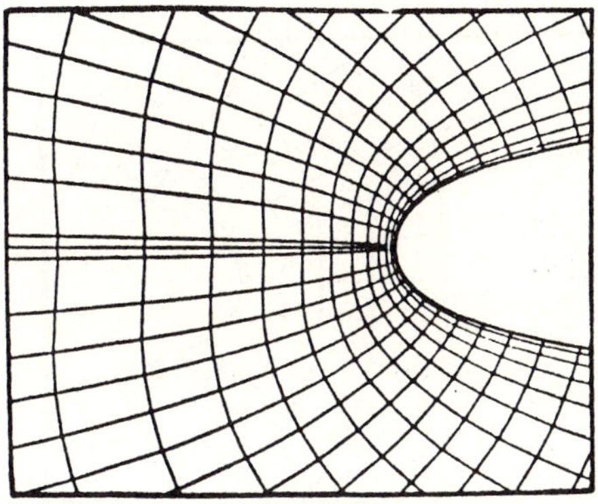

b)

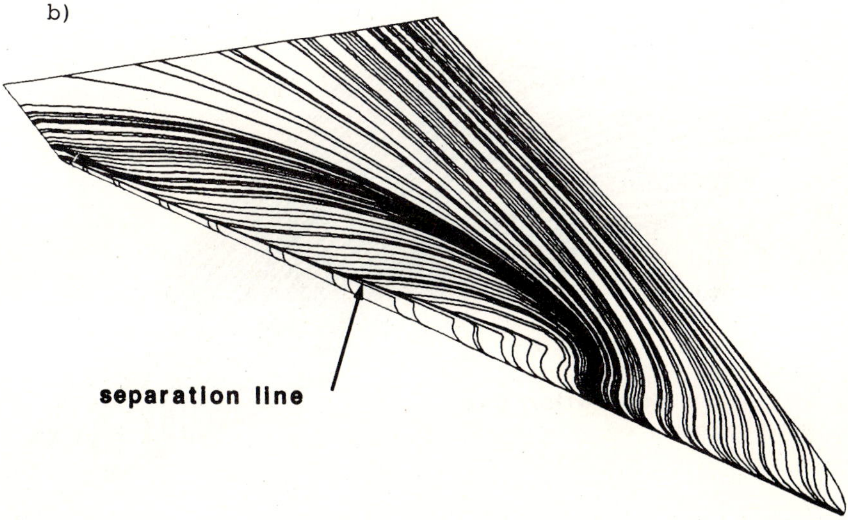

Fig. 11.41 Chance separation (I) on round-edged delta wing at $M_\infty=0.85$ and $\alpha=10^\circ$, Refs. 28 and 29,
a) grid at the leading edge,
b) computed surface-streamline pattern

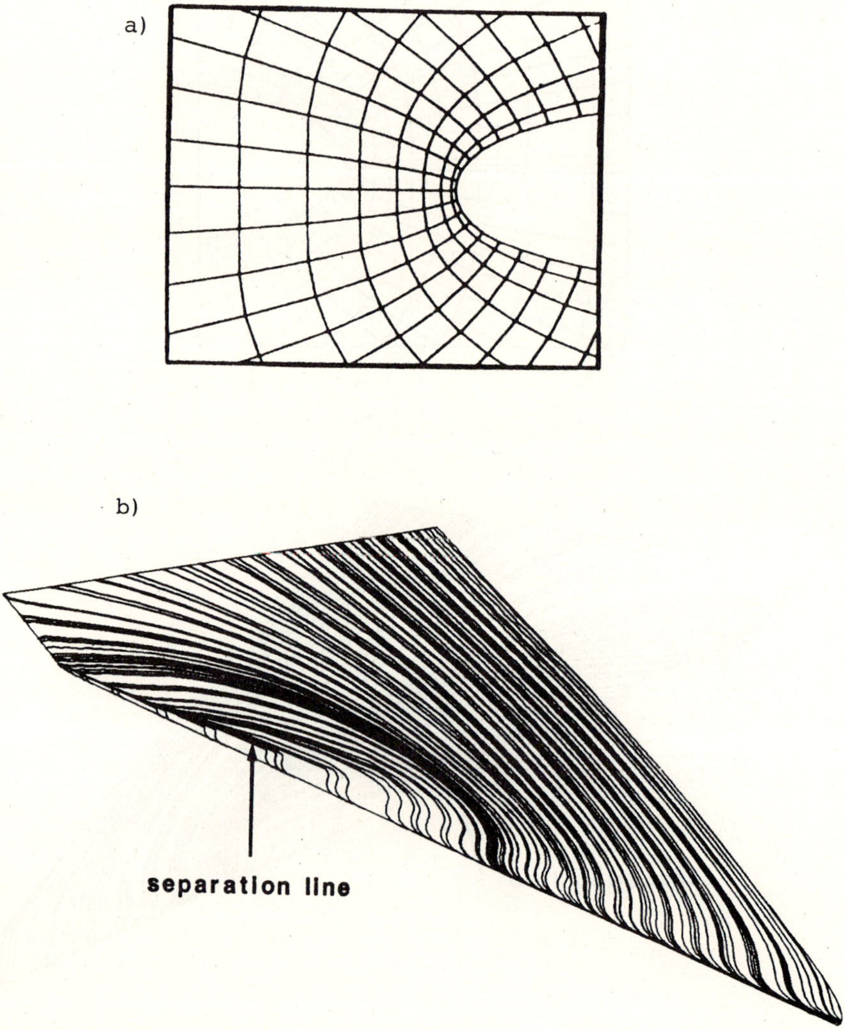

Fig. 11.42 Chance separation (II) on round-edged delta wing at $M_\infty=0.85$ and $\alpha=10^\circ$, Refs. 28 and 29,
a) grid at the leading edge,
b) computed surface-streamline pattern

a)

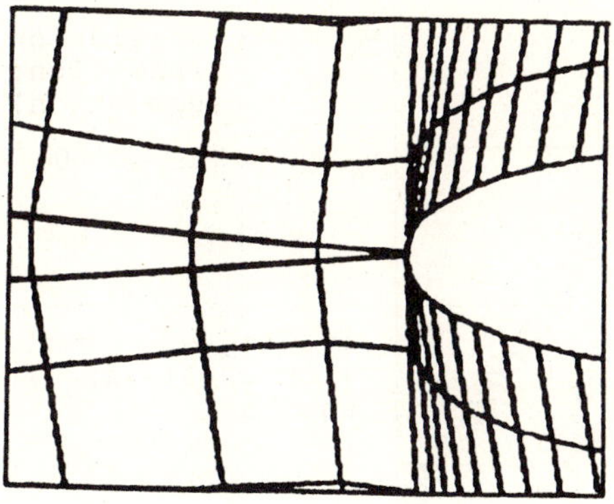

b)

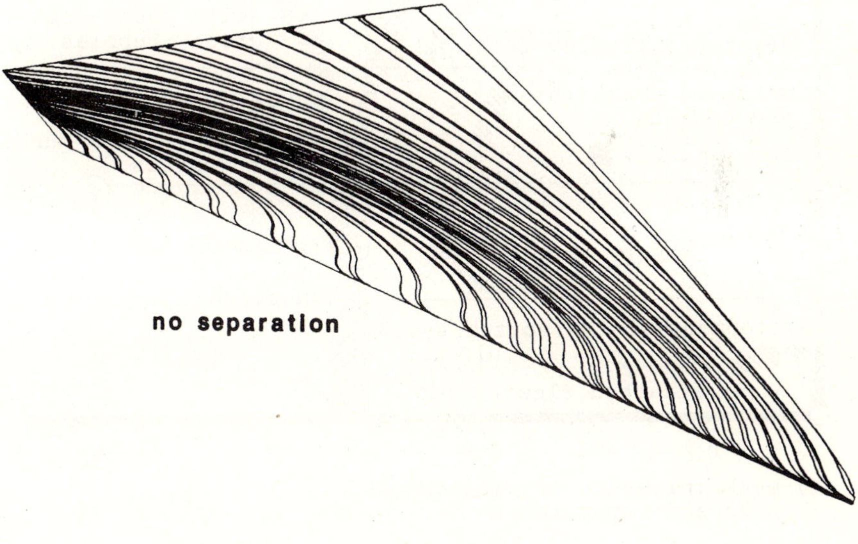

Fig. 11.43 Ideally inviscid flow past round-edged delta wing at $M_\infty=0.85$ and $\alpha=10^\circ$, Refs. 28 and 29,
a) grid at the leading edge,
b) computed surface-streamline pattern

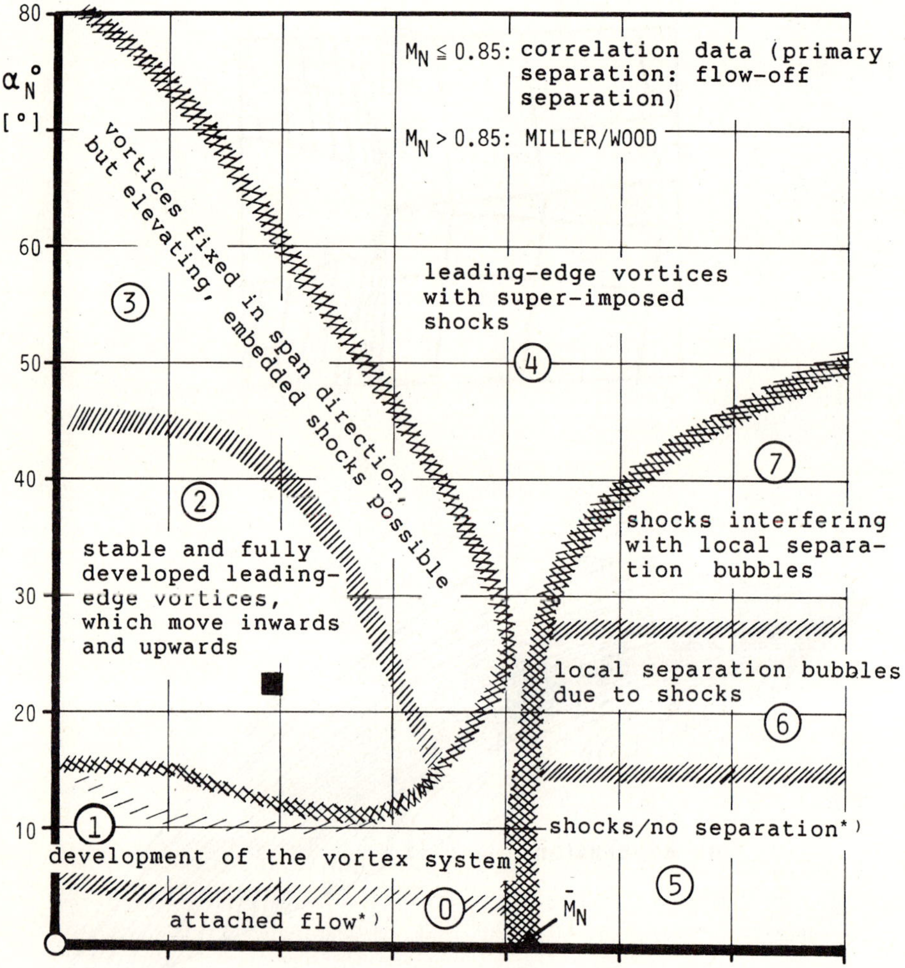

*) weak influence of wing thickness and taper ratio

Fig. 11.44 Flow patterns at slender wings with sharp leading edges and turbulent secondary separation, Ref. 32, ■: data point of vortex-flow model with sharp leading edge at $M_\infty = 0.85$ and $\alpha = 10°$

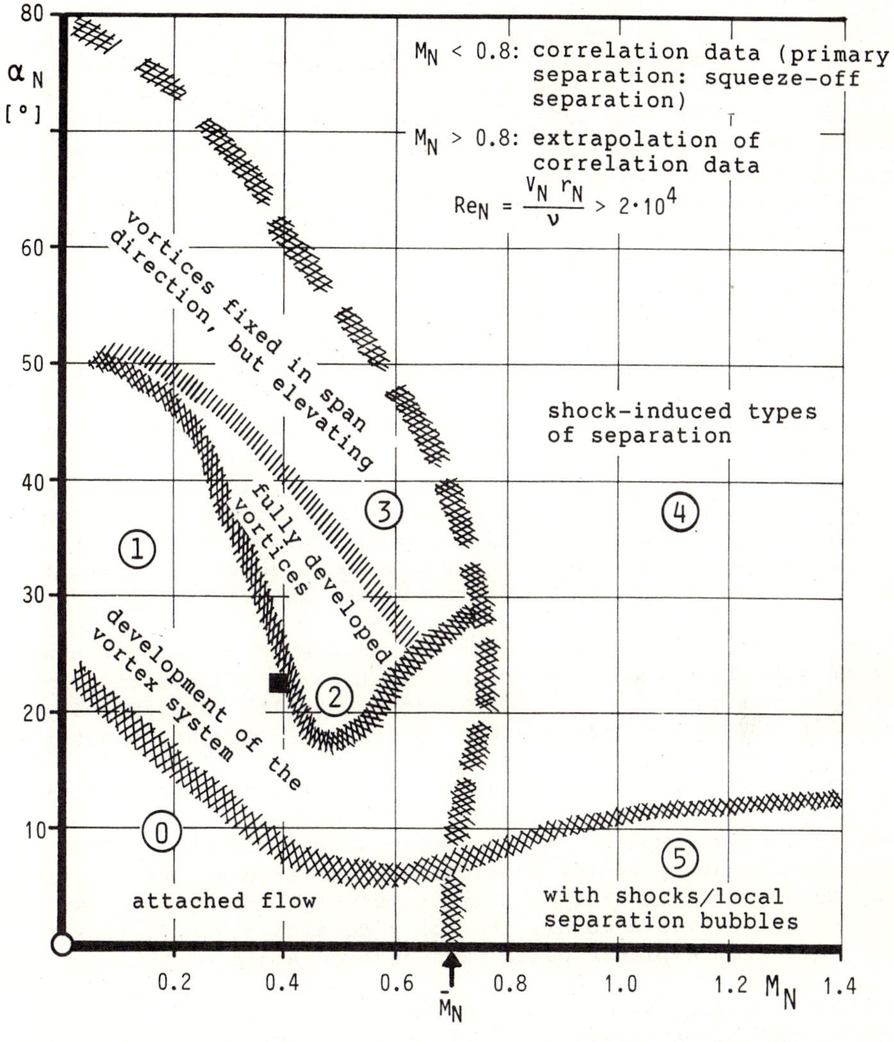

Fig. 11.45 Flow patterns at slender wings with round leading edges and turbulent secondary separation, Ref. 32, ■: data point of vortex-flow model with round leading edge at $M_\infty = 0.85$ and $\alpha = 10°$

11.2.5 Analysis of Flow Around a Project Wing

A cranked delta wing is a suitable one for an advanced military aircraft because it combines the favorable characteristics of a low sweep wing at low speeds with the desirable properties of a slender wing at high speed. But the proper engineering of this concept into a practical airplane wing is a difficult task at every step of the design cycle. Fundamental issues like the vortex pattern over the wing and the steadiness of the flow at off-design conditions are not completely known. The tools that the applied aerodynamics engineer has at his disposal to study these problems are, traditionally, computer simulation by the panel method, measurements made on a scaled model in the wind-tunnel, and, more recently, the Euler equation solver. Each of these three methods, however, provides only limited information. The panel method treats just those flight conditions at which the flow is linear, primarily subsonic and supersonic cruise. Knowledge gained about the character of the flow from wind-tunnel observations is also limited to global force and moment measurements and surface pressure distributions along several chordwise sections. The model typically is small and allows only a small number of pressure orifices along just a few sections. In addition the engineer would like to have more cases at different flow conditions than his budget usually can afford, and with those cases he does obtain he has to worry about lower-than-flight Reynolds-number effects and aeroelastic disturbances.

Keen interest therefore has focused on numerical methods that solve the Euler equations because in the nearly inviscid limit of flight Reynolds numbers this method should, in principle, predict the correct non-viscous non-linear flow throughout the entire envelope of flight conditions, subsonic climb and transonic maneuver. Because the method is relatively new, what is lacking here is full generality in the geometry that can be handled and a complete verification and calibration of the range of accuracy of its prediction. Some attempts at verification do indicate that an Euler simulation does predict with good quantitative accuracy the fully developed vortex flow around a simple planar delta wing at high angle of attack.

Little, however, is known about the simulation for a real project wing, particularly about how well the simulation correlates with the panel method results for attached flow and how well it predicts the transition from linear to non-linear flow as the angle of attack increases.

Fornasier and Rizzi[34] have simulated the flow around the TKF cranked-delta wing of MBB at low and moderate angles of attack in order to study these questions for Mach numbers equal to 0.30, 0.90 and 1.20. They compared the results given by the fully linear higher-order panel method HISSS and the results given by the central difference Euler code[35] with experimental measurements for a range of conditions starting with attached flow and extending to fully developed nonlinear vortex flow.

Globally, good predictions were obtained from both methods at low angle of attack. Euler code results showed a better simulation of the type of flow which develops on the wing upper surface at higher angle of attack and at Mach 0.90. The conclusion reached is that, although a complete agreement with experimental data has not been achieved, the Euler code has the potential of predicting non-linear flow effects. Use of the panel code is, however, still recommendable in the analysis of the linear range of angle of attack because of its still more general geometry modeling capabilities and its lower computational effort requirements.

Here we take a closer look at two of those cases, M=0.9, α=0 and α=6 deg., because they represent the transition from attached flow to the onset of vortex flow. Their analysis sheds some light on the matter of inviscid separation in Euler simulations.

The TKF wing is a interesting test case because it produces a complex flow. The twist and camber distributions of this wing were optimized for best efficiency at attached flow conditions by an inverse panel method technique. Hence, the upper surface flow is expected to evolve smoothly from the design condition to a vortical type flow as the flow separates from the thin leading edge at higher angles of attack. The major objective is to judge whether the Euler results correctly simulate the transition from attached to shed vortex flow.

The Euler results were obtained upon a standard mesh size of 80 cells around the chord, 20 across the span and 20 outward, for a total of 32,000 cells. This relatively coarse model (Fig. 11.46) is deemed to be sufficient for simulating large scale effects. Results on a mesh with increased number of cells in each of the coordinate dimensions (i.e. to 160×80×48) have been obtained on the CYBER 205 eight million word machine but are not presented here. Comparisons of the standard vs. the finer mesh results are reported elsewhere[36-38]. All Euler calculations presented here have been run on the CYBER 205 vector machine located at the CDC facility in Karlruhe, FRG. The final number of iterations at which the calculations were stopped depends on the Mach numbers. Subsonic calculations showed the slowest rate of converence and were therefore iterated up to 3000 cycles. The supersonic case required about 900 cycles to bring the residuals under the prescribed tolerance value. Typical CPU times were 0.785 sec per cylce at subsonic Mach numbers and 0.400 sec/cycle for the supersonic case.

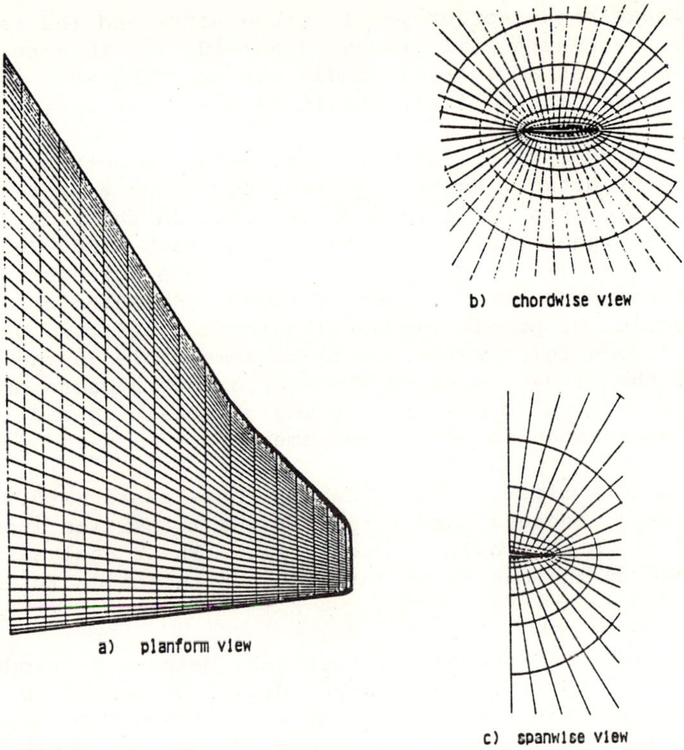

Fig. 11.46 O-O type mesh with 80×20×20 cells used in Euler solutions[34]

Table 11.1 Comparison of global longitudinal characteristics

Mach	α^o	Lift		Drag		Pitching Moment	
		Panel	Euler	Panel	Euler	Panel	Euler
0.90	0	0.021	0.033	0.0003	0.0012	-0.0088	-0.0096
0.90	6	0.347	0.425	0.0304	0.0437	-0.1895	-0.2382

Fornasier and Rizzi[34] compared the predictions given by the two codes for global longitudinal characteristics (Table 11.1). Considering the rather different metric used to define the external surface of the wing (36×10 panels vs. 80×20 cells) the agreement at zero angle of attack is good. As expected, significant variations emerge at angle of attack.

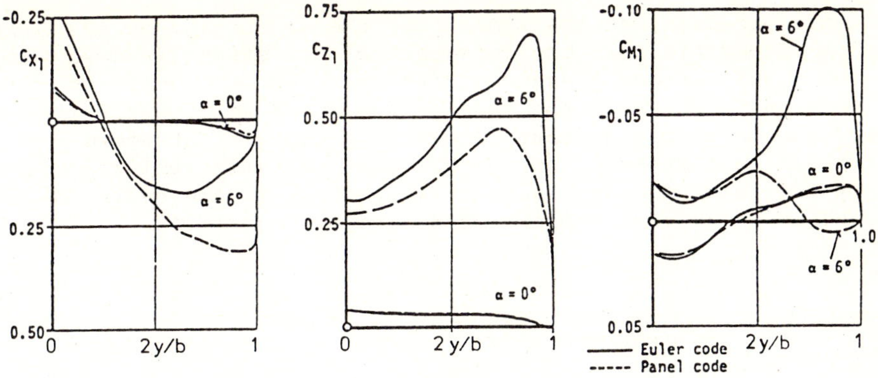

Fig. 11.47 Comparison of local aerodynamic coefficients $M_\infty=0.90$, $\alpha=0^\circ$, $\alpha=6^\circ$, Ref.34

Non-linear lift and pitching moment characteristics are predicted by the Euler code. The Euler code predicts a larger increment in the lift slope than is found in the experimental data. The 17 per cent variation predicted by the panel method is slightly in excess of the value corresponding to the Prandtl-Glauert correction for this low-aspect ratio wing (about 15 per cent). Even more pronounced differences are shown by the comparison of the pitching moment characteristics, where Euler results produces more negative (nose-down) values at angle of attack. The Euler code predicts larger drag values than the panel code.

Spanwise variations of the local aerodynamics coefficients - axial force, normal force and pitching moment about the 25 per cent of the local chord - are compared for all flow conditions in Fig. 11.47. Again, Euler-code and panel-method results show good agreement at zero angle of attack. Differences extend over the whole span at $M_\infty=0.90$, $\alpha=6^\circ$. Most striking discrepancies are the reduction of the suction force and the large downstream shift of the local center of pressure exhibited by Euler results in comparison to panel code predictions.

In order to understand the origins of such discrepancies it is necessary to recall which type of flow develops on this wing at increasing angle of attack and how the two methods simulate this physical flow. When a delta-shaped wing like the present one meets an oncoming stream of air at high angle of attack, the flow separates from the leading edge and under the influence of its own vorticity coils up to form a vortex over the upper surface. The high velocities induced by the vortex create a low pressure region under it which gives the wing a nonlinear lift. At the same time, the separation at the leading edge diminishes the curvature of the streamlines so that the local surface pressure increases. Both these effets contribute to decrease the suction force and to move the center of pressure more downstream over hypothetical attached-flow values. All these phenomena are qualitatively well-reproduced by the Euler code as synthetically shown by the evolution of the upper surface isobars presented in Fig. 11.48.

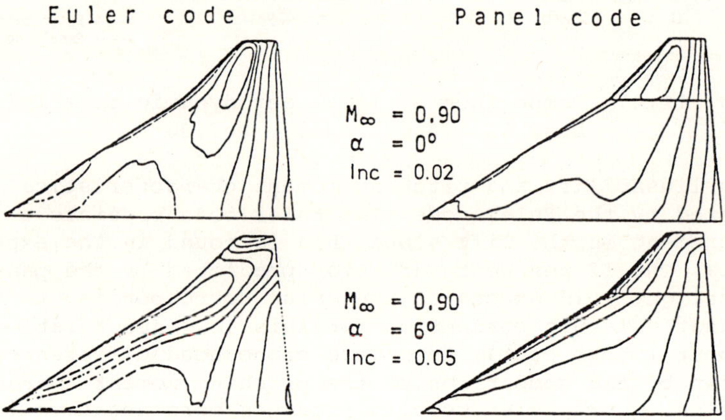

Fig. 11.48 Comparison of upper surface isobars, Ref.34

Already at the intermediate angles of attack $\alpha=6^{\circ}$ at $M_{\infty}=0.90$, the Euler isobars reveal the footprint of a vortex which origins between the apex and the crank. The HISSS-computed isobars - consistent with potential theory - develop the low-pressure peak at the leading edge.

The effects induced by the leading-edge separation vortex are clearly detectable in the comparisons of the longitudinal pressure distributions presented in Fig. 11.49 for the case $M_{\infty}=0.90$, $\alpha=6^{\circ}$. On the lower surface, however, predictions from the two methods are shown to be in good agreement except for a small region close to the wing tip. Here, panel results are influenced by the strong doublet gradient simulating the effect of the tip vortex. Apparently, no trace of this vortex shows up in Euler results. This is probably due to the rounding of the tip geometry which is automatically generated by the mesh construction.

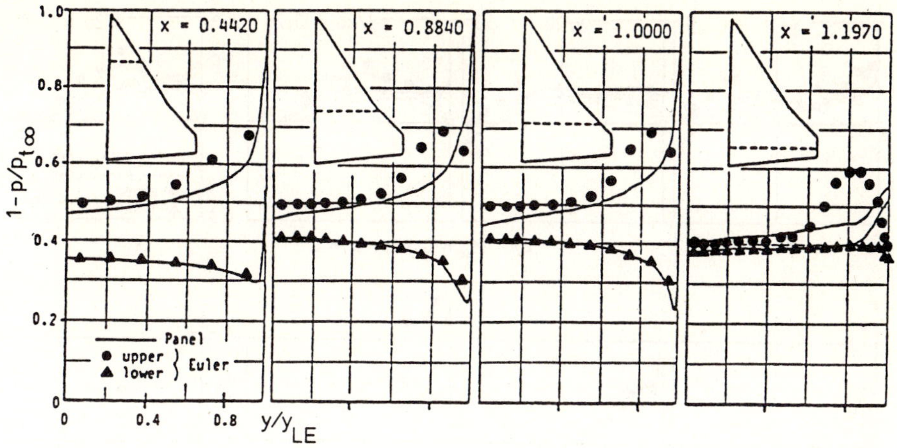

Fig. 11.49 Comparison of spanwise pressure for M_∞=0.90, α=6°, Ref. 34

Experimental pressure data are compared with the calculations at four span stations located at the 25, 50, 75 and 90 per cent of the span, Fig. 11.50. Since the calculations do not simulate the fuselage of the wind tunnel model, body induced effects must be accounted for in the evaluation of the discrepancies between calculations and experiments. The higher velocities induced by blockage and streamline curvature effects due to the presence of the body are more evident at the innermost station at zero angle of attack, but moving out along the span these effects decay rapidly.

At α=0 deg. (Fig. 11.50a), the Euler code follows the experimental trends on both the upper and lower surfaces. The best agreement occurs at y/2b=0.75. The panel code produces values that follow the trend less closely, probably because of the fairly large region of supercritical flow in this problem.

The comparisons at α=6 deg. (Fig. 11.50b) indicate that the discrepancy in the results from all three methods have increased significantly on the upper surface. The lower surface is still reasonable. The Euler results in general agree with the experimental data better than the panel results do. The importance of including non-linear terms directly in the governing equations is rather clear from the comparisons of Fig. 11.50, where HISSS results underpredict the expansion in the supercritical region over the upper surface of the wing at M_∞=0.90.

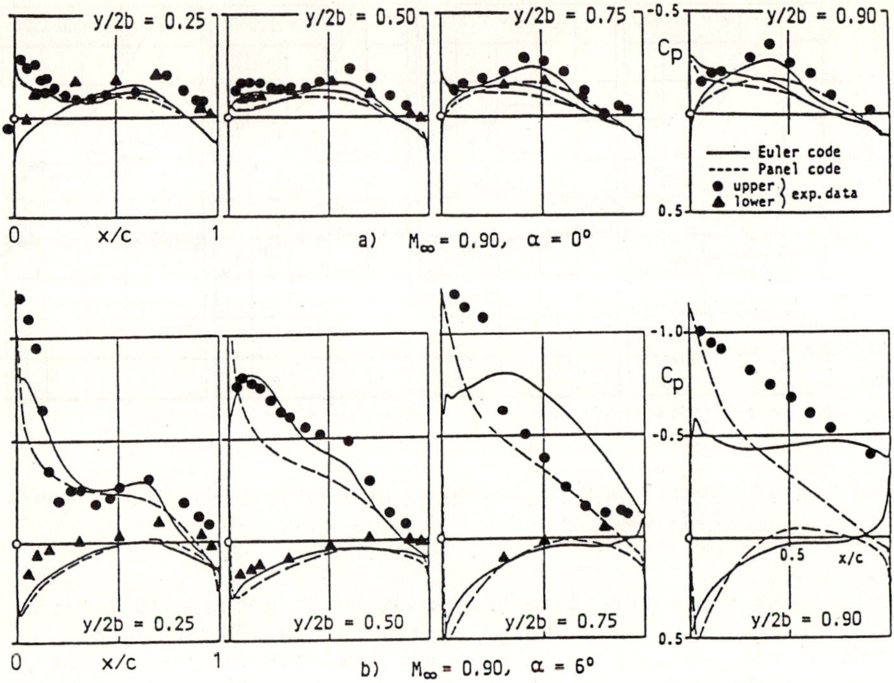

Fig. 11.50 Comparison of chordwise pressure, transonic cases, Ref.34

Predictions from the two codes have been compared with available experimental data. Good agreement has been obtained at low angle of attack. The Euler code thus has been validated at linear flow conditions by an established panel method. Different results were produced by increasing the angle of attack, essentially because of the influence of flow non-linearities over the upper surface of the wing. The panel code, based on linearized potential theory, is unable to simulate the phenomena related to the development of a separated leading edge vortical flow and to the formation of large regions of supercritical flow at transonic speeds. The Euler code proved to be able to predict the essential features of such non-linear effects, but where the flow is in transition from linear to shed vortex flow, the Euler code seems to overpredict the non-linear effect. Possible explanations are either sensitivity to mesh resolution, or the non-realism of inviscid separation at round leading edges.

11.2.6 Flow Through Ducts

The EUFLEX code is applied extensively also to internal flow cases. Here mass-flow balances can easily be made, as well as other, which give an additional check of the accuracy of the solution.

N.C. Bissinger has computed the flow through a typical air duct leading form the supersonic inlet of a fighter to the engine[39]. The axis of the duct is curved. Fig. 11.51 shows isobars in the duct. The quality of the solution is quite good with the relative average mass-flux error $-0.000036 \leq (\dot{m}-\dot{m}_{in})/\dot{m}_{in} \leq 0.000011$, with $-6.9 \cdot 10^{-10}$ at the exit (in ≡ entrance). The total-pressure loss is $-0.0006796 \leq 1-p_t/p_{t_{in}} \leq 0.0003804$, and the total-temperature loss $-0.000179 \leq 1-T_t/T_{t_{in}} \leq 0.000124$.

Another duct flow problem treated by Bissinger[40,41] is that shown in Fig. 11.52 and Fig. 11.53. This scramjet geometry is based on the two-strut design published in Ref.42. It is a perfect test case to demonstrate the capabilities of an Euler code to capture shock surfaces as well as slip surfaces together with their mutual interactions.

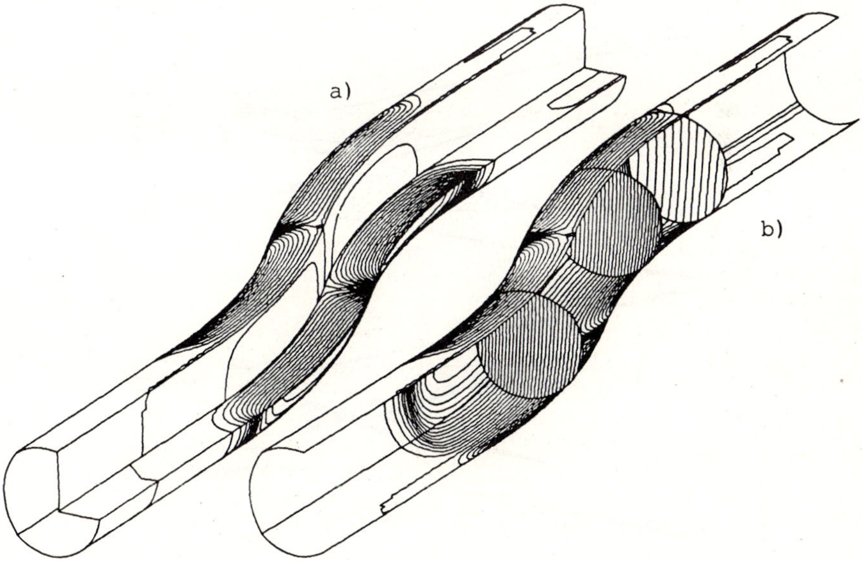

Fig. 11.51 Isobars ($1-p/p_{t_{in}}$) of a duct-flow Euler solution[39], H-grid, 150×15×15 points, $M_{in} \approx 0.6$, $p_{exit}/p_{t_{in}} = 0.784$, a) isobars in longitudinal cuts, b) isobars in cross-sections

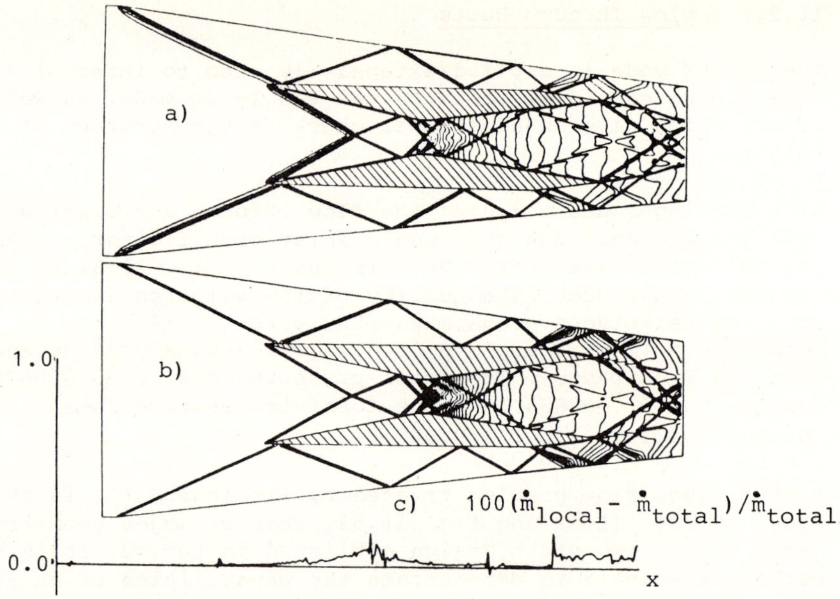

Fig. 11.52 Euler solution[41] for a two-strut scramjet geometry[42] at M_1=3,
a) iso-Mach lines, b) iso-density lines,
c) local relative mass-flux error

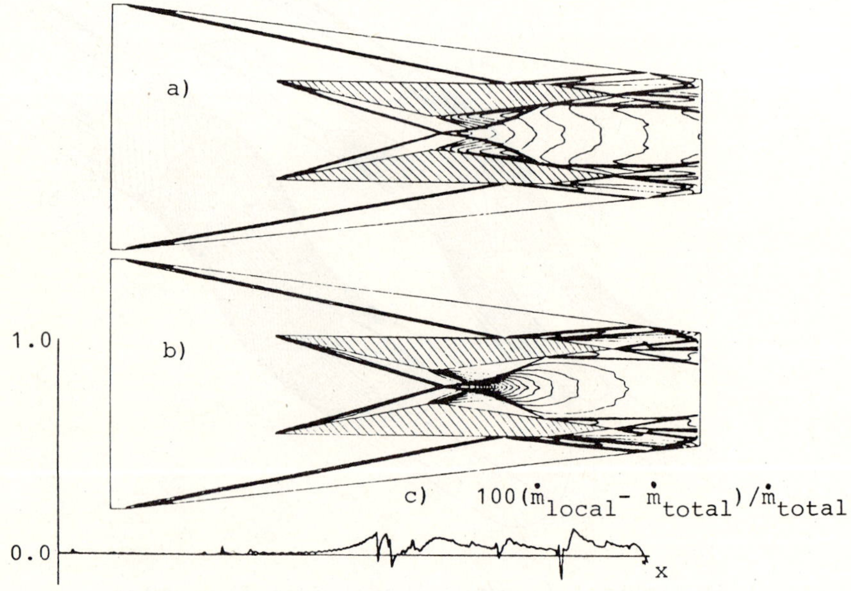

Fig. 11.53 Euler solution[41] like in Fig. 11.52 at M_1=10,
a) iso-Mach lines, b) iso-density lines,
c) local relative mass-flux eror

The results make evident that a careful numerical flux formulation usually leads to much better results than an approach, which takes the code as it is and works, for instance, only on grid refinements, and the like. Although the shocks are far from being aligned with grid lines their capturing by the characteristics based EUFLEX method is very good.

It should be mentioned that in Fig. 11.53 the deformations in the iso-plots are not numerical errors. They always occur when Mach waves develop. In order to make this visible, shaded coloured views of the calculated flow field are necessary, or a tracing of the Mach lines (characteristics) from the computed data.

11.3 Supersonic/Hypersonic Flow

It is generally held that as the Mach number of the flow becomes greater, the shocks become stronger, and a centered scheme with artificial viscosity will no longer work. One must then turn to an upwind scheme. This section presents some results that demonstrate in fact that a centered scheme can function for hypersonic flow. In addition upwind results are presented for some very difficult hypersonic flows.

11.3.1 Leeside Flow Using Centered Scheme

Renewed interest in hypersonic vehicles has motivated the development of methods to numerically simulate such flowfields. The flow about these vehicles has two distinct characters. On the windside the bow shock is intense and the flow is compressed and attached to the body. On the leeside, the bowshock is weak, the flow is thin and generally separated, forming vortex sheets and vortices. Two simple configurations studied here that posess this behaviour are hypersonic flow past a sphere and a delta wing with blunt leading edges (Sub-Chapter 11.3.3). The sphere was choosen as the first suitable test case because a lot of experimental and numerical data are available for comparison. The aim is to predict the bow shock correctly and to model the flow on the leeside where crossflow shocks can induce separation and produce vortices.

The standard approach to inviscid hypersonic flow computations in the 1970's has been the MacCormack scheme together with a bow-shock fitting procedure. Two finite-volume alternatives to this approach are 1) centered flux differences with artificial viscosity and bow-shock fitting, and 2) centered flux differences with artificial viscosity and bow-shock capturing. It has been generally believed that a centered scheme could not capture a strong bow shock satisfactorily. However,

Eliasson and Rizzi[43] show both theoretically and by actual computations that accurate results are obtained for captured shocks that are very strong. Their theoretical argument is based on analysis that shows that the strength of the shock scales with the equations. Hence there should be no strong-shock limitation if the correct artificial viscosity model is implemented. We demonstrate this claim with numerical results for $M_\infty=20$ flow past a sphere which compares well with the shock fitting result. In addition the results on the leeward side of the sphere are discussed in terms of the position of the cross-flow shock and the structure of the wake.

Results are presented for different Mach numbers, M=2,4,8,20. The agreement between numerical and experimental results are very good at the wind side of the sphere. The stand-off distance for the shock and the total pressure loss over the shock are calculated almost exactly compared to theoretical data (Fig. 11.54). We show also that the results on the leeside of the spheroid compare rather well with experiments (Fig. 11.55).

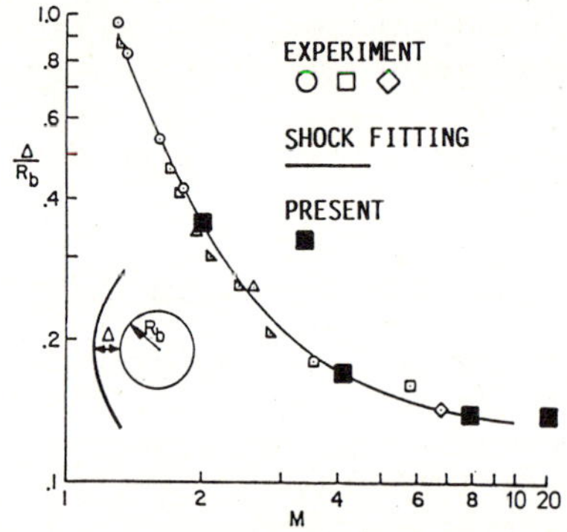

Fig. 11.54 Comparison shock standoff distance experiment vs. computed shock fitting and shock capturing, Ref.43

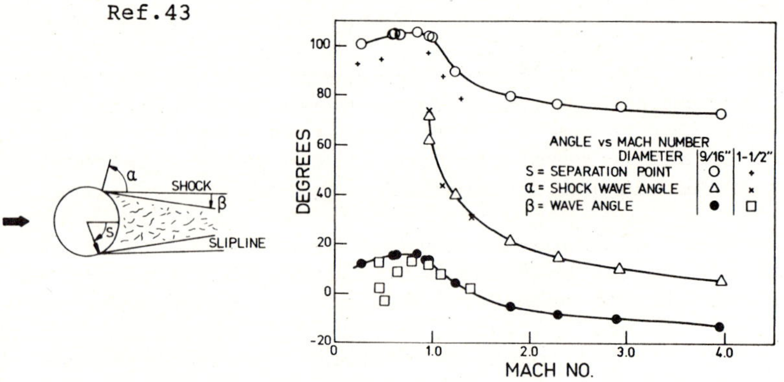

Fig. 11.55 Features of leeside flow. Comparison of measured and computed data, Ref.43

Let us take a closer look at the case $M_\infty=20$. The computed iso-density lines are presented in Fig. 11.56, along with velocity vectors and streamlines in the plane of symmetry. On the windward side these all show a reasonably captured bow shock. Figure 11.54 indicates that the shock-standoff distance is correct. On the leeside the velocity vectors and streamlines show where separation takes place and the extent of the recirculation region. According to Fig. 11.55 above $M_\infty=2$ the point of separation does not change very much with Mach number.

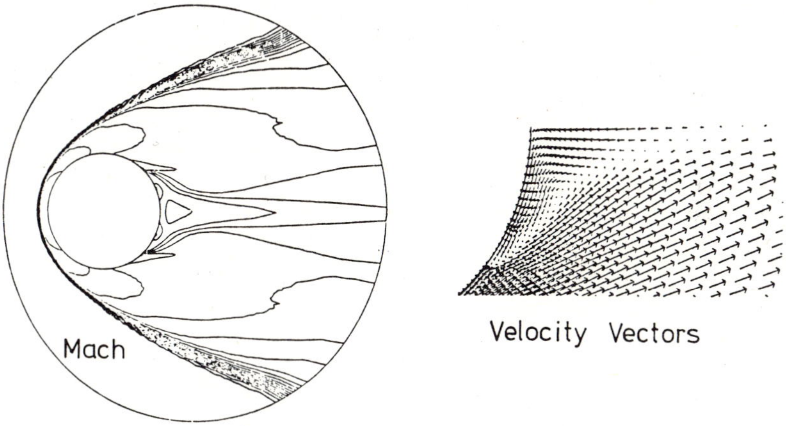

Fig. 11.56 Computed flow past sphere, $M_\infty=20$, $\gamma=1.4$, Ref.43

11.3.2 Computation of Leeside Flow Using an Upwind Scheme

It was mentioned above that upwind schemes often are considered as the only schemes to handle hypersonic flows. In this sub-chapter we don't compare results of an upwind scheme, the EUFLEX scheme with generalized flux vectors[41], with the results given in the previous sub-chapter. Instead we only demonstrate its capabilities to compute leeside flows, and especially its capabilities to compute the very low pressure (and density) there. The ability of this scheme to handle shockes and slip surfaces was demonstrated in Sub-Chapter 11.2.6.

Fig. 11.57 gives results of a computation[41] of the flow past a NACA 0012 airfoil at $M_\infty=30$ and an angle of attack, $\alpha=20^\circ$. The static pressure is nearly zero on the windward side, and the total-temperature conservation (perfect gas assumption) on the windward, and on the leeside is very good.

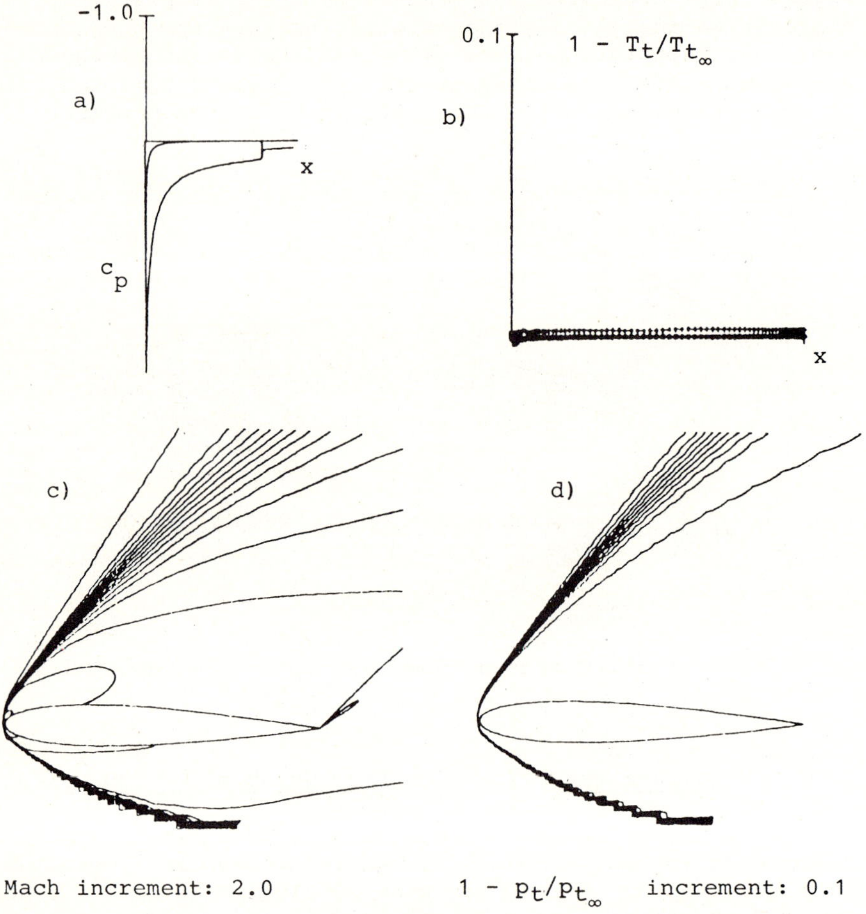

Fig. 11.57 Flow past a NACA 0012 airfoil at M_∞=30 (perfect gas), and α=20° (Ref. 41):

a) static surface pressure,
b) total temperature on surface,
c) iso-Mach lines,
d) iso-total pressure lines

11.3.3 Supersonic Flow Around Delta Wings

A prime motivation to solve the Euler equations is to simulate vorticity-dominated flows. Rizzi[44] used a vectorized center-scheme code for the computation of one such aerodynamic flow of practical interest for the design of a satellite-launch vehicle: high supersonic flow, $M_\infty=3.3$, $\alpha=18^0$ past a sharp-edged 70^0 swept-delta wing. This wing has the same planform as the one discussed in Sub-Chapter 11.2.3, but it now has a maximum thickness of 6 per cent of the local chord.

The mesh used is of O-O type (Fig. 11.58), consisting of 193 nodes around the semispan section, 97 on both the upper and lower chord sections, and 57 from the wing outward to the far boundary, amounting to just over one million. Because of

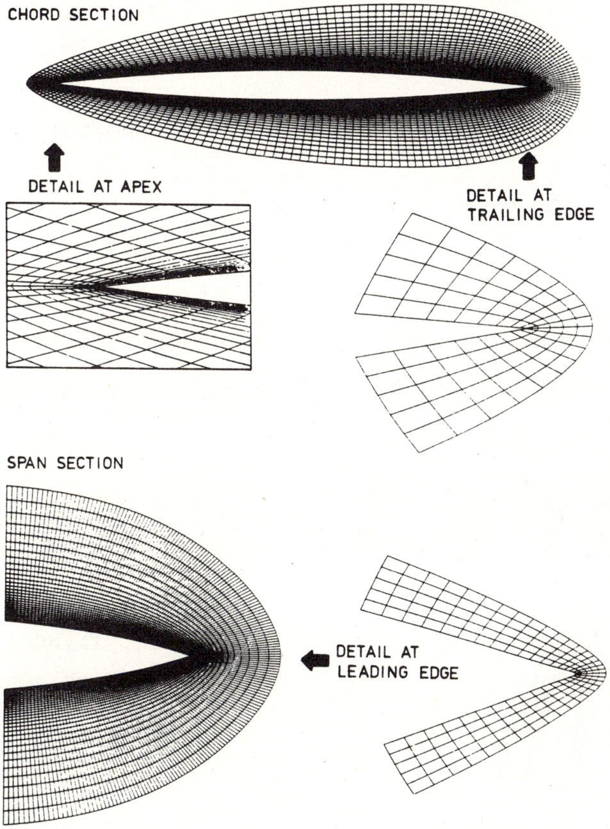

Fig. 11.58 Partial chordwise and spanwise views of the $193\times57\times97$ mesh, Ref.44

the high angle of attack vortex flow is expected, and Fig. 11.59 presents contour lines of computed vorticity magnitude $|\Omega|$ and Mach number in three selected spanwise grid surfaces (nonplanar), that clearly reveal the strong interaction with the leading edge, producing very high shear flow, that then forms cross-flow shocks and vortex structures over the upper surface

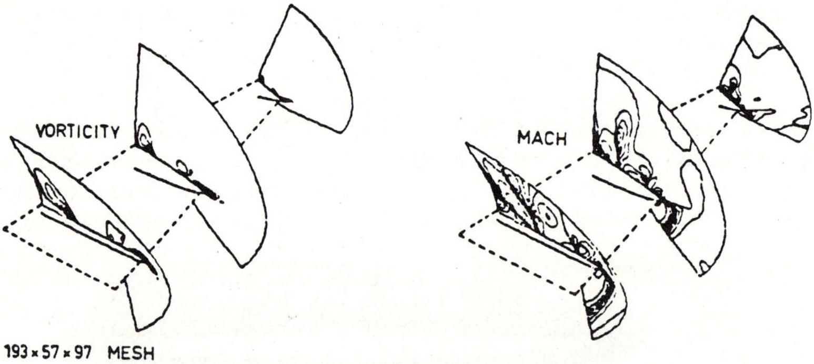

Fig. 11.59 Three-dimensional isograms shown in parallel projection show the bow shock from the windward side interacting strongly at the leading edge with high cross flow velocities, shock waves and vortical flow over the upper surface (Dillner delta-wing $M_\infty=3.3$, $\alpha=18^0$, 197×57×97 mesh), Ref.44

of the wing. Very complex wave interactions take place in the flow over the wing like coalescing waves growing into shocks together with expansion waves that interact with the vorticity produced by the shocks. The iso-Mach lines in Fig. 11.60 on the

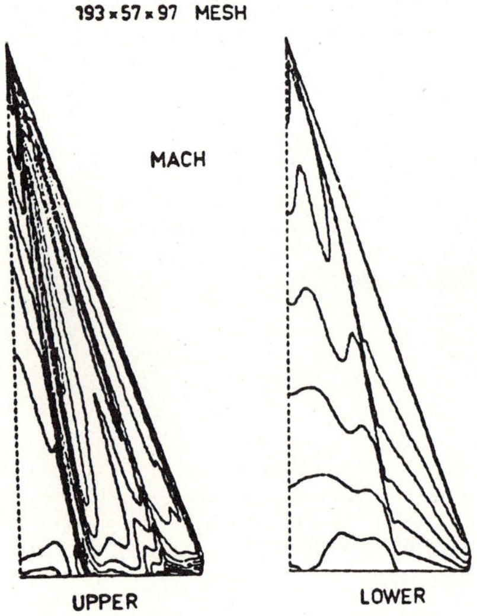

Fig. 11.60

Iso-Mach contours showing the complex wave interactions on the upper and lower surface, Ref. 44

upper surface of the wing show some of these phenomena. A shock wave develops on the lower surface also. Another region worth studying is the wake just behind the trailing edge where a complex interaction between the vortices formed from the leading edges, the shear layers, and the shocks takes place as evidenced by the vorticity and iso-Mach contours just before the trailing edge and in the wake (Fig. 11.61).

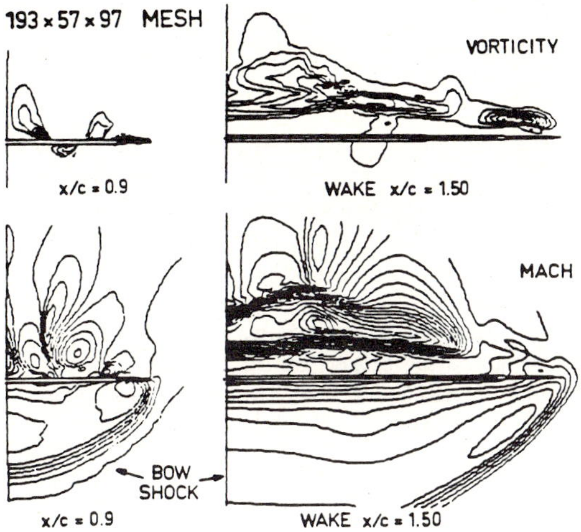

Fig. 11.61

Contours in mesh surface cutting through the wing at x/c=0.9 and the wake at x/c=1.50 show the presence of shocks, shear layers, and vortices, Ref. 44

The convergence history of this case during the course of the computation is given in Fig. 11.62 by plots of log of the residual error, lift and drag all versus the number of time cycles and CPU time on the CYBER 205. The computation begins with the free-stream field on the fine mesh and after 1000 cycles and 8000 seconds the residuals are reduced by practically 3 orders of magnitude for an average reduction rate 0.988. The code executed in 32-bit precision on a two-pipe machine and performed at a rate of 8 CPU microseconds/cycle/ grid point. Since the number of operations per cycle per grid point is something like 1000, this means that the machine sustained a rate of more than 125 megaflops over the entire course of the computation.

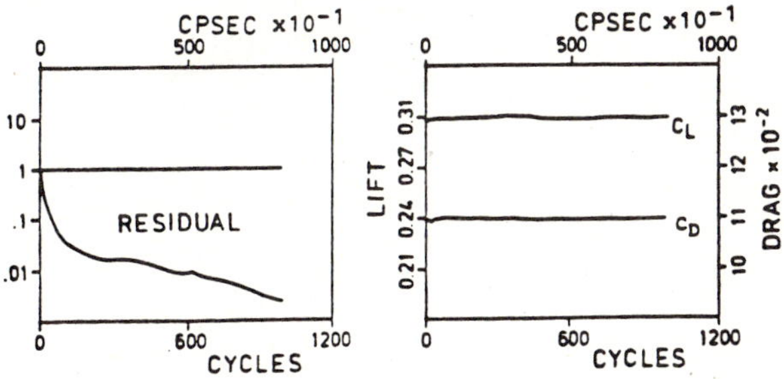

Fig. 11.62

Convergence history versus number of cycles CPU seconds on the CYBER 205. Free stream initial conditions, Ref. 44

Euler solutions are used to compute flow fields where viscous effects can be neglected, or their consequences, vortex sheets and vortices can be modelled. In industrial applications economic reasons are the prime motivation to do so. There are even situations in supersonic flow, where one can step further down in the modelization to the full potential equation, or even further down to the linearized potential equation. In the latter case a panel method, like that prescribed in Ref. 45, is well applicabel, provided the body is sufficiently slender, and no separation effects occur. Even then vortex models are conceivable[9].

To illustrate the use of the full potential equation we show results from Ref.46, where they are compared with results from a solution of the Euler equations. In both cases a space-marching scheme is employed, with shock fitting in the Euler case.

Fig. 11.63 shows the geometry, the Butler wing, which is a circular cone up to 20 per cent, and then is flattened out with elliptical cross sections finally to a straight-lined trailing edge.

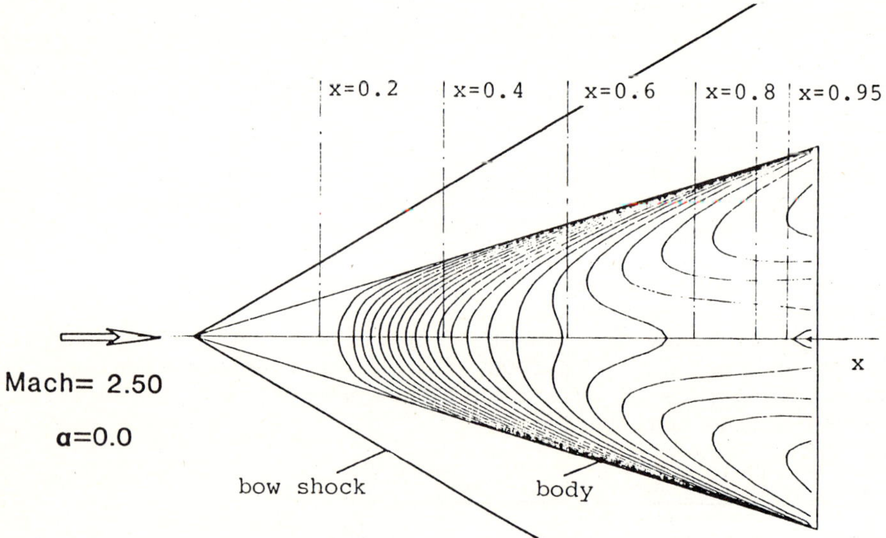

Fig. 11.63 Plan view of the Butler wing with bow shock and surface isobars of Euler solution[46] at $M_\infty=2.5$, $\alpha=0^o$

For the case $M_\infty=2.5$ and $\alpha=5^\circ$ in Fig. 11.64 isobars are given in the cross section x=0.6. Differences are well discernable.

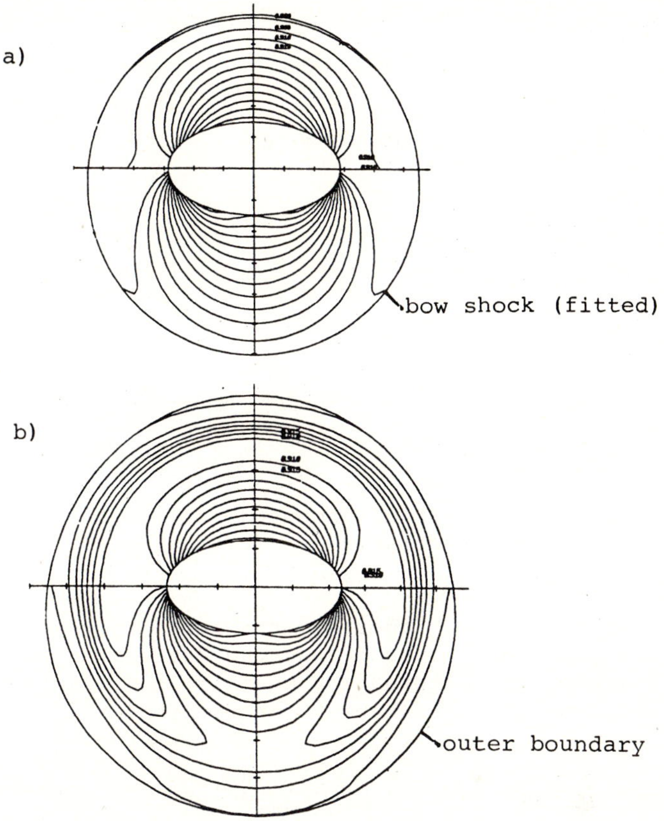

Fig. 11.64 Isobars[46] in the cross section x=0.6 of the Butler wing at $M_\infty=2.5$, $\alpha=5^\circ$,

a) Euler solution,
b) full-potential solution

However, the comparison of the wall pressure in Fig. 11.65 shows a quite good agreement. In both methods 37 grid points were used for the half span in circumferential direction. Away from the surface 17 grid points were used in the Euler solution with bow-shock fitting, and 31 in the potential solution because of the bow-shock capturing.

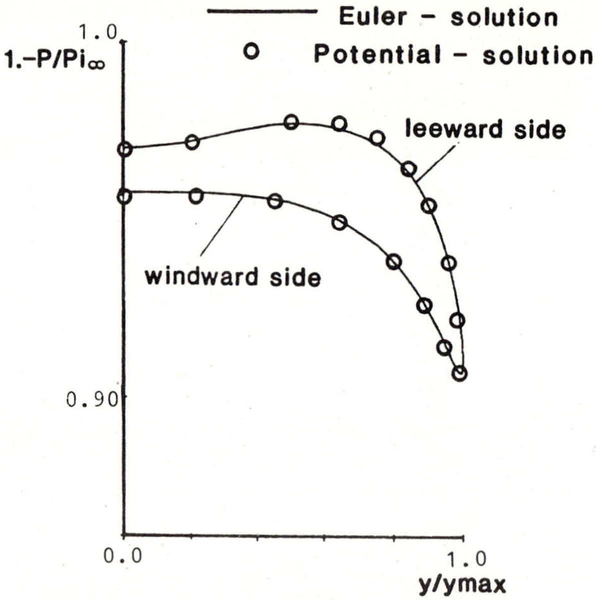

Fig 11.65 Comparison of computed surface pressure[46] at the cross section x=0.6 of the Butler wing at M_∞=2.5, $\alpha=5^\circ$

The Euler method took about four times more computation time than the potential method. Thus the latter can be an interesting economical alternative even if it is limited in application because of the inherent irrotationality.

11.4 Flow Past Complex Configurations

This sub-chapter is intended to give the reader an impression of the power and the possibilities of present-day Euler codes in design work. Details, however, cannot be discussed, because on the one hand too much space would be needed, and on the other hand industrial interests forbid this. Also the cited references are rather laconic in this respect.

11.4.1 Generic Fighter Configuration at Transonic and Supersonic Speed

One of the first large-scale applications of an Euler code to a complete aircraft configuration was made by Eberle, see Refs. 47 to 49. Computed was the flow past a complete configuration with wing, body, canard, V-stabilizer, belly intake, nozzles, and two-engine simulation. Fig. 11.66 gives the half-model with the surface discretization, Fig. 11.66a, and shows the grid extension, Fig. 11.66b, for the lateral motion case. The grid consisted of 57000 cells, which is rather coarse, but which allows the computation of all interesting aerodynamic parameters, except for the moments, with sufficient accuracy for design estimates.

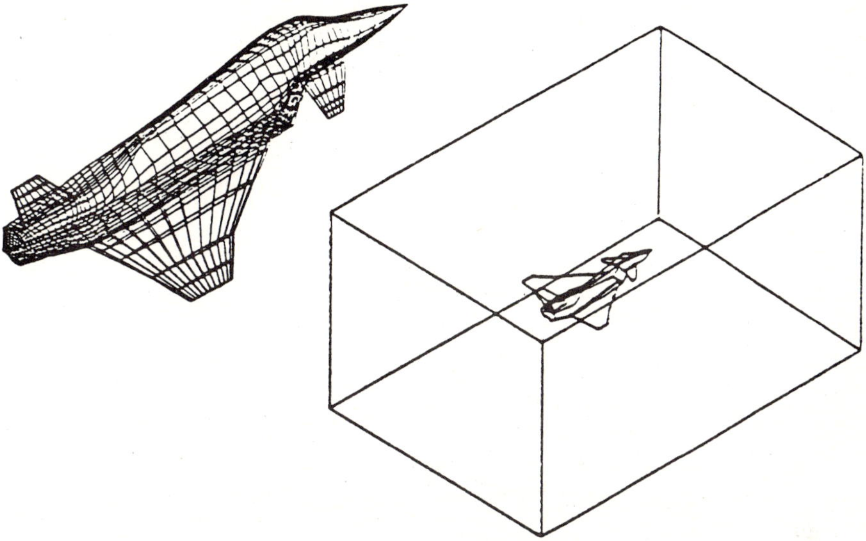

Fig. 11.66 Generic fighter configuration[47],
a) surface grid (half model),
b) grid extension for lateral motion case

Two results from calculations with the then version of EUFLEX are presented in Figs. 11.67 and 11.68 for the transonic case $M_\infty=0.85$ at $\alpha=7.5^\circ$. The intake Mach number was assumed to be $M=0.7$, the static pressure ratio $p_{jet}/p_{intake}=1$, and the total-temperature ratio $T_{jet}/T_{intake}=3$. Other cases computed are with a jaw angle $\beta=5^\circ$, and at supersonic speed $M_\infty=2$. Fig. 11.67 shows the leeside-vortex over the delta wing in a cross-section of the half space and Fig. 11.68 the flow-field vectors in a longitudinal plane cutting the nozzle, with the jet development behind the aircraft.

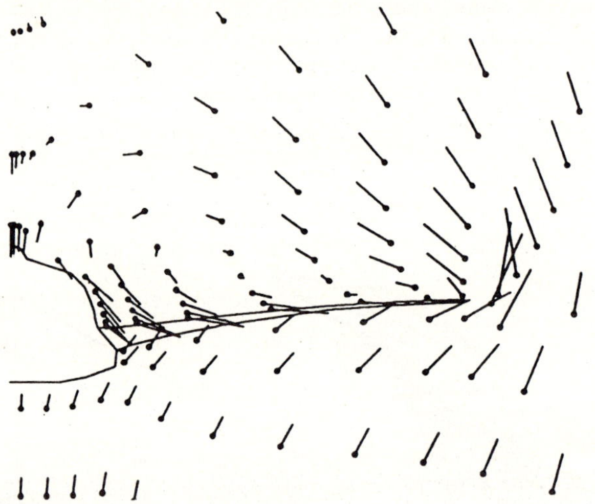

Fig. 11.67 Vortex over the delta wing generic fighter configuration (cross-flow vectors in half space)

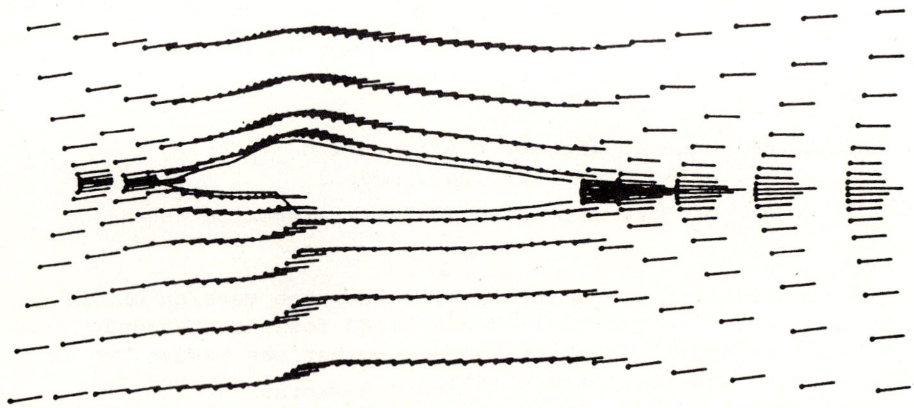

Fig. 11.68 Longitudinal plane cut through flowfield (longitudinal-flow vectors)

11.4.2 Flow Past Hypersonic Generic Aircraft

For the MBB HYPAC, a generic hypersonic configuration, Euler computations were conducted. Fig. 11.69, taken from Ref. 41, shows the surface grid of the configuration, and grid cuts of the 180000 cell grid in several views. For the estimation of moments for bookkeeping purposes, intake and nozzle effects were included, like in the calculations mentioned in Sub-Chapter 11.4.1.

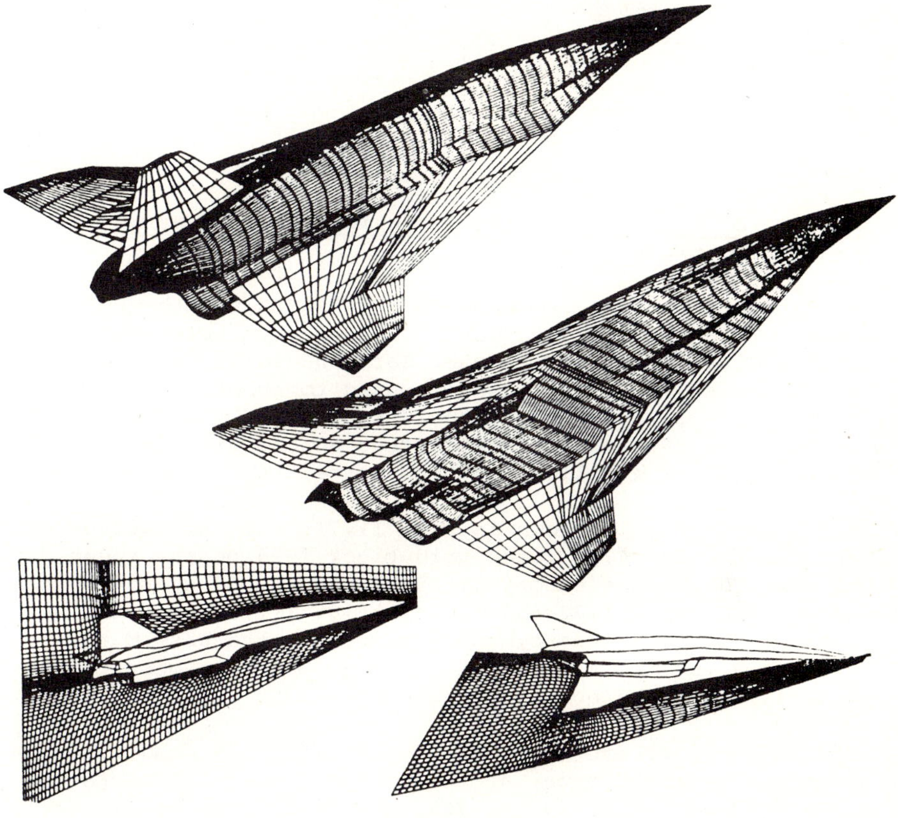

Fig. 11.69 Surface grid and grid cuts of the generic hypersonic aircraft HYPAC[41]

For the case $M_\infty=4.5$ at an angle of attack $\alpha=10^\circ$ only 40 sweeps through the whole computation domain led to convergence because of the very efficient implicit formulation of the EUFLEX-code. In Fig. 11.70 isobars are given on the lower side of the configuration for a combined angle of attack and yaw case. The small jumps of the isobars at the symmetry line are due to pecularities of the grid toplogy, which were not taken into account in the plot routine.

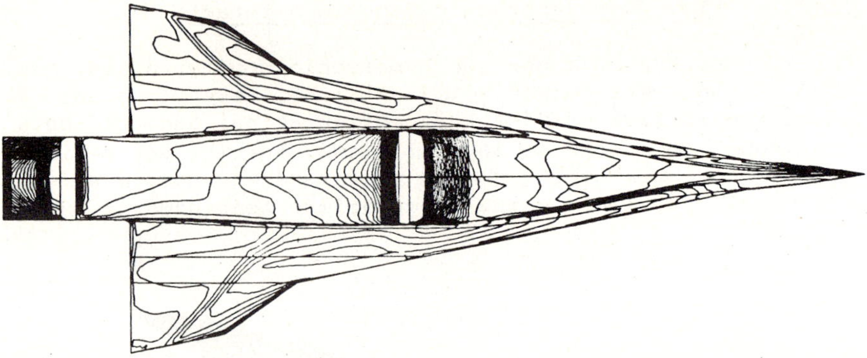

Fig. 11.70 Computed isobars on the lower side of HYPAC, M_∞=5.6, α=10°, β=-5°

11.4.3 Equilibrium Real-Gas Solution for a Reentry Configuration

At MBB contributions are made to the HERMES aerothermodynamics. Illustrating this work, results from Euler calculations with the EULSPLIT-code, Refs. 50 to 52, are given in the following. This code is also used to supply the input for the boundary-layer computations discussed at the end of Sub-Chapter 11.5. Real-gas state-surface approximations, Ref.53 are employed. The following figures are taken from Ref.54.

The surface grid is shown in Fig. 11.71. The off-surface grid fits the bow shock. A fixed grid block lies on the upper part, starting behind the canopy. It does not extend up to the bow shock (see Figs. 11.72 and 11.74). Fig. 11.72 shows the Mach-number contours in the symmetry plane for the case M_∞=8, α=30°, β=0° at the altitude H=42.5 km. Note the very thin shock layer in the nose region.

Fig. 11.73 gives iso-c_p lines on the left side, and iso-Mach lines on the right side, at the windward side of HERMES for M_∞=25, α=30°, β=0°. Fig. 11.73a, and β=5°, Fig. 11.73b, at the altitude H=75 km.

In Fig. 11.74 finally for the yaw case of Fig. 11.73 Mach number isolines are shown in a cross-section close to the base of HERMES.

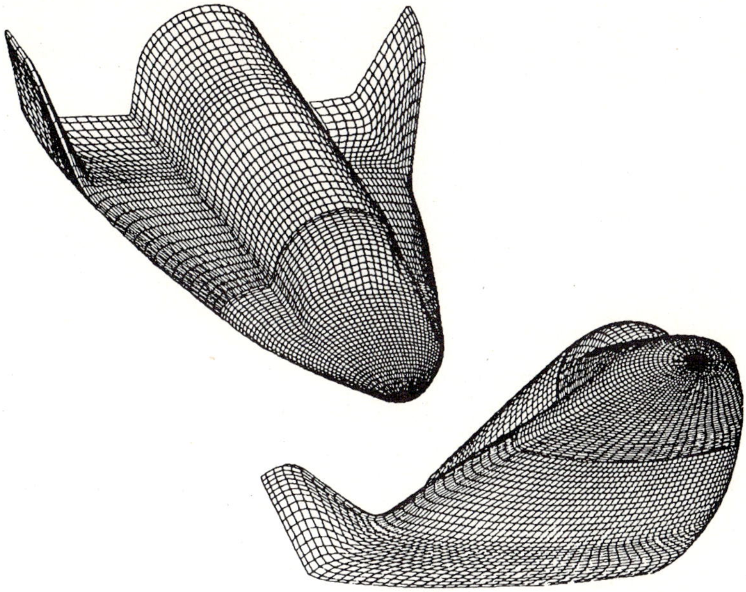

Fig. 11.71 Surface grid views on the HERMES configuration[54]

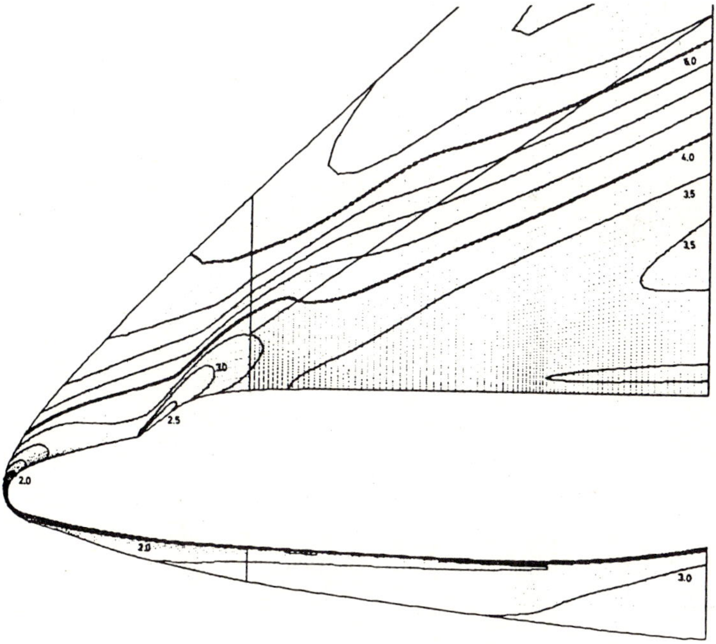

Fig. 11.72 Mach-number contours in the symmetry plane of HERMES[54], $M_\infty=8$, $\alpha=30^\circ$, $\beta=0^\circ$, H=42.5 km, $\Delta M=0.5$

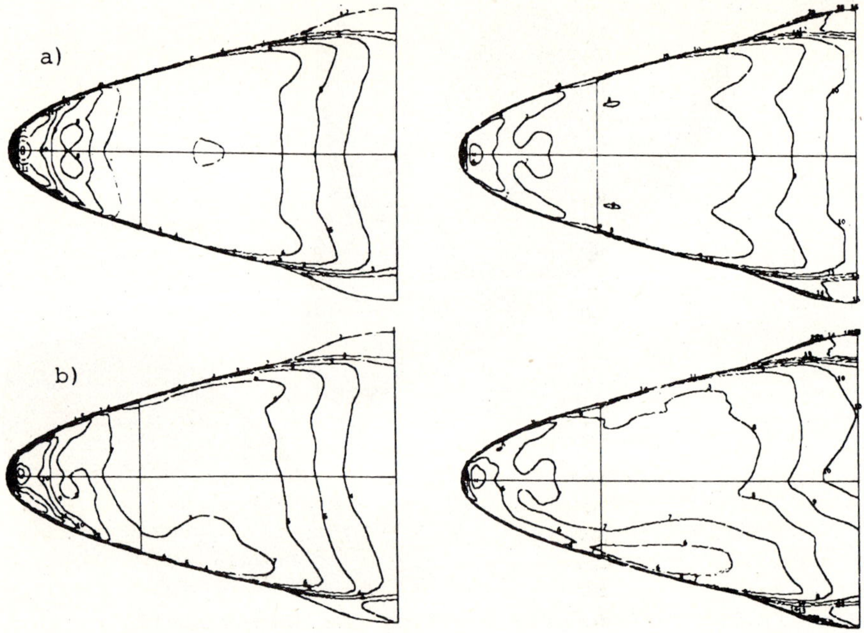

Fig. 11.73 Iso-c_p lines (left side), iso-Mach lines (right side) at the windward side of HERMES[54], $M_\infty = 25$, $\alpha = 30^\circ$, H=75 km, a) $\beta = 0^\circ$, b) $\beta = 5^\circ$, $\Delta c_p = 0.1$, $\Delta M = 0.2$

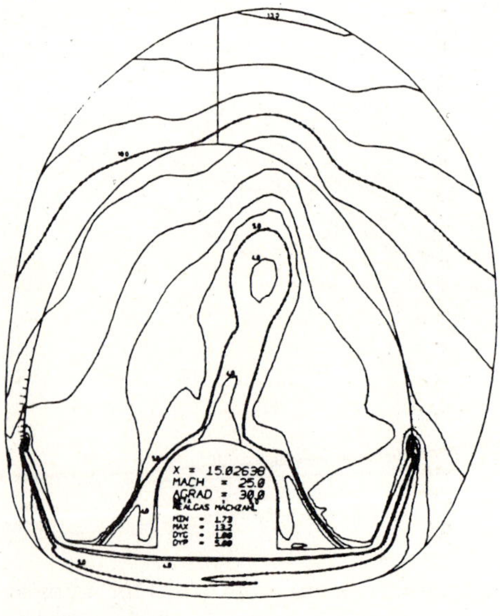

Fig. 11.74 Iso-Mach lines in a cross-section near the base of HERMES[54], yaw case Fig. 11.73b, $\Delta M = 1$

11.5 Coupling with Viscous Models

In Chapter IX the problems of coupling Euler solutions with viscous models were discussed. Here some examples are given. The distinction between weak interaction and strong interaction flow problems leads to two different approaches[55], which are sketched for some basic configurations in Fig. 11.75:

- Coupling of the Euler and the boundary-layer equations where the flow is attached (weak interaction), which alone, for instance on wings, can improve the results by including displacement effects, or which is sufficient if one wants to know, for instance, heat loads in attached flow regions.

- Solution of the Navier-Stokes equations (or other appropriate model) in regions of strong interaction (trailing edge flow, separation, etc.), if the flow past the complete configuration is to be determined. This can be done by zonal coupling with the weak-interaction approach, usually, however, a global Navier-Stokes solution is performed for the whole configuration.

A typical result is shown in Fig. 11.76 for a zonal solution. In the weak-interaction region of the airfoil an Euler solution is performed with the Navier-Stokes code NSFLEX[56,57] in its

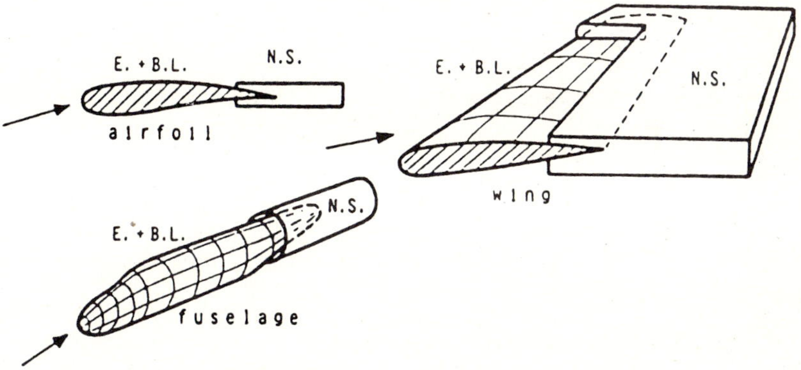

Fig. 11.75 Schematic[55] of solution approaches for some basic configurations (Euler (E.)-, boundary layer (B.L.)-, Navier-Stokes (N.S.) equations)

Euler mode. The boundary layer is computed with the second-order boundary-layer code SOBOL[58]. From the boundary-layer solution the equivalent inviscid source distribution, Sub-Chapter 9.2, is computed and employed in the Euler computation in order to simulate the displacement effect of the boundary layer. Simultaneously the Navier-Stokes solution is coupled in for the strong-interaction region, after boundary data have been constructed at the zonal boundaries.

The complete approach is described in Ref. 59. Except for minor deviations the agreement between the zonal solution and the global solution (NSFLEX-code) is good, and also that with the experimental data. Especially at the zonal boundaries no irregularities appear. This is a good validation of the Euler code, as well as of the other codes, and of the zonal coupling procedure.

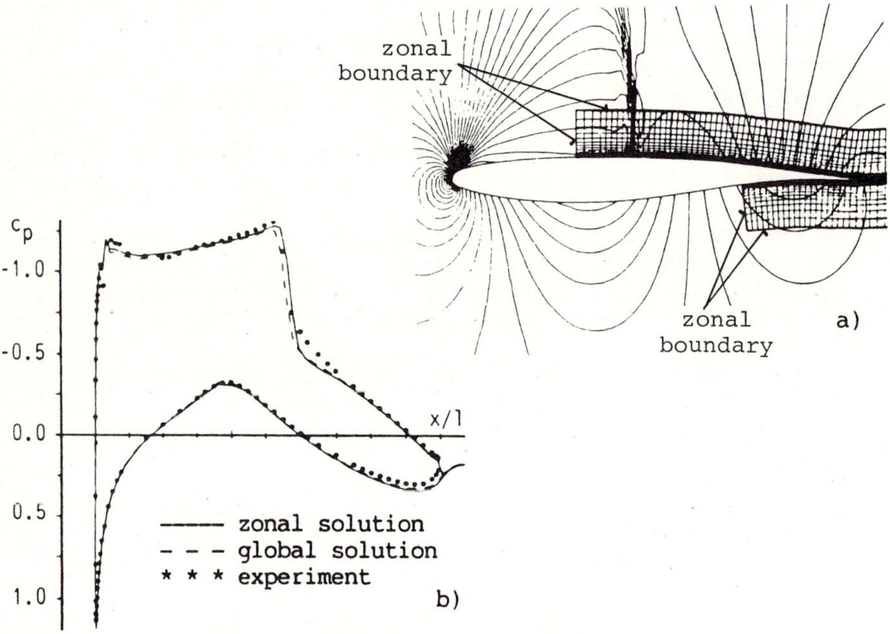

Fig 11.76 Zonal solution[59] for the RAE 2822 airfoil (M_∞=0.73, α=2.79°, Re=6.5·10^6), a) weak and strong interaction regions and computed isobars, b) surface pressure distribution (experimental data Ref. 60)

If should be mentioned, that during the work on the zonal-solution approach it became evident, that first-order boundary-layer theory surprisingly soon starts to loose its validity. Fig. 11.77 shows the static pressure along x^3-coordinate lines in the vicinity of the zonal boundary on an ellipse (1:6), Ref. 59. A first-order boundary-layer solution (first-order mode of SOBOL) of course implies constant pressure in the boundary-layer region, Fig. 11.77b. Although the boundary layer is rather thin at that location, and the surface curvature rather small, second-order effects are appreciable, as the comparison with the second-order result, Fig. 11.77c, and with the global Navier-Stokes result, Fig. 11.77d, shows.

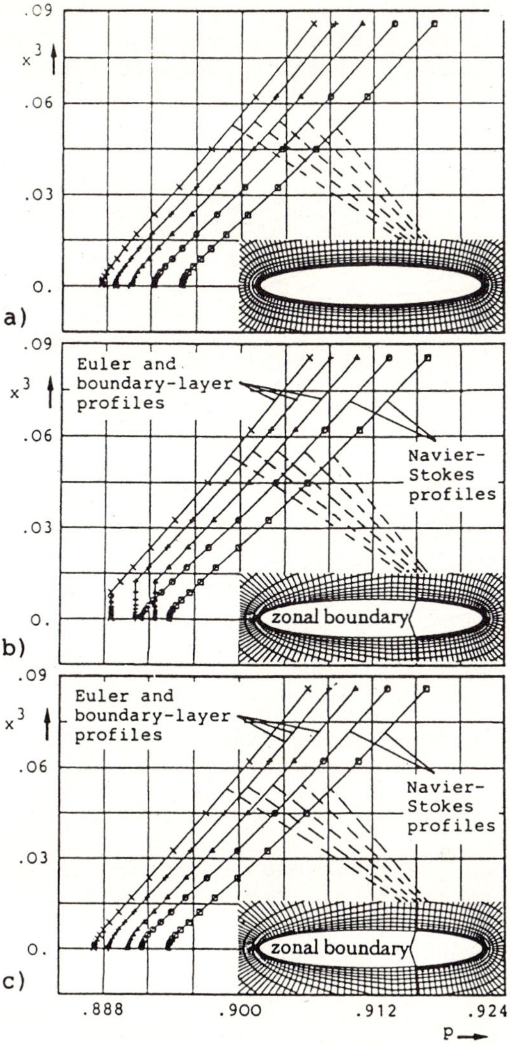

Fig. 11.77 Profiles of static pressure along x^3-coordinate lines in the vicinity of the zonal boundary on a 1:6 ellipse at $M_\infty=0.6$, $Re=10^6$, $\alpha=0^\circ$, transition laminar-turbulent at 10 per cent length (Ref. 61),

a) global Navier-Stokes solution,

zonal solution with:

b) first-order,

c) second-order boundary-layer solution

Without regarding these second-order effects the pressure distribution in the vicinity of the zonal boundary and on the body surface exhibits sizeable irregularities, Fig. 11.78.

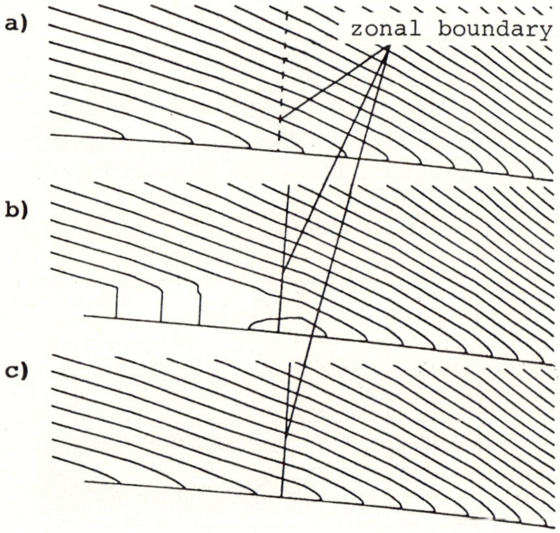

Fig. 11.78 Isobars for the case of Fig. 11.77 in the vicinity of the zonal boundary[61], a) global Navier-Stokes solution; zonal solution with b) first-order, c) second-order boundary-layer solution

Zonal solutions also have been successfully applied to three-dimensional problems[61,62]. Fig. 11.79 shows the zonal geometry for a 1:6 ellipsoid at angle of attack. In Fig. 11.80 the comparison between the zonal solution, the global solution, and the

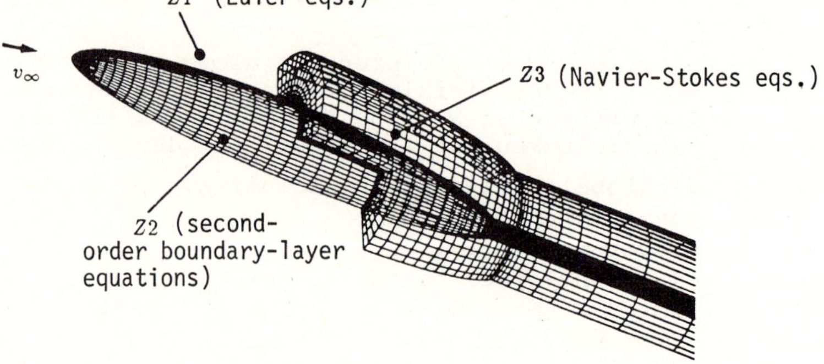

Fig. 11.79 Zonal geometry for a 1:6 ellipsoid at angle of attack, Ref. 61

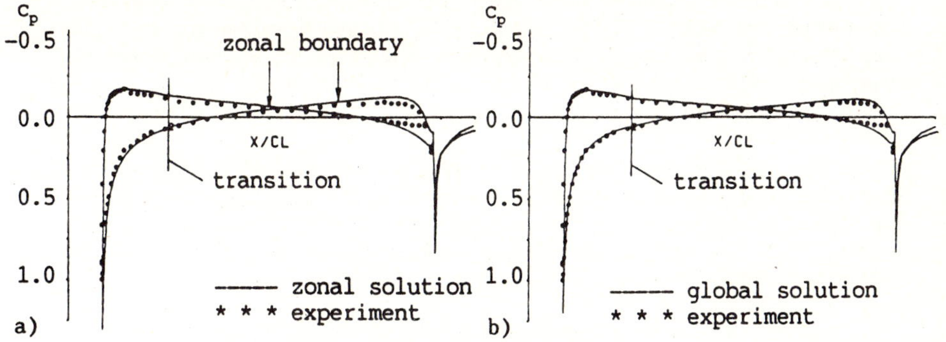

Fig. 11.80 Measured and calculated surface-pressure distribution in the plane of symmetry of a 1:6 ellipsoid at M_∞=0.17, Re=7.7·10^6, α=10°, transition laminar-turbulent at 20 per cent lenght, Ref. 61,

a) zonal solution,
b) global solution, experimental data Ref. 63

experiment exhibits larger differences only at the upper leeward side, where strong three-dimensional separation effects are present. Here of course present-day turbulence models don't allow a better simulation.

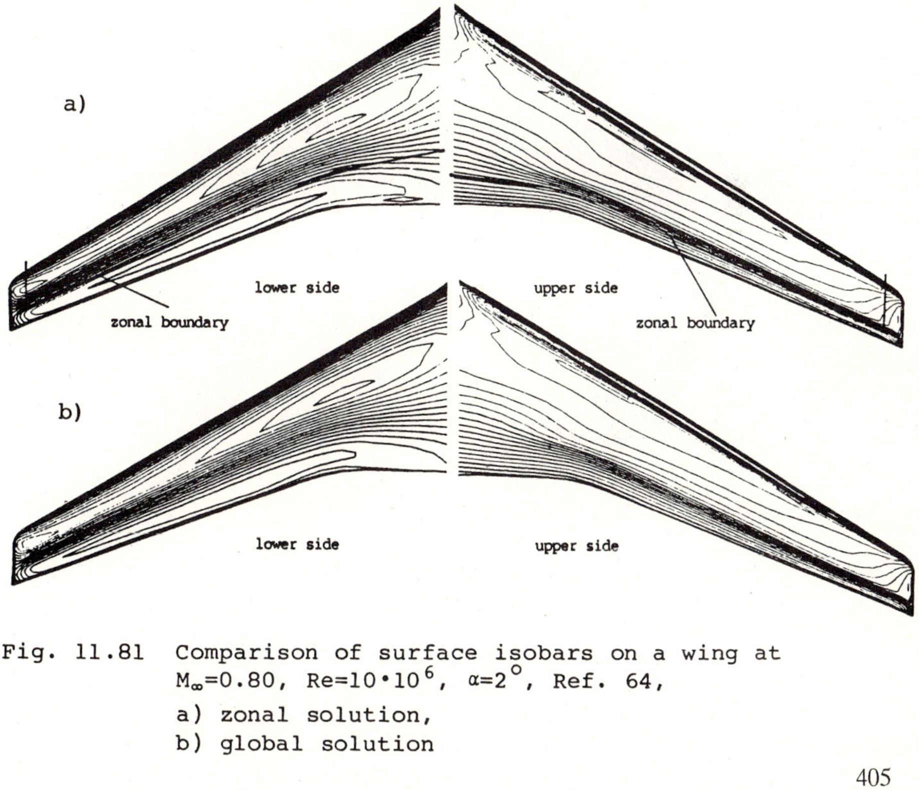

Fig. 11.81 Comparison of surface isobars on a wing at M_∞=0.80, Re=10·10^6, α=2°, Ref. 64,

a) zonal solution,
b) global solution

For a three-dimensional wing case the surface isobars are compared in Fig. 11.81. Again the smooth behaviour of the solution at the zonal boundaries shows that a coupling of viscous models to Euler solutions can be achieved with good results (Ref.64).

Finally two hypersonic flow computation cases are presented. Here second-order boundary-layer theory is mandatory, since because of the low Reynolds numbers thick boundary layers exist, and in addition entropy-layer swallowing must be taken into account.

Fig. 11.82 shows the comparison of an Euler/second-order boundary-layer solution with a Navier-Stokes solution for a hyperbola, taken from Ref. 65. The Euler code employed is a split-matrix algorithm with Runge-Kutta time stepping[50], the boundary-layer code is the SOBOL-code[58], and the Navier-Stokes code the NSFLEX-code[66]. An equilibrium real-gas model was employed with vectorized state-surface approximations[53]. The bow-shock locations (shock fitting in the Euler code, shock capturing in the Navier-Stoke code), as well as the isobars compare well for both the perfect-(ideal-) gas case and the real-gas case. Note that the pressure is not constant in the boundary-layer region. The shock-layer of course is thinner in the real-gas case.

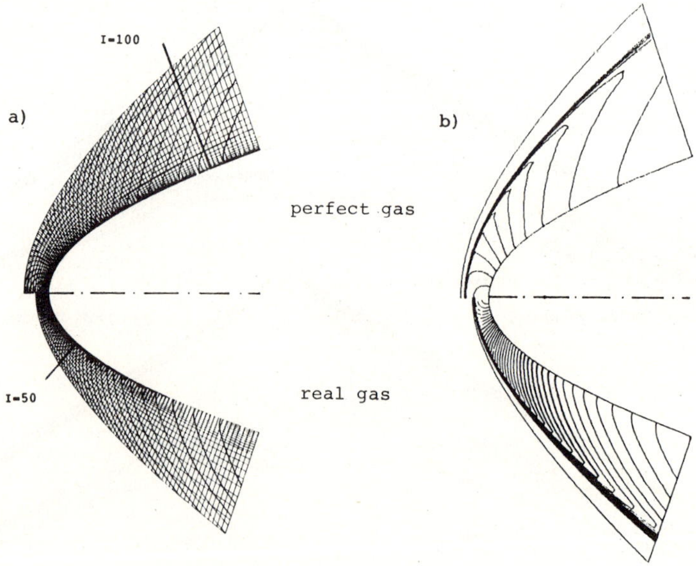

Fig. 11.82 Comparison of a) Euler/second-order boundary-layer result, and b) Navier-Stokes result for a hyperbola, $M_\infty=10$, $Re=2.47 \cdot 10^3$, $T_\infty=220K$, $\rho_\infty=7.7 \cdot 10^4$ kg/m^3, adiabatic wall, laminar flow, $\alpha=0^\circ$ (Ref. 65)

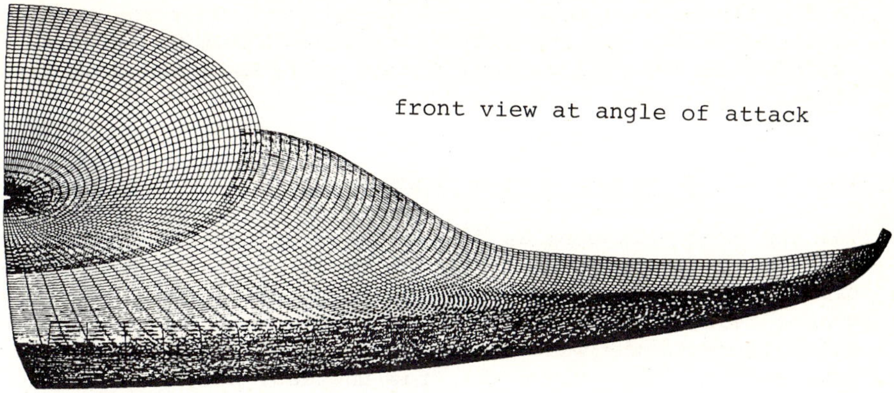

Fig. 11.83 Geometry of the HERMES windward side (Ref. 67)

A three-dimensional result is given for the lower side of the HERMES reentry vehicle, taken from Ref. 67. Fig. 11.83 shows the windward-side geometry. The Euler solution[54] was performed on the locally-monoclinic grid of the boundary-layer solution[58] in order to avoid exessive interpolation work.

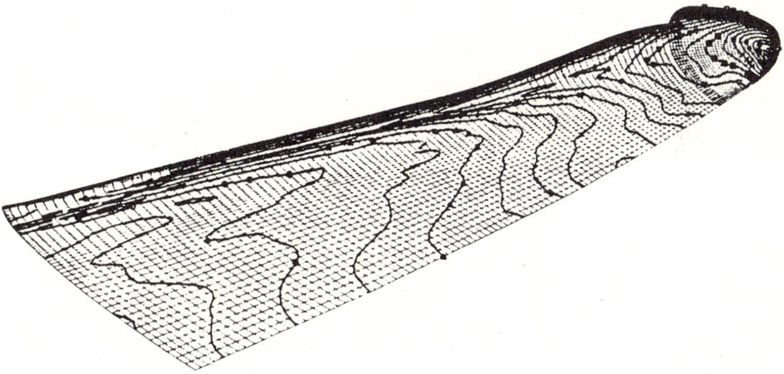

Fig. 11.84 Iso-temperature lines on the HERMES windward side, M_∞=25 at 75 km altitude, laminar flow, radiation-adiabatic wall, ε=0.85, α=30° (Ref. 67)

In Fig. 11.84 the computed iso-temperature lines for the radiation-adiabatic wall condition are given whith an emissivity coefficient $\varepsilon=0.85$. The equilibrium real-gas model from Ref. 53 was employed.

11.6 A Note on Unsteady Applications

The topic of this book is the solution of the Euler equations for steady flow problems. Most of the discussed solution approaches, although quite different from each other, have one common feature: the pseudo-unsteady solution, which asymptotically leads to the wanted steady solution. There are of course many applications, where the true unsteady behaviour of the solution in time is sought. This can be a periodic or a non-periodic behaviour. Again, if viscous effects are negliable, true unsteady Euler solutions are a valiable tool to treat problems in aerodynamics (flutter, control surface motion), in store and weapon separation, and in helicopter aerodynamics.

The work of the first author of this book has spawned developments in this respect. A selection of references is given here in order to allow the reader to get acquainted with these developments. Basic two-dimensional cases were treated in Ref. 68. This work led later to applications to store separation problems. Applications to helicopter problems are reported in Ref. 69, at that time still for the steady hover problem. Refs. 70 and 71 finally give work on two-dimensional and three-dimensional unsteady wing problems, showing very good results as long as viscous effects are negligible.

11.7 References

1. Chorin, A.J.: "A Numerical Method for Solving Incompressible Viscous Problems". J. Comp. Phys., Vol. 2, 1967, pp.12-26.

2. Rizzi, A., Eriksson, L.-E.: "Computation of Inviscid Incompressible Flow with Rotation". J. Fluid Mech., Vol. 153, 1985, pp. 275-312.

3. Rizzi, A.: "Separation Phenomena in 2D and 3D Numerical Solutions of the Euler Equations". In Proc. 7th INRIA Conf. Computation Meth. Appl. Sciences & Engr., R. Glowinski et al. (eds.), North-Holland Publ., Amsterdam 1985, pp. 342-359.

4. Krukow, G.: "Untersuchung des diffusiven Drehungstransportes und Auswirkungen geometrischer Singularitäten auf Lösungen der Eulerschen Bewegungsgleichungen". Diploma Thesis, Technische Universität Münschen, 1985.

5. Eberle, A.: "A New Flux Extrapolating Scheme Solving the Euler Equations for Arbitrary Geometry and Speed". MBB/LKE122/S/PUB 140, 1984.

6. Van Dyke, M.: "An Album of Fluid Motion". The Parabolic Press, Stanford, 1982.

7. Stricker, R., Dick, A.: "Numerical Simulation of Car Aerodynamics and Interior Climate". In: Supercomputer Applications in Automotive Research and Engineering Development, C. Marino (ed.), Cray Research, Minneapolis, 1988, pp. 99-115.

8. Eriksson, L-E. "Calculation of Two-Dimensional Potential Flow Wall Interference for Multicomponent Airfoils in Closed Low Speed Wind Tunnels". FFA TN AU-1116, Part 1, Stockholm, 1975.

9. Hoeijmakers, H.W.M., Vaatstra, W.: "A Higher-Order Panel Method Applied to Vortex-Sheet Roll-Up". AIAA J., Vol. 21, 1973, pp. 516-523.

10. Hoeijmakers, H.W.M., Rizzi, A.: "Vortex-Fitted Potential Solution Compared with Vortex-Captured Euler Solution for Delta Wing with Leading Edge Vortex Separation". AIAA-Paper 84-2144, 1984.

11. Eriksson, L.-E.: Generation of Boundary Conforming Grids Around Wing-Body Configurations using Transfinite Interpolation". AIAA J., Vol. 20, 1982, pp. 1313-1320.

12. Betchov, R.: "On the Curvature and Torsion of an Isolated Vortex Filament". J. Fluid Mech., Vol. 22, 1965, pp. 471-479.

13. Hoeijmakers, H.W.M., Vaatstra, W., Verhaagen, N.G.: On the Vortex Flow over Delta and Double-Delta Wings". J. Aircraft, Vol. 20, 1983, pp. 825-832.

14. Snow, J.T.: "On Inertial Instability as Related to the Multiple-Vortex Phenomenon". J. Atmos. Sci., Vol. 35, 1978, pp. 1660-1677.

15. Rizzi, A.: "Multi-cell Vortices Computed in Large-Scale Difference Solution to the Incompressible Euler Equations". J. Comp. Phys., Vol. 77, No. 1, 1988, pp. 207-220.

16. Powell, K.G., Murman, E., Perez, E., Baron, G.: "Total Pressure Loss in Vortical Solutions of the Conical Euler Equations". AIAA Paper 85-1701, 1985.

17. Saxer, A., Felici, H.: "Etude Numerique d'Ecoulements Internes Incompressibles et Stationnaires par les Equations d'Euler". IMHEF/EPFL Rep. T-87-4, Lausanne, 1987.

18. Eberle, A., Schäfer, O.: "High Order Characteristic Flux Averaging for the Solution of the Euler Equations". Notes on Numerical Fluid Mechanics, Vol. 13, Vieweg Verlag, Braunschweig/Wiesbaden 1986, pp. 78-85.

19. Eberle, A.: "Characteristic Flux Averaging Approach to the Solution of Euler's Equations". VKI Lecture Series 1987-04, 1987.

20. Fluid Dynamics Panel Working Group 07: "Test Cases for Steady Inviscid Transonic and Supersonic Flows". AGARD - AR-211, Paris, 1985.

21. Pandolfi, M.: "Upwind Formulations for the Euler Equations". VKI-Lecture Series 1987-04, 1987.

22. Colella, P., Woodward, P.R.: "The Piecewise-Parabolic Method (PPM) for Gasdynamic Calculations". J. Comp. Physics, Vol. 54, 1984, pp. 174-201.

23. Rizzi, A., Viviand, H. (eds.): "Numerical Methods for the Computation of Inviscid Transonic Flows with Shock Waves". Notes on Numerical Fluid Mechanics, Vol. 3, Vieweg Verlag, Braunschweig/Wiesbaden, 1981.

24. Rizzi, A.: "Spurious Entropy Production and Very Accurate Solutions to the Euler Equations". Aeronautical J., Vol. 31, 1985, pp. 59-71.

25. Rizzi, A., Purcell, C.J.: "On the Computation of Transonic Leading-Edge Vortices Using the Euler Equations". J. Fluid Mech., Vol. 181, 1987, pp. 163-195.

26. Murman, E., Rizzi, A., Powell, K.G.: "High Resolution Solutions of the Euler Equations for Vortex Flows". In: Progr. Supercomputing in CFD (E. Murman, S. Abarbanel, eds.), Birkhäuser, Boston, 1985, pp. 93-114.

27. Elsenaar, A., Eriksson, L.-E. (eds.): "International Vortex-Flow Experiment on Euler-Code Validation". FFA Bromma, 1987, ISBN 91-97, 0914-0-5.

28. Wagner, B., Hitzel, S.M., Schmatz, M.A., Schwarz, W., Hilgenstock, A., Scherr, S.: " Status of CFD Validation on the Vortex Flow Experiment". AGARD CP-437, 1988, pp. 10-1 to 10-10.

29. Schwarz, W.: " IEPG-TA15 Computational Methods in Aerodynamics, MBB Contribution to the Theoretical Program of Phase 2". MBB-FE122-AERO-MT-833, 1989.

30. N.N.: "US/European Vortex-Flow Experiment-Analysis-65° Wing with Sharp Leading Edge, Part II". NLR Memorandum AC-85-006L, Amsterdam, 1985.

31. N.N.: " US/European Vortex-Flow Experiment-Analysis-65° Wing with Round Leading Edge, Test 4026". NLR Memorandum AC-84-025L, Amsterdam, 1984.

32. Staudacher, W.: "Beeinflussung von Vorderkanten-Wirbelsystemen an schlanken Tragflügeln (Influencing Leading-Edge Vortex Systems of Slender Wings)". Doctoral Thesis, University Stuttgart, 1992.

33. Hirschel, E.H.: " Evaluation of Results of Boundary-Layer Calculations with Regard to Design Aerodynamics". In: Computation of Three-Dimensional Boundary Layers Including Separation. AGARD-R-741, 1986, pp. 5-1 to 5-29.

34. Fornasier, L., Rizzi, A.: "Comparisons of Results from a Panel Method and an Euler Code for a Cranked Delta Wing". AIAA-Paper No. 85-4091, 1985.

35. Rizzi, A., Eriksson, L.-E.: "Computation of Flow Around Wings Based on the Euler Equations". J. Fluid Mech., Vol. 148, 1984, pp. 45-71.

36. Rizzi, A., Purcell, C.J., McMurray, J.T.: "Numerical Experiment with Inviscid Vortex-Stretched Flow Around a Cranked Delta Wing, Transonic Speed". In Shear Layer/-Shock-Wave Interactions (J. Delery, ed.), Springer, Berlin, 1986, pp. 283-298.

37. Rizzi, A., Purcell, C.J.: "Numerical Experiment with Inviscid Vortex-Stretched Flow Around a Cranked Delta Wing, Supersonic Speed". Vol. 13, Notes on Numerical Fluid Mechanics, Vieweg, Braunschweig/Wiesbaden, 1989, pp. 310-318.

38. Rizzi, A., Purcell, C.J.: "Numerical Experiment with Inviscid Vortex-Stretched Flow Around a Cranked Delta Wing, Subsonic Speed". AIAA-Paper No. 85-4088, 1985.

39. Bissinger, N.C.: "Berechnung der Strömung in dreidimensionalen Einläufen". MBB-FE122-5-R-1600, 1989.

40. Bissinger, N.C.: "Flow Calculations for Hypersonic Internal Compression Intakes". MBB-FE122-AERO-MT-829, 1988.

41. Eberle, A., Schmatz, M.A., Bissinger, N.C.: "Generalized Flux Vectors for Hypersonic Shock Capturing". AIAA-Paper 90-0390, 1990.

42. Kumar, A.: "Numerical Analysis of the Scramjet-Inlet Flow Field by Using Two-Dimensional Navier-Stokes Equations". NASA Technical Paper 1940, 1981.

43. Eliasson, P., Rizzi, A.: "Hypersonic Leeside Flow Computations Using Centered Schemes for Euler Equations". In Proc. 8th GAMM-Conference on Numerical Methods in Fluid Mechanics, 1989 (P. Wesseling, ed.), Vol. 29 of Notes on Numerical Fluid Mechanics, Vieweg, Braunschweig/Wiesbaden, 1990, pp. 109-118.

44. Rizzi, A.: "Vector Coding the Finite-Volume Procedure for the CYBER 205". Parallel Computing, Vol. 2, 1985, pp. 295-312.

45. Fornasier, L.: "HISSS-a Higher-Order Panel Method for Subsonic and Supersonic Attached Flow About Arbitrary Configurations". In: Panel Methods in Fluid Mechanics with Emphasis on Aerodynamics (J. Ballmann, R. Eppler, W. Hackbusch, eds.), Vol. 21 of Notes on Numerical Fluid Mechanics, Vieweg, Braunschweig/Wiesbaden, 1988, pp. 52-70.

46. Weiland, C.: "A Comparison of Potential- and Euler Methods for the Calculation of 3-D Supersonic Flows Past Wings". In: Proc. of the 5th GAMM-Conference on Numerical Methods in Fluid Mechanics (M. Pandolfi, R. Piva, eds.), Vol. 7 of Notes on Numerical Fluid Mechanics, Vieweg, Braunschweig/Wiesbaden, 1984, pp. 362-369.

47. Eberle, A.: "An Euler Calculation Past a Twin Engine Delta/Canard Fighter Aircraft." MBB-LKE122-AERO-MT-720, 1985.

48. Eberle, A., Misegades, K.: "Euler Solution for a Complete Fighter Aircraft at Sub- and Supersonic Speeds". AGARD CP-412, 1986, pp. 17-1 to 17-12.

49. Eberle, A., Schmatz, M.A., Schäfer, O.: "High-Order Solutions of the Euler Equations by Characteristic Flux Averaging". ICAS-86-1.3.1, 1986.

50. Weiland, C., Pfitzner, M.: "3-D and 2-D Solutions of the Quasi-Conservative Euler Equations". Lecture Notes in Physics No. 264, 1986, pp. 654-659.

51. Pfitzner, M.: "Runge-Kutta Split-Matrix Method for the Simulation of Real-Gas Hypersonic Flows". Vol. 24 of Notes on Numerical Fluid Mechanics, Vieweg, Braunschweig/Wiesbaden, 1986, pp. 489-498.

52. Weiland, C., Pfitzner, M., Hartmann, G.: "Euler Solvers for Hypersonic Aerothermodynamic Problems". Vol. 20 of Notes on Numerical Fluid Mechanics, Vieweg, Braunschweig/ Wiesbaden, 1988, pp. 426-433.

53. Mundt, Ch., Keraus, R., Fischer, J.: "New Accurate Vectorized Approximations of State Surfaces for the Thermodynamic and Transport Properties for Equilibrium Air". ZFW, Vol. 15, No. 3, 1991, pp. 173-184.

54. Pfitzner, M., Hartmann, G.: "HERMES Euler Calculations, Equilibrium, Phase C.1". H-NT-1-0045-MBB, 1989.

55. Hirschel, E.H., Schmatz, M.A.: "Zonal Solutions for Viscous Flow Problems". Vol. 14 of Notes on Numerical Fluid Mechanics, Vieweg, Braunschweig/Wiesbaden, 1986, pp. 118-131.

56. Schmatz, M.A.: "NSFLEX – a Computer Program for the Solution of the Compressible Navier-Stokes Equations". MBB-LKE122-AERO-MT-778, 1987.

57. Schmatz, M.A.: "Three-Dimensional Viscous-Flow Simulations Using an Implicit Relaxation Scheme". Vol. 22 of Notes on Numerical Fluid Mechanics, Vieweg, Braunschweig/Wiesbaden, 1988, pp. 226-243.

58. Monnoyer, F.: "Calculation of Three-Dimensional Attached Viscous Flow on General Configurations with Second-Order Boundary-Layer Theory". ZFW, Vol. 14, No. 1/2, 1990, pp. 95-108.

59. Wanie, K.M., Schmatz, M.A., Monnoyer, F.: "A Close Coupling Procedure for Zonal Solutions of Navier-Stokes, Euler and Boundary-Layer Equations". ZFW, Vol. 11, No. 6, 1987, pp. 347-359.

60. Cook, P.H., McDonald, M.A., Firmin, M.C.P.: "Aerofoil RAE 2822 Pressure Distributions, and Boundary Layer and Wake Measurements". AGARD-AR-138, 1979, A6-1 to A6-77.

61. Wanie, K.M.: "Simulation dreidimensionaler viskoser Strömungen durch lokale Lösung der Navier-Stokesschen Gleichungen". Doctoral Thesis, Technical University München, 1988.

62. Hirschel, E.H., Wanie, K.M.: "Close-Coupled Zonal Solution for Viscous Flow Problems". Vol. 25 of Notes on Numerical Fluid Mechanics, Vieweg, Braunschweig/Wiesbaden, 1989, pp. 197-215.

63. Meier, H.U., Kreplin, H.-P., Landhäusser, A.: "Wall-Pressure Measurements on a 1:6 Prolate Spheroid in the DFVLR 3m×3m Low-Speed Wind Tunnel ($\alpha=10^{\circ}$, $u_{\infty}=55$ m/s, Artificial Transition) - Data Report". DFVLR IB222-86 A04, 1986.

64. Schmatz, M.A., Monnoyer, F., Wanie, K.M.: "Numerical Simulation of Transonic Wing Flows Using a Zonal Euler/Boundary Layer/Navier-Stokes Approach". ICAS-88.4.6.3, 1988.

65. Mundt, Ch., Pfitzner, M., Schmatz, M.A.: "Calculation of Viscous Hypersonic Flows Using a Coupled Euler/Second Order Boundary-Layer Method". Vol. 29 of Notes on Numerical Fluid Mechanics, Vieweg, Braunschweig/Wiesbaden, 1990, pp. 422-433.

66. Schmatz, M.A.: "Hypersonic Three-Dimensional Navier-Stokes Calculations for Equilibrium Gas". AIAA-Paper 89-2183, 1989.

67. Monnoyer, F., Mundt, Ch., Pfitzner, M.: "Calculation of the Hypersonic Viscous Flow Past Reentry Vehicles with an Euler-Boundary Layer Coupling Method". AIAA-Paper 90-0417, 1990.

68. Deslandes, R.: "Eine explizite Methode zur Lösung der Eulergleichungen angewandt auf instationäre ebene Strömung". Doctoral Thesis, Technical University Braunschweig, 1986.

69. Krämer, E., Hertel, J., Wagner, S.: "A Study of the Influence of a Helicopter Rotor Blade on the Following Blades Using Euler Equations". Paper No. 2.6, Fourteenth European Rotorcraft Forum, Milano, 1988.

70. Brenneis, A.: "Berechnung instationärer zwei- und dreidimensionaler Strömungen um Tragflügel mittels eines impliziten Relaxationsverfahrens zur Lösung der Eulergleichungen". Doctoral Thesis, University of the Armed Forces, München. VDI-Fortschrittberichte, Reihe 7, Strömungstechnik, Nr. 165, 1989.

71. Brenneis, A., Eberle, A.: "Application of an Implicit Relaxation Method Solving the Euler Equations for Time-Accurate Unsteady Problems". Journal of Fluids Engineering, Vol. 112, No. 4, 1990, pp. 510-520.

XII FUTURE PROSPECTS

This chapter closes the book with some considerations of future prospects of the numerical solution of the Euler equations. It is of course not only academic interest in Euler solutions to do so. It is also and especially the use of Euler solutions in industrial design work. The development of Navier-Stokes solvers is very welcome, but those do not meet every demand in design work. Because of cost considerations in aerodynamic design work, always a whole palette of modelization levels and hence computation methods will be employed. There the cheapest method will always be selected which just gives the requested answer with the desired accuracy. The answers sought are very different in the different stages of a design process.

In the following sub-chapters first some objectives of future research on numerical Euler solutions are sketched. Then possible developments, which lead beyond dimensional splitting, are discussed, followed by notes on finite-element formulations. The chapter is closed by thoughts about geometric complexity and possible ways to handle it, and a consideration of interdisciplinary problems.

12.1 General Considerations

The discrete numerical methods for the solution of the Euler equations have reached a considerable level within the last decade. They have matured to a degree that allows their application in industrial aerodynamic design work. Certainly, however, they have not reached a stage where future work on them has become superfluous.

Experience especially from industrial applications teaches the necessity of extended future research work in at least the following main areas:

- General understanding of discrete numerical solutions of the Euler equations. The classical theory of inviscid flows with embedded vortex sheets, vortices, and shock surfaces considers these phenomena as discontinuities. Discrete numerical solutions due to the numerical viscosity at best handle them in a way which resembles reality, and in many cases it can be shown that in principle this is right (see for instance Chapter X with regard to vortex sheets and vortices). What is needed in much more depth is what can be called a Theory of Discrete Euler Flows, which bridges the gap between the ideal Euler flow, which certainly borrows much from potential-flow theory, and viscous reality.

We have seen how both upwind and central-difference schemes contain numerical viscosity. It is the details of the numerical viscosity that distinguishes one method from another.

Because of the decomposition process in upwind schemes, each component field of the solution receives a carefully prescribed amount of artificial viscosity. The common practice with central schemes is to apply some higher difference formula to each equation of the system.

In general more research needs to be done in order to advance the understanding of the effect of these differences beyond the current level (e.g. Ref. 1).

A word is also in order here about terminology. It is common, and we do so also, to refer to these models as "artificial viscosity". Strictly speaking the term "viscosity" is somewhat ambiguous. It is more precise mathematically to call it "artificial diffusion" because the second (and higher, even,) differences allow the mass, momentum, and energy to diffuse across streamlines, something which does not occur in the Euler equations. It is, for example, only by this diffusion mechanism that a wake in an Euler simulation can thicken up as we have seen in Chapter X. It is also this process that helps transients, particularly highly spatially varying ones, to diffuse out of the solution, i.e. the solution displays a parabolic behavior.

Further developments in the "artificial diffusion" models might be motivated by the physical viscous process. For example, models might be developed with diffusion terms that truly mimic the effect of dissipation in the energy equation, which has a physical basis, and none at all in the mass equation. Only time will tell if such considerations lead to improved models.

- Reliability and effectiveness of Euler solvers. Still much too often Euler solutions exhibit unwanted, and even outright wrong characteristics. The reasons are partly to be found among the above discussed phenomena, which are present in the flow field, but which cannot be described sufficiently with the given code, or with the given, not suitable discretization of the computation domain. It can be imagined that more flexible codes together with self-adaptive discretization techniques, which identify the appearance of discontinuities, can help to overcome such problems.

In supersonic and hypersonic flow fields combinations of time-marching and space-marching techniques will enhance effectiveness. It should be mentioned that the question when to use bow-shock fitting or bow-shock capturing in such cases has not been answered definitely. Of course embedded shock surfaces should be captured, especially when self-adaptive discretization is employed. Similarly it has not been answered whether multi-grid approaches have the potential with general hyperbolic problems, which they have with scalar elliptic problems in regular domains.

- Validation efforts. Because "inviscid" experiments cannot be conducted, no inviscid experimental data base can be imagined for validation purposes. Validation efforts therefore must be conducted with help of other means. Comparisons of codes on common grids, check of flow properties (conservation of total pressure (global, piecewise), total enthalpy, mass, momentum and energy flux balances, balances of kinematically active vorticity or circulation, etc.), and check of compatibility conditions can help to validate Euler solutions. Of course, results of high Reynolds number experiments, with negligible kinematically inactive vorticity, have been employed successfully in validation efforts. Such approaches should be pursued further, because they also provide knowledge about aerodynamic design problems, where Euler solutions effectively can be used. Zonal solutions, where boundary-layer and Navier-Stokes codes are coupled in appropriate flow regions to an Euler code, indirectly also help to validate Euler codes. Such techniques, which have a high cost-saving potential, certainly deserve further development.

12.2 Beyond Dimensional Splitting

The concept of splitting a multidimensional difference operator into a series of single-direction operators has a long history. Perhaps its oldest usage was by Peaceman and Rachford[2] in 1955 when they used it to accelerate the convergence of matrix inversion.

An implicit difference scheme in two space dimensions leads to a matrix equation to be solved at the advanced time level. If, for example, it is a second-order-accurate method, the matrix is penta-diagonal, and the equation is cumbersome to solve.

Accordingly alternating-direction-implicit (ADI) methods were introduced which (in this case) are two-step methods involving the solution of tridiagonal matrix equations along lines parallel to the two coordinate directions at the first and second steps respectively. It can be seen as a factorization of the right-hand side operator in a product of one-dimensional operators. A class of schemes, called approximate-factorization methods, have evolved from this concept for treating parabolic and hyperbolic equations.

A related idea was advanced by Russian numerical analysts[3,4], called locally one-dimensional (LOD) or time-splitting[5] methods for solving time dependent partial differential equations in several space dimensions. It appears most commonly with explicit difference schemes and replaces the original difference equation with a series of time-dependent equations in one space dimension. Originally the goal of matching the numerical and physical domains of dependence more accurately is what motivated MacCormack to adopt this approach[5]. While this is an advantage, its major drawback is that boundary conditions have to be specified on the variables during the intermediate fractional steps, and these are in general difficult to determine accurately. Today the method by and large is not used in this form.

Currently the concept is invoked in upwind schemes where the state at a cell face at the new time has to be determined from the known values at the old time level. One way of doing it is to solve the Riemann problem for the new level given the two old states on each side of the cell face. Multidimensional Riemann problems are virtually intractable, so the standard recourse is to approximate the process as one-dimensional wave propagation normal to the cell face. This introduces a mesh dependency in the definition of the upwind properties which can produce errors of unknown magnitude.

Recently several researchers are working towards the definition of convection algorithms for multidimensional flows which are based on the local flow structure and independent of the local mesh orientation. Hirsch et al.[6] analyze the structure of the multidimensional Euler equations and define particular propagation directions which lead to maximal decoupling of the system of equations.

Roe[7] instead decomposes a 2D flow disturbance into eight simple waves that propagate in specific directions defined by certain combinations of velocity gradients. These simple waves then are discretized in an upwind manner observing the associated wave speed.

It remains to be seen whether the improved accuracy warrants the increased work[8].

12.3 Finite-Element Formulations

Traditionally the approach in computational fluid dynamics has been with finite difference or finite volume methods. During the past several years much progress has been made with finite element methods.

Correspondences and relationships have been found linking the discretizations schemes in each of the three methods. The main distinction lies in the way the mesh is ordered, and one can actually classify the methods accordingly. Finite difference methods are usually associated with structured grids that are based on some coordinate mapping (also called tensor-product grids or Cartesian grids). On the other hand finite element methods are associated with unstructured grids that have no underlying coordinate system. The grids are a set of unordered points that then are linked as triangles in 2D and tetrahedra in 3D. With sufficient methods for linking the set of points, this method has a powerful advantage for tackling complex geometry. It also has the further advantage that the grid can be refined adaptively in a very natural manner. The prospects look very bright for this approach in the future.

The current status of development in this area can be glimpsed in the references below. They are by no means a survey of the field, because this is not the intention here.

Jameson and Baker start with the finite-volume formulation, apply the Galerkin technique, and obtain a scheme that loops over cells faces to calculate the flux[9,10]. Combined with a triangulation procedure[11] for the point set, this approach has seen good success for treating very realistic airplane configurations.

Another approach is the Taylor-Galerkin scheme implemented and applied to a large class of problems by Löhner and Coworkers[12,13]. Included in these are transient problems, and Löhner[14] has developed a very powerful adaptive mesh technique based on the advancing front idea.

A third approach differs somewhat from these two in that Shapiro and Murman[15,16] work with quadrilaterals instead of triangles, and the Galerkin algorithm uses biquadratic elements. It also allows adaptive meshing, and follows the idea of directional embedding[17]. Jameson[18], and Shapiro and Murman[16] show that the Galerkin finite-element discretization in space is equivalent to a finite-volume method.

Progress on these three fronts, as well as on others, is sure to develop rapidly in the near future.

12.4 Geometric Complexity

If one of the ultimate goals of the development of Euler solvers is their application in the aerodynamic design process, the flow past complex configurations, at least in airplane design, has to stand in the center of attention.

Geometric complexity, however, also means complex flow patterns, which appear especially when lifting configurations are considered. On delta-wing aircraft with canards, for instance, or on missiles, appear complicated arrangements of vortex sheets and concentrated vortices, and at transonic or higher Mach numbers in addition shock surfaces (bow shocks, imbedded shocks), and slip surfaces, which all may interact with each other, and with the aircraft configuration.

As long as pure viscous effects are small, such flows in principle can be simulated by Euler solutions, in some cases by modelling for instance primary separation lines, if they are not fixed by sufficiently sharp edges.

The main problem with geometric complexity is of course the discretization of the computation domain. On the one hand the aircraft geometry forces the discretization, on the other hand the mentioned phenomena. As long as computer size and costs are limiting factors, an uniformly fine discretization of the whole computation domain is not possible. The only way out is to make a fine discretization where it is needed, and the question is how to achive this.

From the above discussion two main problem classes can be distinguished, which, of course, may overlap to a certain degree, depending on the particular computation object under consideration:

- <u>weak grid forcing:</u> grid refinement because of aircraft geometric properties (surface curvature), which lead locally to continuous changes of the flow direction, accompanied by continuous acceleration or decelaration of the flow, and the subsequent changes of the scalar flow properties,

- <u>strong grid forcing:</u> grid refinement, or even adaptation (fitting), in order to capture discontinuity surfaces (shock surfaces, slip surfaces, vortex sheets, discrete vortices, which are all discontinuities in the inviscid limit), including, if it applies, edges where these originate, for example also regions with explicit Kutta condition.

Weak grid forcing, e.g. clustering in part of the computation domain is made routinely, however, not necessarily with automatic processes, because real aircraft geometries partly are very complicated, especially fighter aircraft or missile geometries. Of course, automatic surface-grid distribution routines exist, enhanced by interactive tools, but the weak grid-forcing problem as such is waiting for a general solution.

Strong grid forcing has been discussed almost for decades with regard to shock surfaces. The question "shock fitting" or "shock capturing" has only partly been answered. Embedded shocks certainly should be captured. For bow shocks, shock-fitting certainly is superior, if the body geometry is sufficiently smooth. However, shock fitting means larger formulation and computation effort, and ways are explored to achive instead a grid refinement controlled by sensor functions[19,20], especially for hypersonic Mach numbers. Automatic grid adaptation techniques of course should also apply to embedded shocks, and to all kind of discontinuity surfaces as well. However, sensors must work on quite different flow properties, depending on the kind of discontinuity surface present.

A complete grid adaptation to a discontinuity surface certainly is not desirable, if this means to make the discontinuity to a grid surface. This would be a "fitting" process, which is much to complicated, at least regarding the experience with shock fitting. Conceivable is an adaptation in form of a local clustering, more or less strongly following the discontinuity surface. Even embedded finer grids, which are not fully aligned with the base grid, are a possibility[21].

In general it appears that the question of self-adaptive grid generation presently is gaining momentum, and hopefully will lead soon to realistic and convenient means to enhance the applicability and accuracy of Euler solutions. A good overview over grid-generation problems in general is found in the proceedings Ref. 22.

12.5 Interdisciplinary Problems

In classical aerodynamics the aircraft is considered as a rigid body, which in addition does not answer in its attitude on a change for instance in lift. In general this view is sufficient for aircraft design, although there are fields like flutter and buffeting, where another view is adopted.

A coupling of aerodynamics with flight mechanics, i.e., the motion of the hole aircraft, is a typical interdisciplinary problem, which can be simulated in the wind tunnel only in certain aspects. However, the numerical simulation of such problem is possible, and even to a large extend in the frame of the Euler modelling of the flow, if the computer power would be available.

Out of the interdisciplinary topics discussed by Hankey[23] we list

- flow - rigid body motion (stall, wing, rock, etc.),
- flow - elastic body motion (as above),
- flow - structure interactions (aeroelasticity in general),
- flow - moving control surfaces.

These topics, if they are to be tackled with regard to realistic cases, need much larger computer power as is general available today[24]. Other interdisciplinary fields, like

- internal flow and propulsion systems,
- hypersonic aerothermodynamics,

are already well developed, and especially equilibrium and non-equilibrium real gas models can be handled in Euler computations past reentry and other hypersonic vehicles, as well as in propulsion and nozzle problems.

While Euler solvers are employed in order to study for instance drag properties of delta wings, see for example Ref. 25, the problem of improving the aerodynamic performance of a given design by means of an Euler solver is treated by Jameson[26]. This leads to the two very important interdisciplinary topics, which have the potential to be treated in a very general way in the Euler frame:

- inverse calculations,
- configuration optimization.

The first of these topics deals with the question to find for example to a desired aerodynamic behaviour the configuration of the wing or aircraft, while, of course, several constraints must be fulfilled. The second topic is closely related to the first one, but starts with an already given configuration and tries to improve it within certain limits. Both could be summarized under the heading "design problems".

Another field, where methods developed for the solution of the Euler equations proved to be successful, are Maxwell's equations, which are the governing equations of electromagnetics. They are hyperbolic-type first-order partial differential equations with coefficients, which can be taken as constants in a wide class of applications. In Reference 27 an elaborate description of an upwind finite-volume scheme is found which emanated from a CFD code. Reference 28 proposes a scheme based on a centered difference approach, which uses a heuristic coupling term composed of Laplacians and cross derivatives as artificial viscosity. In both references the highly oscillatory unsteady dynamics of high-frequency radar scattering are considered. In principle the schemes also could be used in the steady state mode, say, the evaluation of electromagnetic fields in electric propulsion, magnetoplasmahydrodynamics, etc..

This short discussion certainly shows that Euler solutions in the present application frame have by far not exploited their potential. Research and development has started on some of the topics, industrial use, however, will only be possible after further strong improvement of computer power, and strong decrease of computation costs.

12.6 References

1. Pulliam, T.H.: "Artificial Dissipation Models for the Euler Equations". AIAA J., VOl. 24, No. 12, 1986, pp. 1931-1940.

2. Peaceman, D.W., Rachford, H.H., Jr.: "The Numerical Solution of Parabolic and Elliptic Differential Equations". SIAM Journal of Applied Mathematics, Vol. 3, 1955, pp. 28-41.

3. Yanenko, N.N.: "The Method of Fractional Steps for Numerical Solution of the Problems of Mechanics of Continuous Media". Fluid Dynamics Transactions, Vol. 4, 1969, pp. 135-147. Institute of Fundamental Technical Research, Polish Academy of Science, Warsaw.

4. Samaskii, A.A.: "Local One Dimensional Difference Schemes for Multidimensional Hyperbolic Equations in an Arbirary Region". Vycisl. Mat. i Mat. Fiz., 4 21-35, 1964.

5. MacCormack, R.W., Paullay, A.J.: "Computational Efficiency Achieved by Time Splitting of Finite Difference Operators". AIAA-Paper 72-154, 1972.

6. Hirsch, C., Lacor, C., Deconinck, H.: "Convection Algorithms Based on a Diagonalization Procedure for the Multidimensional Euler Equations". AIAA-Paper 87-1163, 1987.

7. Roe, P.L.: "Discrete Models for the Numerical Analysis of Time Dependent Multidimensional Gas Dynamics". J. Comp. Phys., Vol. 63, 1986, pp. 458-476.

8. Powell, K.G., van Leer, B.: "A Genuinely Multidimensional Upwind Cell-Vertex Scheme for the Euler Equations". AIAA-Paper 89-0095, 1989.

9. Jameson, A., Baker, T.J.: "Improvements to the Aircraft Euler Method". AIAA-Paper 87-0452, 1987.

10. Mavriplis, D., Jameson, A.: "Multigrid Solution of the Two-Dimensional Euler Equations on Unstructured Triangular Meshes". AIAA-Paper 87-0353, 1987.

11. Baker, T.J.: " Three Dimensional Mesh Generation by Triangulation of Arbitrary Point Sets". AIAA Paper 87-1124, 1987.

12. Löhner, R., Morgan, K., Zienkiewicz, O.C.: "The Solution of Non-Linear Hyperbolic Equation Systems by the Finite Element Method". Intl. J. Num. Meth. Fluids, Vol. 4, 1984, pp. 1043-1063.

13. Morgan, K., Peraire, J.: "Finite-Element Methods for Compressible Flows". VKI Lecture Notes, Brussels, 1987.

14. Löhner, R.: "Adaptive Remeshing for Transient Problems with Moving Bodies". AIAA-Paper 88-3737, 1988.

15. Shapiro, R.A., Murman, E.M.: "Adaptive Finite Element Methods for the Euler Equations". AIAA-Paper 88-0034, 1988.

16. Shapiro, R.A.: "Adaptive Finite Element Solution Algorithm for the Euler Equations". NNFM 32, Vieweg, Braunschweig/-Wiesbaden, 1991.

17. Kallinderis, J., Baron, J.R.: "Adaptation Methods for a New Navier Stokes Algorithm". AIAA-Paper 87-1167, 1987.

18. Jameson, A.: "Computational Transonics". Comm. Pure Appl. Math., Vol. XLI, 1988, pp. 507-549.

19. Brackbill, J.U., Saltzmann, J.S.: "Adaptive Zoning for Singular Problems in Two Dimensions". J. of Computational Physics, Vol. 46, 1982, pp. 342-368.

20. Kim, H.J., Thompson, J.F.: "Three-Dimensional Adaptive Grid Generation on a Composite Block Grid". AIAA-Paper 88-0311, 1988.

21. Reggio, M., Trepanier, J.-Y., Camarero, R.: " A Composite Grid Approach for the Euler Equations". Int. J. for Numerical Methods in Fluids, Vol. 10, 1990, pp. 161-178.

22. Sengupta, S., Häuser, J., Eisemann, P.R., Thompson, J.F. (eds.): "Numerical Grid Generation in Computational Fluid Mechanics '88". Pineridge Press, Swansea, 1988.

23. Hankey, W.L.: "ICFD-Interdisciplinary Computational Fluid Dynamics". AIAA-Paper 85-1522, 1985.

24. Hirschel, E.H.: "Super Computers Today: Sufficient for Aircraft Design? — Experiences and Demands". In: H.W. Meuer (ed.): Super-Computer '88. Hanser, München/Wien, 1988, pp. 110-150.

25. Drougge, G.: "The Wave Drag of Delta Wings at Supersonic Speeds: a Recent Study". Aeronautical Journal, August/September 1990, pp. 225-230.

26. Jameson, A.: "Aerodynamic Design via Control Theory". AGARD-CP-463, 1990, pp. 22-1 to 22-32.

27. Shankar, V., Hall, W. Mohammadian, A.H.: "A CFD-based Finite Volume Procedure for Computational Electromagnetics-Interdisciplinary Applications of CFD Methods". AIAA-Paper 89-1987-CP, 1989.

28. Goorjian, P.M.: "Algorithm Development for Maxwell's Equations for Computational Electromagnetism". AIAA-Paper 90-0251, 1990.

XIII LIST OF SYMBOLS

The numbers in parantheses are the page numbers where the symbols are defined, or where they appear first. If a symbol appears "sufficiently local", and is there completely defined, it may not appear in this list.

1. Latin Letters

A	matrix (48)
A	cell-face area (160)
A	constant (28, 180)
A	stream-tube cross section (190)
A	parameter (220)
A_{ij}	matrix element (248)
a	edge length of control volume (32)
a	constant (61)
a	weighting parameter (147)
a	determinant of covariant metric tensor of surface (286)
a_{ij}	matrix element (49)
$a_{\alpha\beta}$	coordinate of covariant metric tensor of surface (295)
B	Bernoulli constant (30)
B	matrix (48)
B	function (169)
B	parameter (220)
B	boundary-condition function (256)
B_r	rth-order extrapolation operator (260)
b	edge length of control volume (32)
b	constant (61)
b	parameter (148)
b	span of wing (379)
$b_{\alpha\beta}$	coordinate of covariant curvature tensor of surface (295)
b_{α}^{β}	coordinate of mixed-variant curvature tensor of surface (295)
C	matrix (48)
C	parameter (220)

CFL	Courant-Friedrichs-Lewy number (129)
CN	Courant number (85)
$C^{(n)}$	family of real curves (51)
C_D	drag coefficient (350)
C_L	lift coefficient (350)
C_{M_1}	local pitching moment (25 per cent chord)(379)
C_{X_1}	local axial force (379)
C_{Z_1}	local normal force (379)
c	constant (5)
c	edge length of control volume (32)
c	speed of sound (49)
c	chord length (314)
c_p	specific heat at constant pressure (310)
c_v	specific heat at constant volume
c_p	$=(p-p_\infty)/0.5\ \rho_\infty u_\infty{}^2$, coefficient of static pressure (325)
c_R	chord length of wing root (358)
D	drag (10)
D	=abc, control volume (33)
D	matrix (48)
D	Jacobian mapping determinant (146)
D_o	central difference (72)
D_+	forward difference (72)
D_-	backward difference (72)
D_ξ, D_η	differenc operators (126)
d	distribution function
d_i	normalizing coefficient (103)
E	flux vector (32)
E_M	$=A\rho s$, split-flux derivative (190)
e	specific energy (28)
$\underline{e}_x$, $\underline{e}_y$, $\underline{e}_z$	base vectors (unit vectors) of Cartesian coordinate system (89)
$\underline{e}_x{}^{i'}$	base vector (unit vector) of Cartesian reference coordinate system (294)
F	flux vector (32)
F	function (65)
F	difference operator (132)
F_c	convective part of difference operator (132)

F_D	artificial-viscosity part of difference operator (132)
F_j^n	difference operator at point j and time level n (73)
$F^{*\alpha}$	physical vector coordinate (295)
F^{α}	$= F^{*\alpha}/\sqrt{a_{(\alpha\alpha)}}$, contravariant vector coordinate (295)
f	function (6)
f	Riemann invariant (56)
G	flux vector (32)
G	difference operator (71)
G	growth coefficient (213)
g	coordinate (152)
g	determinant of covariant metric tensor (general)(294)
$\underline{g}^k$	contravariant base vector of general coordinates (97)
g_{ij}	coordinate of covariant metric tensor (general)(294)
H	total enthalpy (30)
H	function (84)
$\underline{H}$	flux-vector function (90)
h	mesh spacing (116)
I	unit matrix (47)
I	direction in computational space (91)
Im	imaginary part of complex number
i	$= \sqrt{-1}$, imaginary unit
i	counting integer (51)
J	direction in computational space (91)
J	Jacobian mapping determinant (97)
j	counting integer (71)
K	direction in computational space (91)
k	counting integer (71)
k	wave number (274)
L	shock trace (66)
L	flow state to left of cell face (85)
L, L_ℓ, L_r	limiter functions (226)
L	Laplace transform operator (258)

L	characteristic (body) length (282)
l	counting integer (71)
ℓ	$=\rho u$, mass flux in x-direction (37)
$\underline{\ell}$	left eigenvector (52)
$\underline{\ell}_{ij}$	line-segment vector (93)
M	parameter (187)
M	Mach number (190)
M_e	Mach number at edge of boundary layer or shear layer (311)
m	$=\rho v$, mass flux in y-direction (44)
$\dot{m}$	mass flux (383)
N_I, N_J, N_K	number of cells in I, J, K
n	$=\rho w$, mass flux in z-direction (44)
n	counting integer (time level)(72)
$\underline{n}$	surface normal vector (42)
n_x, n_y, n_z	components of surface normal vector (42)
$\underline{n}_I$, $\underline{n}_J$ $\underline{n}_K$	unit normal vectors (91)
P	projection operator (268)
p	static pressure (7)
p	parameter (227)
p_∞	free-stream static pressure (10)
p_t	total pressure (311)
p_{t_∞}	total pressure of free stream (329)
Q	lower-order term (256)
q	$=(u^2+v^2+w^2)^{1/2}$, magnitude of velocity (28)
q	solution vector (48)
q^n	q at time level n (123)
q^2	energy quantity (134)
q_{ijk}	volumetric cell average of q (94)
R	Riemann invariant (56)
R	region (66)
R	flow state to right of cell face (85)
R	residual (268)
R	gas constant (311)
Re	real part of complex number
Re	Reynolds number (282)
R_1, R_2	Riemann invariants (176)
r	Riemann invariant (55)

$\underline{r}$	displacement vector (29)
$\underline{r}$	eigenvector (51)
$\underline{r}$	right eigenvector (52)
r_1, r_2	Riemann invariants (166)
S	velocity potential (7)
S	surface (42)
S	entropy (44)
S	Riemann invariant (56)
S	matrix (63)
S	averaging operator (272)
$\underline{S}, \underline{S}_I, \underline{S}_J, \underline{S}_K$	cell surface-area vectors (92)
s	speed of sound (36)
s	time coefficient (47)
s	Riemann invariant (55)
s	switching function (222)
s	eigenvalue (258)
s	entropy (310)
s	span of wing (313)
s	half span of delta wing (364)
s_p	eigenvalue (129)
sign	signum, sign $q = q/\|q\|$ (198)
sup	limit superior (70)
T	time function (212)
T	eigenvector (246)
T	static temperature (310)
T_t	total temperature (310)
T_x, T_u, T_v	(local) truncation errors (108)
T_j^n	truncation error at point j and time level n (73)
t	time (5)
U	solution vector (32)
U	contravariant velocity component (90)
U_1, U_2	dimensionless velocities ahead and behind shock (35)
U_j^n	value of U at point j and time level n (118)
$\underline{U}$	column vector (49)
u	Cartesian velocity component (7)
u_∞	free-stream velocity (10)

u	function (107)
u_i	element of column vector (52)
u_1, u_2	velocities ahead and behind shock (35)
u^n	u at time level n (71)
u_{jkl}	volumetric cell average of u (71)
u_j^n	value of u at point j and time level n (118)
V	volume (6)
V	contravariant velocity component (90)
V	cell volume (267)
$\underline{V}$	velocity vector (89)
v	Cartesian velocity component (7)
v^α	contravariant surface-tangential velocity coordinate (286)
v_e^α	contravariant surface-tangential velocity at edge of boundary layer or shear layer (286)
v_o^3	equivalent inviscid outflow velocity (286)
v^i	contravariant velocity coordinate (general) (294)
v^3	contravariant surface-normal velocity coordinate (294)
$\tilde{v}^3$	$=v^3\sqrt{Re_{ref}}$, contravariant surface-normal velocity coordinate with boundary-layer stretching (294)
$\underline{v}$	velocity vector (310)
$\underline{v}_{e_u}$	velocity vector at upper edge of boundary layer or shear layer (294)
$\underline{v}_{e_\ell}$	velocity vector at lower edge of shear layer (294)
$v^{*\alpha}$	physical surface-tangential velocity coordinate (295)
$v_{e_u}^{*\alpha}$	physical surface-tangential velocity coordinate at upper edge of boundary layer or shear layer (296)
$v_{e_\ell}^{*\alpha}$	physical surface-tangential velocity coordinate at lower edge of shear layer (296)
$\underline{v}_e$	velocity vector at edge of boundary layer or shear layer (298)
v_e	$=\lvert\underline{v}_e\rvert$, magnitude of velocity vector at edge of boundary layer or shear layer (298)
var	variation of function (70)

W	contravariant velocity component (90)
w	Cartesian velocity component (7)
w	test function (66)
$\underline{w}$	vector function (62)
X	external force
X_I, X_J, X_K	curvilinear coordinates (91)
x	Cartesian coordinate (5)
x_j	grid point (71)
x^3	surface-normal locally monoclinic coordinate (286)
$\tilde{x}^3$	$=x^3\sqrt{Re_{ref}}$, surface-normal locally monoclinic coordinate with boundary-layer stretching (294)
x^i	general coordinate parameter (294)
x^α	surface-tangential coordinate parameter (Gaussian parameter) (294)
$x^{i'}$	Cartesian reference coordinate (294)
Y	external force (7)
y	Cartesian coordinate (5)
y_k	grid point (71)
Z	external force (7)
Z	complex scalar
Z_j^n	error at grid point j and time level n (77)
z	Cartesian coordinate (6)
z_l	grid point (71)

2. Greek Letters

α	coefficient in wave ansatz (47)
α	quantity (51)
α	constant (220)
α	angle of attack (297)
β	coefficient in wave ansatz (47)
β	quantity (51)
β	constant (220)
β	yaw angle (395)
β_m	frequency (78)
Γ	non-characteristic curve (58)
γ	$=cp/c_v$, ratio of specific heats (27)
γ	constant (133)

δ	difference operator (94)
δ	parameter (235)
δ	boundary-layer thickness, edge of boundary layer (286)
δ_u	upper edge of boundary layer or shear layer (294)
δ_ℓ	lower edge of shear layer (294)
δ_1	boundary-layer thickness (283)
δ_1^α	local displacement thickness (three-dimensional boundary layer) (286)
δ^2	second difference operator (274)
ε	coefficient in wave ansatz (47)
ε	coefficient 4th difference artificial viscosity (135)
ε	smoothing parameter (272)
ε	vortex-line angle (inclination of local wake-coordinate system) (306)
ε	radiation emissivity coefficient of body surface (406)
εx, εy	smoothing parameters (274)
ζ	local Cartesian shock coordinate (32)
ζ	general coordinate (39)
ζx, ζy, ζz	components of grid-surface (ζ=const.) normal vector (40)
$\bar{\zeta}x$, $\bar{\zeta}y$, $\bar{\zeta}z$	components of metric tensor (39)
$\dot{\zeta}$	grid velocity (41)
η	local Cartesian shock coordinate (32)
η	general coordinate (39)
η	variable (ratio of slopes) (226)
ηx, ηy, ηz	components of grid-surface (η=const.) normal vector (40)
$\bar{\eta}x$, $\bar{\eta}y$, $\bar{\eta}z$	components of metric tensor (39)
$\dot{\eta}$	grid velocity (41)
θ	angle between coordinates (109)
θ	weighting function (137)
θ	phase angle (211)
κ_1, κ_2, κ_3	exponents (265)
Λ	eigenvalue matrix (246)
Λ	cell aspect ratio (248)

Λ_1, Λ_2	matrices (62)
λ	eigenvalue (51)
λ	$=\Delta t/\Delta x$, coefficient in leap-frog scheme (123)
λ	wave speed (161)
λ	viscosity coefficient (268)
λ_o	$= \xi_o^* A$, cell-face normal velocity (160)
λ_i, λ_D	eigenvalues (48)
$\lambda^{(i)}$	eigenvalue (51)
$\tilde{\lambda}$	shock-wave propagation speed (67)
μ	viscosity coefficient (268)
μ_∞	free-stream viscosity (294)
μ_I, μ_J, μ_K	averaging operators (95)
μ_ξ, μ_η	operators (126)
ν	viscosity (69)
ν	iteration level (243)
ν	Courant number (268)
$\underline{\nu}$	direction of wave propagation (48)
ξ	local Cartesian shock coordinate (32)
ξ	general coordinate (39)
ξ	amplification factor (78)
$\xi x, \xi y, \xi z$	components of grid-surface (ξ=const.) normal vector (40)
$\bar{\xi} x, \bar{\xi} y, \bar{\xi} z$	components of metric tensor (39)
$\dot{\xi}$	grid velocity (41)
ρ	density (6)
ρ_∞	free-stream density (294)
ρ_o	equivalent inviscid outflow density (286)
ρ_e	density at edge of boundary layer or shear layer (286)
$\rho_{j,k}$	local spectral radius at j,k (129)
σ	piston path (65)
$\sigma_1, \sigma_2, \sigma_3$	coefficients (265)
τ	time (90)
ϕ	potential (29)
ϕ	coordinate line (50)
ϕ	variable (240)
χ	constant (132)
ψ	function (52)

ϕ	general variable (94)
ϕ_e	angle between x^2=constant coordinate line and flow direction at edge of a three-dimensional boundary layer (304)
ϕ_{e_u}	angle between x^1-direction and flow direction at upper edge of shear layer in local wake-coordinate system (304)
ϕ_{e_ℓ}	angle between x^1-direction and flow direction at lower edge of shear layer in local wake-coordinate system (304)
Ω, $\partial\Omega$	volume, surface (90)
Ω^{*i}	physical vorticity-content coordinate (295)
Ω^i	$=\Omega^{*i}/\sqrt{a_{(\alpha\alpha)}}$, contravariant vorticity-content coordinate (295)
Ω_{ijk}	cell volume at i,j,k (94)
$\underline{\Omega}$	vorticity-content vector (295)
$\underline{\Omega_s}$	shock vorticity-content vector (297)
ω	frequency (264)
ω	interpolation coordinate (235)
ω^i	contravariant vorticity coordinate (294)
$\underline{\omega}$	vorticity vector (294)

3. Indices

3.1 Upper Indices

(i)	counting integer (51)
I,II	counter of partitioned system (63)
n	discrete time level (71)
*	steady state (137)
*	physical quantity (295)
+	positive flux (186)
-	negative flux (186)
3	surface-normal vector coordinate (286)
'	quantity in Cartesian reference coordinate system (294)

3.2 Lower Indices

b	bottom (32)
c	ceiling (32)
e	edge value (286)
I	counting integer (91)
i	inboard (32)
i	counting integer (94)
in	initial value (161)
J	counting integer (91)
j	counting integer (71)
K	counting integer (91)
k	counting integer (71)
L	characteristic (body) reference length (283)
l	counting in layer (71)
ℓ	left face of cell volume (31)
ℓ	lower edge of shear layer (294)
o	outboard (32)
o	cell-face normal speed
o	equivalent inviscid outflow quantity (286)
r	right face of cell volume (31)
t	total value (310)
u	upper edge of shear layer (294)
x	partial derivative, e.g. $u_x \equiv \partial u/\partial x$ (27)
y	partial derivative, e.g. $u_y \equiv \partial u/\partial y$ (27)
z	partial derivative, e.g. $u_z \equiv \partial u/\partial z$ (27)
x	coordinate (90)
y	coordinate (90)
z	coordinate (90)
xx, xxx	second, third derivative, e.g. $f_{xx} \equiv \partial^2 f/\partial x^2$ (107)
ref	reference value (294)
1	upstream of shock surface (33)
2	downstream of shock surface (33)
∞	freestream infinity (10)

3.3 Upper and Lower Indices

$i,j,k,\ldots$	=1,2,3, denotes vector and tensor quantities in general coordinates (294)
$\alpha,\beta,\gamma\ldots$	=1,2, denotes surface-tangential vector and tensor quantities in locally monoclinic coordinates (286)

3.4 Other Symbols

$(a^1;\ a^2;\ a^3)$	$\equiv \underline{a}$, vector (294)
$,_i$	partial derivative, e.g. $u,_i \equiv \frac{\partial u}{\partial x^i}$ (286)
$\bullet$	$\equiv \frac{\partial}{\partial t}$, time derivative, e.g. $\dot{u} \equiv \frac{\partial u}{\partial t}$ (27)
o	time derivative in moving coordinate system (32)
–	component of metric tensor (39)
*	reference state (179)
~	quantity with boundary-layer stretching (294)
$v^\alpha{}_{,\alpha}$	$\equiv \partial v^1/\partial x^1 + \partial v^2/\partial x^2$, Einstein summation convention (286)
(α)	no summation over α (286)

XIV INDEX OF AUTHORS

XV SUBJECT INDEX

Addresses of the Editors of the Series "Notes on Numerical Fluid Mechanics"

Prof. Dr. Ernst Heinrich Hirschel (General Editor)
Herzog-Heinrich-Weg 6
D-8011 Zorneding
Federal Republic of Germany

Prof. Dr. Kozo Fujii
High-Speed Aerodynamics Div.
The ISAS
Yoshinodai 3-1-1, Sagamihara
Kanagawa 229
Japan

Prof. Dr. Bram van Leer
Department of Aerospace Engineering
The University of Michigan
Ann Arbor, MI 48109-2140
USA

Prof. Dr. Keith William Morton
Oxford University Computing Laboratory
Numerical Analysis Group
8-11 Keble Road
Oxford OX1 3QD
Great Britain

Prof. Dr. Maurizio Pandolfi
Dipartimento di Ingegneria Aeronautica e Spaziale
Politecnico di Torino
Corso Duca Degli Abruzzi, 24
I-10129 Torino
Italy

Prof. Dr. Arthur Rizzi
FFA Stockholm
Box 11021
S-16111 Bromma 11
Sweden

Dr. Bernard Roux
Institut de Mécanique des Fluides
Laboratoire Associé au C. R. N. S. LA 03
1, Rue Honnorat
F-13003 Marseille
France

Brief Instruction for Authors

Manuscripts should have well over 100 pages. As they will be reproduced photomechanically they should be typed with utmost care on special stationary which will be supplied on request.
In print, the size will be reduced linearly to approximately 75 per cent. Figures and diagrams should be lettered accordingly so as to produce letters not smaller than 2 mm in print. The same is valid for handwritten formulae. Manuscripts (in English) or proposals should be sent to the general editor, Prof. Dr. E. H. Hirschel, Herzog-Heinrich-Weg 6, D-8011 Zorneding.